Behaviour and Management
of European Ungulates

Behaviour and Management of European Ungulates

edited by

RORY PUTMAN AND MARCO APOLLONIO

Whittles Publishing

Published by
Whittles Publishing,
Dunbeath,
Caithness KW6 6EG,
Scotland, UK
www.whittlespublishing.com

Printed by Gomer Press Ltd.

Contents

List of Contributors

Professor Marco Apollonio
Department of Science for Nature and Environmental Resources, University of
Sassari, via Muroni 25, 07100 Sassari, Italy

Seán Cahill
Estació Biològica del Parc Natural de la Serra de Collserola, Ctra. de
l'Església, 92 08017 Barcelona, SPAIN

Professor Juan Carranza
Ungulate Research Unit, CRCP Game and Fish Research Centre, University of
Cordoba, Cordoba, Spain

Dr Roberta Chirichella
Department of Science for Nature and Environmental Resources, University of
Sassari, via Muroni 25, 07100 Sassari, Italy

Professor Sándor Csányi
Institute for Wildlife Conservation, Szent István University, Gödöllő, H-2100,
Hungary

Dr Dave Cowan
National Wildlife Management Centre, Animal Health and Veterinary
Laboratories Agency (AHVLA), Sand Hutton, York YO41 1LZ, UK

Dr Douglas Eckery
USDA APHIS National Wildlife Research Center, 4101 Laporte Avenue, Fort
Collins, CO 80521, USA

Dr Francesco Ferretti
Research Unit of Behavioural Ecology, Ethology and Wildlife Management,
Department of Life Sciences, University of Siena, via P.A. Mattioli 4, 53100,
Siena, Italy

Professor Carlos Fonseca
Department of Biology and Centro des Estudos do Ambiente e do Mar (CESAM),
University of Aveiro, Campus de Santiago, 3810-193 Aveiro, Portugal

Peter Green, BVSc, MRCVS
South Woolley Farm, Shirwell, Barnstaple, Devon, EX31 4JZ, United Kingdom

Dr Stefano Grignolio
Department of Science for Nature and Environmental Resources, University of
Sassari, via Muroni 25, 07100 Sassari, Italy

Dr Marco Heurich
Bavarian Forest National Park, Department for Conservation and Research,
Freyunger Straße 2, 94481 Grafenau, Germany

Dr Jochen Langbein
Langbein Wildlife Associates, Chapel Cleeve, Minehead, Somerset, TA24 6HY,
United Kingdom

Professor Sandro Lovari
Research Unit of Behavioural Ecology, Ethology and Wildlife Management
Department of Life Sciences, University of Siena, via P.A. Mattioli 4, 53100,
Siena, Italy

Dr Giovanna Massei
National Wildlife Management Centre, Animal Health and Veterinary Laboratories
Agency (AHVLA), Sand Hutton, York YO41 1LZ, UK

Professor Atle Mysterud
Centre for Ecological and Evolutionary Synthesis (CEES), Department of
Biology, University of Oslo, P.O. Box 1066 Blindern, N-0316 Oslo, Norway

Professor Frauke Ohl
Department of Animals in Science and Society, Division of Animal Welfare and
Laboratory Animal Science, Faculty of Veterinary Medicine, Utrecht University,
Yalelaan 2, 3584 CM Utrecht, The Netherlands

Dr Boštjan Pokorny
ERICo Velenje, Ecological Research and Industrial Co-operation,
Koroška 58, 3320 Velenje, and Environmental Protection College,
Trg mladosti 7, 3320 Velenje, Slovenia

Professor Rory Putman
Department of Animals in Science and Society, Utrecht University and Institute
of Biodiversity, Animal Health and Comparative Medicine, University of
Glasgow, UK

Mark Ryan
International Council for Game and Wildlife Conservation (CIC), CIC Headquarters, P.O. Box 82, H-2092 Budakeszi, Hungary

João P. V. Santos
Department of Biology Centro des Estudos do Ambiente e do Mar (CESAM), University of Aveiro, Campus de Santiago, 3810-193 Aveiro, Portugal.

Dr Massimo Scandura
Department of Science for Nature and Environmental Resources, University of Sassari, via Muroni 25, 07100 Sassari, Italy

Dr Nikica Šprem
Department for Fisheries, Beekeeping and Special Zoology, University of Zagreb, Svetošimunska 25, 10000, Zagreb, Croatia

Dr Naomi Sykes
Department of Archaeology, University of Nottingham, Nottingham NG7 2RD, UK

Dr Rita Torres
Department of Biology and Centro des Estudos do Ambiente e do Mar (CESAM), University of Aveiro, Campus de Santiago, 3810-193 Aveiro, Portugal

Dr José Vingada
Department of Biology and Centro des Estudos do Ambiente e do Mar (CESAM), University of Minho, Campus de Gualtar, 4710-057 Braga, Portugal

Peter Watson
The Deer Initiative, The Carriage House, Brynkinalt Business Centre, Chirk, Wrexham LL14 5NS, UK

Scientific Names of Species Referred to in this Text

In this book, common names are used through the text. Species implied are:

Cervidae:

Axis (or chital)	*Axis axis*
Chinese, or Reeves' muntjac	*Muntiacus reevesi*
Chinese water deer	*Hydropotes inermis*
Fallow deer	*Dama dama*
Moose	*Alces alces*
Reindeer	*Rangifer tarandus*
Roe deer	*Capreolus capreolus*
Red deer	*Cervus elaphus*
Sika deer	*Cervus nippon*
Wapiti	*Cervus canadensis*
White-tailed deer	*Odocoileus virginianus*
	and their subspecies

Bovidae/Ovidae:

Alpine chamois	*Rupicapra rupicapra*
Pyrenean chamois	*Rupicapra pyrenaica*
Alpine ibex	*Capra ibex*
Spanish or Iberian ibex	*Capra pyrenaica*
Barbary sheep	*Ammotragus lervia*
European bison (or wisent)	*Bison bonasus*
Mouflon	*Ovis orientalis musimon*
Musk ox	*Ovibos moschatus*
Wild goat	*Capra aegagrus*
	and their subspecies

Suidae:

Wild boar	*Sus scrofa*

Common names are used throughout the text in place of formal scientific names, except where the latter are used in text or in subheadings to identify a particular subspecies.

Chapter 1

Behaviour and Management of European Ungulates

Rory Putman and Marco Apollonio

Ungulates are an extraordinarily important group of animals within a European context, at a multiplicity of different levels. Ecologically, they represent a remarkable biodiversity in their own right; in addition, as key species within many of the communities of which they are a part, they may have a very significant effect on the functioning of wider ecological systems (through the impacts of grazing and browsing on plant communities and habitats, and the relocation of nutrients within nutrient-limited ecosystems); they provide a prey base for populations of large carnivores – many of which within Europe may be endangered or only recently showing signs of recovery.

Deer and other ungulates are also of a disproportionate significance both culturally and economically, as a major source of protein and because of their wide exploitation in recreational hunting, which is still a major form of land-use across Europe. Finally, they may have a tremendous impact on other human land-use interests through damage to agriculture or forestry, through competition with domestic livestock, the spread of diseases to humans or livestock and through involvement in collisions with vehicles.

In November 2004, a seminal meeting was held in Erice (Sicily) where representatives from a wide range of European countries were asked to come and offer presentations on ungulate populations and how those populations are managed in their respective countries. The overall idea was to learn from each other's experiences (and each other's mistakes), in the hope of developing improved management strategies for the future. Speakers were asked to review the status of populations of wild ungulates in their countries, describe current legislation and management philosophy, and review problems and actual practice with day-to-day management.

From that meeting two major books have already arisen. The first was conceived primarily as a reference work, establishing for some 28 countries (all EU countries except Malta, plus Norway and Switzerland) what species of ungulates were present, the history and current distribution and status of the different species within each country, approaches to management and issues associated both with the ungulates themselves and their management. Published in 2010 (Apollonio *et al.*, 2010a, *European Ungulates and their Management in the 21st Century*) we believe that this is the first time anyone has attempted to try and draw together information on wild ungulates and their management across Europe. Such was the scale of the project that we did not at that time attempt to offer a detailed synthesis of this diverse body of material. Our main aim was, explicitly, to draw together in one place a convenient single source of reference for the primary information itself – and with the book already extending to some 30-odd chapters, an equivalent 'weight' of synthesis would render the work so large as to be virtually unusable.

However, one of the main themes rehearsed again and again by authors in the first volume, and highlighted by the editors in conclusion, was a need for science-based management (Apollonio *et al.*, 2010b). In a second volume, therefore (Putman *et al.*, *Ungulate Management in Europe: Problems and Practices*) we adopted a more 'functional' or 'process-based' approach, identifying a number of types of impact, or management issues which clearly recurred across most, or all of the different countries of Europe and then inviting experts in each of those different biological areas to offer a review of the management issues themselves and of the different management approaches adopted in different countries of Europe to try to resolve those issues, to prepare some sort of synthesis of that experience.

While inevitably each chapter in this second book was to an extent *informed* by the material summarised in that earlier volume, we asked each author to use their own research experience and expertise to essay further development of the material in particular topical areas.

Topics included an overview of the basic resource and the administrative structures within which management is carried out, while other chapters addressed consideration of particular management issues. Thus topics included:

• An overview of the ungulate species present: native species, problems associated with the introduction of non-native genotypes of native species or other exotic taxa and the implications for management of this varied genetic resource
• Management context: a consideration of European legislation and legal or administrative structures facilitating or constraining effective management
• Other constraints on management: hunting seasons in relation to biological breeding seasons and the implications for the control or regulation of ungulate populations
• Impact of large ungulates on agriculture, forestry and conservation habitats in Europe
• Road traffic accidents involving ungulates and available measures for mitigation.
• The role of diseases in limiting or regulating large ungulate populations and the role of wild ungulates as vectors of disease
• Large carnivores and the impact of predation on populations of wild ungulates
• Large herbivores as 'environmental engineers', or agents of deliberate habitat change

- Climate change and implications for future distribution and management of ungulates in Europe.

These two volumes were awarded a Prize by the International Council for Game and Wildlife Conservation (CIC) in the technical section of their annual Literary Award, so have clearly been well received; in response to this and given a sense of 'unfinished business', we are encouraged to present here a third volume. Like the second book, this adopts a 'cross-cutting' approach: taking a series of issues or topics and offering an overview of these issues and how they might best be addressed by management; the remit of this text is however somewhat broader, in that we also seek to explore the wider economic or welfare consequences of alternative approaches which may be adopted to our management of ungulate populations.

In this case we consider:

- both the positive and negative issues associated with introductions or reintroductions of ungulates as a management tool (Chapters 3 and 4)
- problems of competition between introduced exotics and native species of ungulates and problems of competition between wild ungulates and domestic livestock (Chapters 4 and 5)
- effects of selective harvesting on dynamics and evolution of ungulate populations (Chapter 6)
- problems associated with the colonisation of urban and peri-urban areas by wild ungulates and the difficulties of management of ungulate populations within such urban landscapes (Chapter 7)
- special problems associated with the management of ungulates in National Parks or other protected areas (Chapter 8)
- the problems of management of populations of ungulates whose distribution crosses national or provincial boundaries where legislation or even management objectives may differ (Chapter 9)
- the advantages and disadvantages of non-lethal management methods such as translocation and immunocontraception (Chapter 10)
- welfare issues in the management of wild ungulates (Chapter 11).

As with the previous volumes, this book is directed at practising wildlife managers and stalkers; policy makers in local, regional or national administrations, responsible for formulating policies affecting management of different wildlife species and those who may be actively involved in research into improving methods of wildlife management.

1.1 Behaviour and management of European ungulates: The effects of management on behaviour

This book takes its title in a nod towards an earlier work, *The Behaviour of Ungulates and its Relation to Management*, edited by Valerius Geist and Fritz Walther in 1974: in recognition not only of the fact that an understanding of the behaviour of different species may assist efforts at managing their impacts (or at least save managers from trying to work against the grain of natural behavioural responses), but also in acknowledgement of the fact that our management activities may themselves have a significant impact on the resultant behaviour of the animals we seek to control.

We know for example that culling may have a significant effect on behaviour – both immediately, through the effects of disturbance, but also through its effect on population size and structure. Removal of particular individuals (loss of a lead or matriarchal hind in more herding species; loss of a close group member) may lead to disorientation or fragmentation of that group; alteration of the age and sex-structure may also result in a disruption of normal behaviours.

One of the most obvious and immediate effects of disturbance by culling is that animals tend to group together in larger aggregations and become much less faithful to a defined home range. Rather than abandon familiar territory altogether, many species also show a tendency to become more nocturnal, concentrating activity in the hours of darkness, when levels of disturbance are reduced. First reported for red deer by van der Veen (1979), this tendency towards a nocturnal habit has also been reported, for example, for both sika (Putman and Mann, 1989) and roe (Bonnot *et al.*, 2013). Heavy culling pressure may also have a profound effect on animal distribution patterns and use of space, with a tendency ultimately to avoid areas of high risk, and clear evidence of immigration into areas of lower risk (e.g. Grignolio *et al.*, 2010; Putman, 2012; Wäber *et al.*, 2013; Goldberg *et al.*, 2014).

Within the more immediate social context, little is known about the effects on fitness or welfare of loss from a social group of lead animals. Very few species in a European context show overt evidence of dependence on an accumulated group memory – of a familiar range or routes to localised resources which may be used only rarely – although like many other ungulates they do make use of mineralised licks which may be present only in discrete areas of a range and some species and populations may traverse considerable distances between discrete winter and summer ranges, whose routes must be learned, as in the case of red deer on the Alps (Georgii, 1980; Luccarini *et al.*, 2006). How important may be this social dependency on established lead animals is largely unknown.

For many species, again, individuals within a group show closer associations with some group members than they do with others, spending significantly more time feeding or resting in association with certain other individuals than would be predicted by purely random patterns of association. Such closer 'allegiances' (among both males and females) are well-established, for example in groups of both fallow and red deer (e.g. Hall, 1983; Carlin, 2002). To what extent disruption of this bond by death of one of those associates may impact upon others within that 'clique' through some sense of 'emotional' loss, or through actual loss of mutual advantages in shared foraging experience, shared competitive ability, etc., is again unknown and is not readily quantified. Amongst European ungulates, focus has tended to concentrate on the potential implications of disruption of the rut (Apollonio *et al.*, 2011), or orphaning of juveniles by culling of the dam during a vulnerable period of juvenile dependency (Apollonio *et al.*, 2011; Holand *et al.*, 2012).

The longer-term effect of more general disruption of population or social structure resulting from deliberate or unconscious selection in culling is explored in detail in this volume by Atle Mysterud (Chapter 6). And even non-lethal approaches to management of populations may have significant effects on behaviour. As Massei *et al.* explore in Chapter 10, suppression of the reproductive cycle of females, or measures adopted to prevent the implantation of fertilised embryos may prolong the period of the rut and lead to increased aggression between males or increased harassment of females, which repeatedly return to oestrus.

Control of population numbers by lethal or non-lethal means is only one of the management interventions we may impose upon ungulate populations. We may affect their local distribution, their time budgets and foraging behaviours in provision of supplementary feeds (see, for example, Wiersema 1974; Linn, 1986, 1987; Schmidt, 1992; Seivwright, 1996; Luccarini *et al.*, 2006; review by Putman and Staines, 2004). Management actions like the reintroduction of large carnivores can also have a marked impact on spatial behaviour and habitat selection by ungulate prey (e.g. Fortin *et al.*, 2005; Goldberg *et al.*, 2014). But many actions which may be taken by managers are not explicitly targeted at wildlife populations directly in this way: we may also impinge upon animals and their behaviour inadvertently, simply as a consequence of effects of habitat change due to agricultural, forestry or other activity, through effects of habitat fragmentation (by roads, fences and other human infrastructure) or due to the effects of specific developments within a management area such as wind farms (for this last, see Walter *et al.*, 2006; Colman *et al.*, 2013).

1.2 The effects of behaviour on management

While management decisions, whether directed specifically at wildlife populations or directed towards some totally independent land-use objective, may have a profound implication on the behaviour of wild ungulates, so conversely, the behaviour of the animals themselves may have significant impact in terms of choice of management methods adopted to control populations or their impacts – or in determining which of a variety of alternative management options is likely to be more or less effective in any given instance. The effectiveness of any given strategy is likely to be influenced by the social structure of different species and the level of social organisation: thus more mobile, herding species such as red, fallow and sika deer are likely to have a rather different 'style' of impact and require different management strategies from those effective for more solitary 'territorial' species with small home ranges such as roe or muntjac (see, for example, Putman *et al.*, 2011b); required management strategies are likely to be rather different for animals which have extensive home-ranges, because of the difficulty of coordinating management activity across the entire home range, and there may be particular issues associated with migratory species. The huge difficulties that are faced in some European countries in managing large ungulates are in fact the direct consequence of the adoption of small management units that include only a part of the effective range of any given population (Apollonio *et al.*, 2010b; and for further discussion, see Chapter 9). Much may also depend on reproductive strategies or seasonality of breeding (Apollonio *et al.*, 2011). More solitary species are less impacted by culling during the rut than more social ones as in the latter case many individuals are disrupted by a single hunting effort.

But while behavioural considerations may thus constrain the effectiveness of different management strategies, an understanding of behaviour may also enable managers to develop more effective management policies based on active exploitation of particular behavioural responses. Many studies now agree that impacts from deer or damage to human interests (agriculture, forestry, etc.) are only weakly related to density (see, for example, Putman, 2004; Palmer *et al.*, 2004; Ward *et al.*, 2008; Gill and Morgan, 2010; reviews by Putman *et al.*, 2011b; Reimoser and Putman, 2011). Such a weak relationship between impacts and density implies that control of impacts through traditional management efforts directed at reductions in population size is not necessarily especially effective. Putman (2004) and Reimoser and Putman (2011) argue that there is therefore a need to adopt different (if complementary) approaches to management of population size and management of impacts. Perhaps more to the point, the fact that so many additional factors contribute to levels of impact observed even at a constant density – and that in many cases we can clearly identify those factors – opens up the possibility of addressing problems of game damage through manipulation of those other environmental factors known to have an influence on damage level. If we know, for example, that damage caused to forest crops is in large part a function of forest structure and the lack of understorey which in turn restricts the amount of alternative forage available to forest ungulates (e.g. Reimoser and Gossow, 1996; Gill and Morgan, 2010), then alteration to forest structure, or to forest 'hygiene' practices (in terms of removal of understorey 'weeds'), may have a significant impact on future levels of damage sustained (e.g. Putman, 1998). If we know that ungulates are causing damage to vulnerable species and habitats in conservation areas – and can identify what resources are attracting those animals into the sensitive area in the first place, then by providing alternative sources of that commodity elsewhere within their range, we may be able to alter the balance of distribution within, or patterns of usage of the home-range, reducing time spent within the sensitive site and thus reducing damaging impacts (Putman, 2004).

To offer a further example: it has been clearly established by a number of authors that the frequency of deer–vehicle collisions is not simply related to deer density but also road density, traffic volume and traffic speed (e.g. Langbein, 2007; Langbein *et al.*, 2011) as well as a number of other environmental factors (e.g. Bashore *et al.*, 1985; Finder *et al.*, 1999; Hubbard *et al.*, 2000; Malo *et al.*, 2004; Putman *et al.*, 2004; Seiler, 2004). In all these studies certain consistent features emerge as characteristic of sites likely to suffer a high frequency of deer-related road traffic accidents (Putman *et al.*, 2004); these include: number of lanes of traffic (width of road); traffic volume and speed; presence or absence of a central barrier; but also, close association with woodland or forest cover close to the carriageway; landscape diversity (variability and patch size); the presence of obvious travel corridors across the roadway, such as rivers, dry gullies or other linear structures leading down at an angle to, or perpendicular to the roadway.

In a recent analysis in different sites within the UK, Uzal (2013) has demonstrated that such environmental or traffic-related factors account for up to 70% of recorded variation in road traffic accident incidence at a landscape (10 km^2) scale, leaving comparatively little to be explained by variations in ungulate density. If we can pay attention to such features in planning, or alter their disposition in relation to the carriageway even retrospectively, we are likely to achieve a far greater and far longer-lasting reduction in risk of wildlife-related traffic accidents than in any attempts at local reduction of density of deer or other larger wildlife. Lavsund and Sandergren (1991) report that clearance of a 20 m strip either side of the highway decreased moose collisions by near 20%, and in experimental manipulations to test effectiveness along a railway in reducing the frequency of collisions between trains and moose, Jaren *et al.* (1991) found that removal of vegetation again from a 20–30 m strip on either side of the railway line caused a 56% reduction in the number of recorded accidents (see also Langbein *et al.*, 2011).

An additional component within such management approaches is offered by the deliberate exploitation of behavioural responses to disturbance considered above in relation to the effects of management on behaviour, whereby animals may positively avoid areas of higher disturbance or high risk (above Wäber *et al.*, 2013; Goldberg *et al.*, 2014). Thus manipulation of 'the landscape of fear' (Brown *et al.*, 1999) of ungulate populations, by targeting disturbance or intense hunting pressure within certain areas of the home range, may be a very effective way of causing changes in the distribution of animals within their range and altering usage of more vulnerable areas, since long-lived ungulates may learn to avoid such high risk locations. Hunting for fear (Cromsigt *et al.*, 2012) may thus offer a key element in management practices designed to persuade ungulates to avoid cultivated areas or other places where reduced impacts may be desired (Ciuti *et al.*, 2012).

All such management interventions are also likely to have a longer-term effectiveness than localised reduction in population density of the animals themselves, since such local reductions are, inevitably, rapidly reversed through reproductive recruitment or immigration into what are clearly preferred areas (Putman, 2004, 2012; Wäber *et al.*, 2013).

Such considerations imply that the most effective management strategies for reducing ungulate impacts in the future will require integration of a number of different approaches of both population control and habitat management. We believe that there is a need for a much more holistic approach to the integration of ungulate species into cultivated landscapes, with proper, landscape level planning to ensure adequate habitat structures available for plants and animals, thereby reducing conflicts. Such ideas have been rehearsed extensively by (e.g.) Putman (2004); Putman *et al.* (2011b); Kenward and Putman (2011); Davies and White (2012); Austin *et al.* (2013); Skonhoft *et al.* (2013). While the regulation of numbers of ungulates within any area may contribute partially to regulation of impact, it is perhaps best seen as directed primarily to regulating the numbers of the ungulates themselves in relation to the land's capacity to support healthy stocks – while separate consideration may need to be given to complementary strategies which will help to control their impact on conservation, forestry or agricultural interests. While the latter (attempts to control damage) may indeed include elements of population control, they should also explore alternative approaches such as fencing and other physical barriers; habitat manipulations to increase the availability of alternative forages or alterations to forest management and culture methods which may change the balance between forage availability and the availability of cover habitats; diversionary feeding, etc. (Putman, 1998, 2004; Reimoser and Putman, 2011).

We fully accept that none of these approaches is likely to be entirely effective on its own (any more than are simple attempts at population control). Kuiters *et al.* (1996) for example point out that ungulate impact on forest development and its dependence on spatial and temporal patterns of forest regeneration are still poorly understood. This makes it very difficult to control ungulate herbivory effectively in commercial and conservation forests by habitat manipulation alone. Much further work is required on all the factors 'predisposing' different cultural ecosystems to ungulate damage, if such manipulations are to be effective, and Putman (1996, 2004) recognised that such methods would usually need to be deployed within a wider management package involving at least some element of direct population control. Nonetheless, a greater focus on such methods is likely to be of much greater *long-term* effectiveness in controlling impacts in many situations (Putman, 2004; Reimoser and Putman, 2011).

1.3 Conclusions

As implied by the title of Geist and Walther's compilation, effective management of ungulate populations needs to take into account the behaviour of the different species concerned – and the likely response, in terms of changes in behaviour, to management actions taken (whether targeted directly on the animals themselves or on their wider environment). A proper understanding of behaviour can help facilitate effective management (even on occasion by deliberate manipulation of behaviour to deliver some management goal), while at the same time enabling managers to avoid management decisions which might prove counterproductive. But behaviour is in itself not a fixed, or species-specific characteristic. Because behaviour constitutes one of the main ways in which individuals interact with their environment, behavioural responses must be flexible and adaptable in order to be adaptive (Putman and Flueck, 2011).

Significantly, the social and sexual behaviour of any one species changes markedly with environmental context and character (Emlen and Oring, 1977; Lott, 1991; review for cervids by Putman and Flueck, 2011). This has, amongst other things, profound implications for the manager: if a species such as roe is likely to be territorial only under some circumstances, while in other situations territoriality is suppressed or not apparent (e.g. Lamberti *et al.*, 2006), then the whole approach to managing the roe population will be affected in the two distinct sets of conditions. Further, most managers seek to plan their cull in such a way as to maintain the natural social structure of the population targeted (not least because disruption of the natural sex- or age-structure of the population, or distortion of the social structure may in itself lead to an increase in damage sustained by agricultural or forest crops through an increase in aggression). But if social structure itself changes with local conditions (Putman, 1988; Putman and Flueck, 2011), then the manager cannot simply aim for the 'typical' structure reported in the standard texts: for this may itself be entirely *in*appropriate in his particular environmental conditions and not represent the social structure that would naturally be adopted in his area at all. That is why it is so important to understand what is responsible for the underlying variation and what factors in the environment do have an effect on the social structure expressed.

If we recognise that group sizes are generally larger in more open habitats, that sexual segregation is less marked in areas of low buck density, territoriality of roe less pronounced at low density or in open landscapes, rutting behaviour affected by both density of females and density of mature bucks etc., and if we know which direction the behaviour changes in response to variations in these environmental cues, then from a knowledge of local environmental conditions we can hope to predict what should be 'normal', or at least expected, in any given area or for any given population – and aim our management towards that endpoint rather than some textbook norm.

This book then explores some of the issues surrounding management of wild (and feral) ungulate populations in Europe – and the effect that management actions currently taken may have on the ecology, behaviour and welfare of the animals themselves. It focuses attention specifically on issues, and the management options available for their solution but implicit, in each case, is this same consideration of the need to take the animals' behaviour into account in developing effective solutions, and the likely implications of management interventions on subsequent behaviour.

References

Apollonio, M., Andersen, R. and Putman, R.J. (eds) (2010a) *European Ungulates and Their Management in the 21st Century.* Cambridge, UK: Cambridge University Press.

Apollonio, M., Andersen, R. and Putman, R.J. (2010b) Present status and future challenges for European ungulate management. In M. Apollonio, R. Andersen and R.J. Putman (eds), *European Ungulates and their Management in the 21st century.* Cambridge, UK: Cambridge University Press, pp. 578–604.

Apollonio, M., Putman, R.J, Grignolio, S. and Bartos, L. (2011) Hunting seasons for ungulates in relation to biological breeding seasons and the implications for control of population size. In R.J. Putman, M. Apollonio and R. Andersen (eds), *Ungulate Management in Europe: Problems and Practices.* Cambridge, UK: Cambridge University Press, pp. 80–105.

Austin, Z., Raffaelli, D.G. and White, P.C.L. (2013) Interactions between ecological and social drivers in determining and managing biodiversity impacts of deer. *Biological Conservation* **158**, 214–222.

Bashore, T.L., Tzilkowski, W.M. and Bellis, E.D. (1985). Analysis of deer-vehicle collision sites in Pennsylvania. *Journal of Wildlife Management* **49**, 769–774.

Bonnot, N., Morellet, N., Verheyden, H., Cargnelutti, B., Lourtet, B., Klein, F. and Hewison, A.J.M. (2013) Habitat use under predation risk: hunting, roads and human dwellings influence the spatial behaviour of roe deer. *European Journal of Wildlife Research* **59**, 185–193.

Brown, J.S., Laundre, J.W. and Gurung, M. (1999) The ecology of fear: optimal foraging, game theory, and trophic interactions. *Journal of Mammalogy* **80**, 385–399.

Carlin, C.M. (2002) *Social organisation in female fallow deer (Dama dama L.): ecology of grouping patterns, association indices, mother-offspring bonds and synchrony of oestrus.* PhD thesis, University College Dublin.

Ciuti, S., Muhly, T.B., Paton, D.G., McDevitt, A.D., Musiani, M. and Boyce, M.S. (2012) Human selection of elk behavioural traits in a landscape of fear. *Proceedings of the Royal Society, B, Biological Sciences* **279**, 4407–4416.

Colman, J.E., Sindre Eftestøl, S., Tsegaye, D., Flydal, K. and Mysterud, A. (2013) Summer distribution of semi-domesticated reindeer relative to a new wind-power plant. *European Journal of Wildlife Research* **59**, 359–370.

Cromsigt, J.P.G.M., Kuijper, D.P.J., Adam, M., *et al.* (2013) Hunting for fear: innovating management of human–wildlife conflicts. *Journal of Applied Ecology* **50**, 544–549.

Davies, A.L. and White M.R. (2012) Collaboration in natural resource governance: reconciling stakeholder expectations in deer management in Scotland. *Journal of Environmental Management* **12,** 160–169.

Emlen, S.T. and Oring, L.W. (1977) Ecology, sexual selection, and the evolution of mating systems. *Science* **197**, 215–223.

Finder, R.A., Roseberry, J.L. and Woolf, A. (1999). Site and landscape conditions at white-tailed deer collision locations in Illinois. *Landscape and Urban Planning* **44**, 77–85.

Fortin, D., Beyer, H.L., Boyce, M.S., Smith, D.W., Duchesne, T. and Mao, J.S. (2005) Wolves influence elk movements: behavior shapes a trophic cascade in Yellowstone National Park. *Ecology* **86**, 1320–1330.

Geist, V. and Walther, F.R. (1974) *The Behaviour of Ungulates and its Relation to Management.* Morges, Switzerland: International Union for Conservation of Nature and Natural Resources.

Georgii, B. (1980) Home range patterns of female red deer (*Cervus elaphus*) on the Alps. *Oecologia* **58,** 238–248.

Gill, R.M.A. and Morgan, G. (2010) The effects of varying deer density on natural regeneration in woodlands in lowland Britain. *Forestry* **83**, 53–63.

Goldberg, J.F., Hebblewhite, M. and Bardsley, J. (2014) Consequences of a refuge for the predator–prey dynamics of a wolf–elk system in Banff National Park, Alberta, Canada. *PlosOne*, in press.

Grignolio, S., Merli, E., Bongi, P., Ciuti, S. and Apollonio, M. (2010) Effect of hunting with hounds on a non target species living on the edge of a protected areas. *Biological Conservation* **144,** 641–649.

Hall, M.J. (1983) Social organisation in an enclosed group of red deer (*Cervus elaphus* L.) on Rhum: 1. The dominance hierarchy of females and their offspring. *Zeitschrift fur Tierpsycholgie* **61**, 250–262.

Holand, O., Weladji, R.B., Mysterud, A., Røed, K., Reimers, E. and Nieminen, M. (2012) Induced orphaning reveals post-weaning maternal care in reindeer. *European Journal of Wildlife Research* **58**, 589–596.

Hubbard, M.W., Danielson, B.J. and Schmitz, R.A. (2000) Factors influencing the location of deer–vehicle accidents in Iowa. *Journal of Wildlife Management* **64**, 707–713.

Jaren, V., Andersen, R., Ulleberg, M., Pedersen, P.H. and Wiseth, B. (1991). Moose–train collisions: the effects of vegetation removal with a cost–benefit analysis. *Alces* **27**, 93–99.

Kuiters, A.T., Mohren. G.M.J. and van Wieren, S.E. (1996) Ungulates in temperate forest ecosystems. *Forest Ecology and Management* **88**, 1–5.

Lamberti, P., Mauri, L., Merli, E., Dusi, S. and Apollonio, M. (2006) Use of space and habitat selection by roe deer *Capreolus capreolus* in a Mediterranean costal area: how does wood landscape affect home range? *Journal of Ethology* **24**, 181–188.

Langbein, J. (2007) *National Deer–Vehicle Collisions Project: England 2003–2005.* Final Report to the Highways Agency. The Deer Initiative, Wrexham UK.

Langbein, J., Putman, R.J. and Pokorny, B. (2011). Road traffic accidents involving ungulates and available measures for mitigation. In R.J. Putman, M. Apollonio and R. Andersen (eds), *Ungulate Management in Europe: Problems and Practices.* Cambridge, UK: Cambridge University Press, pp. 215–259.

Lavsund, S. and Sandegren, F. (1991) Moose–vehicle relations in Sweden: a review. *Alces* **27**, 118–126.

Linn, S. (1986) The social behaviour of a red deer herd (*Cervus elaphus*) at the winter feeding place: spatial distribution at the feeding place. *Zeitschrift für Jagdwissenschaft* **32**, 13–21.

Linn, S. (1987) *Zum sozialen verhalten eines rotwildrudels (*Cervus elaphus*) am winter futterungsplatz unter besonderer berucksichtigung soziobiologischer hypothesen.* PhD thesis, University of Geneva.

Lott, D. (1991) *Intraspecific Variation in the Social Systems of Wild Vertebrates.* Cambridge, UK: Cambridge University Press.

Luccarini, S., Mauri, L., Ciuti, S., Lamberti, P. and Apollonio, M. (2006) Red deer (*Cervus elaphus*) spatial use in the Italian Alps: home range patterns, seasonal migrations, and effect of snow and winter feeding. *Ethology, Ecology and Evolution* **18**, 127–145.

Malo, J.E., Suarez, F. and Diaz, A. (2004) Can we mitigate animal–vehicle accidents using predictive models? *Journal of Applied Ecology* **41**, 701–710.

Palmer, S.C.F., Mitchell, R.J., Truscott, A.-M. and Welch, D. (2004) Regeneration failure in Atlantic oakwoods: the roles of ungulate grazing and invertebrates. *Forest Ecology and Management* **192**, 251–265.

Putman R.J. (1988) *The Natural History of Deer.* Beckenham, Kent, UK: Christopher Helm.

Putman, R.J. (1998) The potential role of habitat manipulation in reducing deer impact. In C.R. Goldspink, S. King and R.J. Putman (eds), *Population Ecology, Management and Welfare of Deer.* Manchester, UK: British Deer Society/Manchester Metropolitan University, pp. 95–101.

Putman, R.J. (2004) *The Deer Manager's Companion: A Guide to Deer Management in the Wild and in Parks.* Shrewsbury, UK: Swan Hill Press.

Putman, R.J. (2012) Effects of heavy localised culling on population distribution of red deer at a landscape scale: an analytical modelling approach. *European Journal of Wildlife Research* **58**, 781–796.

Putman, R.J. and Flueck, W.T. (2011) Intraspecific variation in biology and ecology of deer: magnitude and causation. *Animal Production Science* **51**, 277–291.

Putman, R.J. and Mann J.C.E. (1990) Social organisation and behaviour of British sika deer in contrasting environments. *Deer,* **8**, 90–94.

Putman, R.J. and Staines, B.W. (2004) Supplementary winter feeding of wild red deer *Cervus elaphus* in Europe and North America: justifications, feeding practice and effectiveness. *Mammal Review* **34**, 285–306.

Putman. R.J, Langbein, J. and Staines, B.W (2004): *Deer and road traffic accidents: a review of mitigation measures: costs and cost-effectiveness.* Contract report RP23A. Deer Commission, Inverness, Scotland.

Putman, R.J., Apollonio, M. and Andersen, R. (eds) (2011a) *Ungulate Management in Europe: Problems and Practices.* Cambridge, UK: Cambridge University Press.

Putman, R.J., Langbein, J., Green, P. and Watson, P. (2011b) Identifying threshold densities for wild deer in the UK above which negative impacts may occur. *Mammal Review* **41**, 175–196.

Reimoser, F. and Gossow, H. (1996) Impact of ungulates on forest vegetation and its dependence on the silvicultural system. *Forest Ecology and Management* **88**, 107–119.

Reimoser, F. and Putman, R.J. (2011) Impact of large ungulates on agriculture, forestry and conservation habitats in Europe. In R.J. Putman, M. Apollonio and R. Andersen (eds), *Ungulate Management in Europe: Problems and Practices.* Cambridge, UK: Cambridge University Press, pp. 144–191.

Schmidt, K.T. (1992) Uber den einfluss von futterung und jagd auf das raum-zeit-verhalten von rotwild. *Zeitschrift fur Jagdwissenschaft* **38**, 88–100.

Seiler, A.(2004) Trends and spatial pattern in ungulate–vehicle collisions in Sweden. *Wildlife Biology* **10**, 301–313.

Seivwright, L.J. (1996) *The influence of supplementary winter feeding on the social behaviour of red deer (*Cervus elaphus*).* BSc (Hons) thesis, Environmental Biology, University of St Andrews, UK.

Skonhoft, A., Veiberg, V., Gautepass, A., Olaussen, G.O., Meisigngset, G.L. and Mysterud, A. (2013) Balancing income and cost in red deer management. *Journal of Environmental Management* **115,** 179–188.

Uzal, A. (2013) *Reported deer road casualties and related accidents in England 2003–2010: their potential to develop an index of deer density.* Report to the Deer Initiative, Wrexham, UK.

van der Veen, H.E. (1979). *Food selection and habitat use in the red deer (Cervus elaphus L.).* PhD thesis, Rijksuniversiteit te Groningen, The Netherlands.

Wäber, K., Spencer, J. and Dolman, P.M. (2013) Achieving landscape-scale deer management for biodiversity conservation: The need to consider sources and sinks. *Journal of Wildlife Management* **77**, 726–736.

Walter, W.D., Leslie Jr., D.M. and Jenks, J.A. (2006) Response of Rocky Mountain elk (*Cervus elaphus*) to wind-power development. *American Midland Naturalist* **156**(2), 363–375.

Ward, A.I., White, P.C.L., Walker, N.J. and Critchley, C.H. (2008) Conifer leader browsing by roe deer in English upland forests: Effects of deer density and understorey vegetation. *Forest Ecology and Management* **256**, 1333–1338.

Wiersema, G.J. (1974) *Observations on the supplementary winter feeding of red deer on an estate in the Central Highlands of Scotland.* MSc thesis, University of Wageningen, The Netherlands.

Chapter 2

Valuing Ungulates in Europe

Sándor Csányi, Juan Carranza, Boštjan Pokorny,
Rory Putman and Mark Ryan

The European Parliament adopted a resolution on the EU 2020 Biodiversity Strategy on 20 April 2012. Part of this resolution urged a transition towards a resource-efficient and green economy, one which values natural resources, and their sustainable exploitation, as tabled previously by the European Commission (2011). Such natural resources include ungulates, which must be recognised as a core part of our natural capital.

While assessments vary, there are some 20 species of ungulates across Europe, representing an immense potential resource, whether considered in terms of biodiversity or economics (Apollonio *et al.*, 2010a). The more than 5.2 million animals harvested annually represent >120,000 tonnes of meat, and a potential hunting revenue of 100 million euros (Apollonio *et al.*, 2010a; Kenward and Putman, 2011). In addition, these ungulates have invaluable aesthetic, cultural, or even sacred, values depending on the diversity of cultural, historical and hunting traditions of European countries (see Chapter 12).

Wild ungulates are key elements for ecosystem processes (e.g. Smit and Putman, 2011; Murray *et al.*, 2013) as well as important assets for ecosystem services (Kenward and Putman, 2011; SEEA, 2013). However, the total value of ungulates may be difficult to capture. Economic or cultural impacts are not just associated with exploitation of different ungulate species for food or recreation; ungulates may also impose negative impacts through damage to forests and agriculture, damaging natural habitats (for review, see Reimoser and Putman, 2011; Barrios-Garcia and Ballari, 2012), as vectors of diseases (e.g. Ferroglio *et al.*, 2011), and through implication in wildlife–vehicle collisions (e.g. Langbein *et al.*, 2011). These negative impacts, whether ecological or economic, are easily understood; against those, however, there are several and significant economic and ecological benefits connected with the presence of ungulates in the environment at appropriate densities (again for review, see Reimoser and Putman, 2011; specifically for wild boar, see Pokorny and Jelenko, 2013). It becomes increasingly evident that a significant business associated with ungulates, for instance through big game hunting or tourism, provides a large contribution to the economy in many rural areas in Europe (PACEC, 2006a, b; Carranza, 2010; Putman and Watson, 2010; WFSA, 2010; Feuereisel, 2012).

Wildlife, when managed sustainably, is a part of a functioning renewable system characterised by a symbiotic relationship between natural resources and its users (Gilbert and Dodds, 1992; Decker *et al.*, 2001; Kawata, 2010). Classically, economists distinguish between:

1. *Consumptive uses* of wildlife (sometimes termed *extractive uses*), i.e. activities where the wildlife resource is exploited by removing a certain amount of either live or dead animals. The most important form of this type of use is hunting, while the reasons for that hunting vary from subsistence to commercial to leisure (recreation).
2. *Non-consumptive uses* of wildlife (sometimes termed *non-extractive uses*), which give value to wildlife without removing the resource. The non-consumptive use of wildlife is mostly based on the aesthetic value of animals. Wildlife becomes the support of the 'close-to-nature tourism' or ecotourism industry, as beaches are the support of the seaside tourism industry. This category of tourism is essentially based on wildlife viewing/watching and is almost entirely part of the service sector (Newsome *et al.*, 2005).

Together, consumptive and non-consumptive uses of wildlife generate billions of euros each year.

Wildlife-related recreation, whether consumptive or non-consumptive, is not only a widespread leisure activity but can also be a catalyst for economic growth, contributing to local economies through employment, increased economic output (money circulation) and tax revenue generation, all of which can act as motivating forces for conservation initiatives. The value of such contributions can be computed using scientific methodology from environmental economics (Hanley *et al.*, 2007; Edens and Hein, 2013). However, rigorous studies are lacking and the economic significance of ungulates is often only represented by numbers of animals harvested, the amount of meat produced, or financial values collected in hunting statistics (e.g. PACEC, 2006a, b; Carranza, 2010; WFSA, 2010; Putman and Watson, 2010; Feuereisel, 2012; Garrido-Martin, 2012; Vajai, 2012; Macaulay *et al.*, 2013). These studies tend to be rather localised/country-specific and in general do not employ consistent methodologies. In order better to understand the full range of economic benefits and the costs of wild ungulates and their management, more data are required, and crucially a more comprehensive system of evaluation is needed.

According to the TEEB approach (TEEB, 2010), valuation of ecosystem services is undertaken at three levels: recognising, demonstrating and capturing value. Values such as the benefits provided by ungulates must first be recognised by people, which depends on their understanding of, and attitude to, what ungulates may bring to their well-being. Having established these attitudes, the role of ungulates may then be demonstrated by an analysis of their benefits and costs to human well-being, which is important to inform subsequent decisions on management, conservation and sustainable use of populations. But finally, modern economics requires that environmental values be assessed under standardised scientific procedures which permit the integration of environmental accounting into the general economy in monetary terms.

Much work has been done during the last decades on environmental accounting (see e.g. SEEA, 2013). However, research on the valuation of environmental assets and incomes in monetary terms is very recent and our knowledge of the role of ungulates in this economy is still very rudimentary. As an initial step, it is necessary to summarise what information is currently available; it is particularly important to identify gaps in our knowledge of the benefits and costs of ungulates and their management. The purpose of this chapter is to identify what information is currently available, what are the main gaps in our knowledge and to provide some initial scoping of values at the three levels mentioned above: recognising, demonstrating and capturing value for ungulates under the framework of ecosystem services and environmental economics.

2.1 Recognizing value: Human attitudes toward large mammals/ungulates

Large mammals represent far more than simply biological entities and people continue to attach great meaning and importance to these animals. The cultural values attributed to large mammals are usually quite stable and deeply held features of human personality and society (Kellert and Smith, 2000). They greatly influence attitudes and behaviours affecting the conservation and exploitation of these animals but these values often depend on experience, learning and culture. Management of ungulates and their habitats frequently involves choosing among different values that may affect contrasting interests and needs. Information on human values can be relevant for a wide range of management objectives including allocation decisions, policy choices, damage assessments, mitigation strategies, conflict resolutions, and educational programmes, respectively (Kellert and Smith, 2000; Decker *et al.*, 2001). Various schemes have been developed to classify people's values of nature and wildlife. Concepts represented by each of these values can be summarised briefly as follows (Kellert and Smith, 2000):

- *Naturalistic value* focuses on the personal pleasure and satisfaction derived from direct experience and contact with large mammals in their natural habitats. This experience occurs through a variety of outdoor recreational activities, such as hunting (consumptive use) and wildlife observation (non-consumptive use). Hunting in different European countries is pursued by 0.5–5% of the human population, with motivations closely connected to the activities or challenges in natural conditions (Lecocq and Meine, 1995). Outdoor recreation is often associated with the chance to see large animals and the presence of ungulates can be a major motivating force and add significantly to the pleasure derived from the outdoors experience, even in the absence of actually observing them (Kellert and Smith, 2000).

• *Scientific value* emphasises the empirical study and understanding of these animals, and a scientific perspective can foster intellectual growth and development. Ungulates and other large mammals have been particularly instrumental in strengthening a scientific perspective of nature.

• *Aesthetic value*: Large mammals are among the most aesthetically valued wild creatures; reflecting this, many ungulates are perhaps the most prominently featured animals in pictures, posters, toys, paintings, sculptures, film, cartoons, and other visual media. Since the times of the cave paintings, in their idealised form or state, these species can suggest a model of harmony, majesty and power. In the modern world, the emotional significance of them has been transformed, especially among urban dwellers and higher socio-economic groups, into a highly positive aesthetic view (Kellert and Smith, 2000).

• *Utilitarian value* emphasises the practical and material importance of nature. Ungulates have long provided people with a wide variety of tangible benefits. Food is among the most important of these benefits and, even today, a significant proportion of hunters chase these animals primarily for their meat value. Some of the utilitarian values are less known but wild ungulates can even provide various medical benefits. Antlers and studies of the genetic regulation of antler growth may open new perspectives for pharmaceutical development (Bubenik, 1990; Borsy *et al.*, 2009; Stéger, 2011; Landete-Castillejos *et al.*, 2012). Ungulates are also exploited for their skins, pelts, horns and other body parts for use in clothing and decorative products.

• *Humanistic value* underlies the emotional affinity people have for nature, particularly for other creatures. Large ungulates are often the subject of strong affection and emotional interest. Identification with the presumed emotional and mental experience of individual animals represents a critical basis for a humanistic perspective of nature. However, this emotional identification can also result in dysfunctional anthropomorphism (e.g. the so-called 'Bambi' syndrome) (Kellert and Smith, 2000).

• *Dominionistic value* emphasises the human inclination to subdue, dominate and master nature (Kellert, 1978; Kellert and Smith, 2000). From a dominionistic perspective, nature is valued as an arena of contest and control. People achieve physical strength, mental fortitude and associated feelings of self-reliance and self-confidence by demonstrating an ability to function effectively under difficult and trying circumstances. Historically, hunting has provided this opportunity for contest and challenge in the wild (Hull, 1964; Myrberget, 1990; de Ferrieres, 2006), but today this experience is increasingly achieved through non-consumptive pursuits.

• *Moralistic values* emphasise a sense of ethical and moral responsibility for conserving, protecting, and properly treating nature and animals (Kellert and Smith, 2000). For example, a code of sportsmanship and ethical restraint in hunting wild animals serves to protect and restore the health of associated habitats and ecosystems. This moral code has encouraged many hunters to act with restraint and respect despite immediate practical advantages (Belházy, 1903; Békés, 2012). This sense of moral and spiritual connection has been facilitated by a strong affinity and respect for large ungulates, like red deer (Kowalsky, 2010).

• *Negativistic values*: Nature also serves as a powerful source of human fears and anxieties. Fear and avoidance of wildlife can sometimes assume irrational proportions, the avoidance of injury, harm, and death is a basic characteristic of all organisms (Kellert and Smith, 2000). Humans often behave naively, ignore their continuing vulnerability and the lack of fear or of maintaining a healthy distance from wild animals can result in injuries of visitors to game parks or in dangerous encounters with wild boar or deer in urban environments (see Chapter 7).

• *Symbolic value*: Large mammals have been especially prominent in our symbolic uses of nature, for example in historical tales, ancient myths, stories, legends, images, children's books, as well as in modern advertising and marketing (Kellert and Smith, 2000).

The above typology shows that human values toward large mammals/ungulates are diverse and their quantification is problematic. However, we can to an extent simplify this diversity and consider wildlife values and services under six general categories (according to Gilbert and Dodds, 1992):

• *Commercial values*: incomes derived from the sales of wild animals or their products or from direct and controlled use of wild animals and their progeny.

• *Recreational values*: values associated with the pursuit of wildlife in connection with sport and recreation, whether this is associated with consumptive or non-consumptive use (direct hunting, but also wildlife watching, wildlife photography, etc.).

• *Biological values*: the worth of services rendered to humans by wild animals (sometimes called ecosystem services).

• *Social values*: values accruing to human communities from the use of wildlife and values associated with structures that exist because of the common interest in wildlife.

• *Aesthetic values*: values of animals and places (with animals) possessing beauty, affording inspiration, contributing to arts, etc.

• *Scientific values*: values realised through the use of wildlife as a means of investigating certain natural phenomena.

Chardonnet *et al.* (2002) give an overview of wildlife values in a system where these values are grouped into *direct* and *indirect value* categories:

• Direct values:
  • consumptive use value: non-market value of game
  • productive use value: commercial value of game.

- Indirect values:
  - non-consumptive use value: scientific research, wildlife watching, etc.
  - option value: value of maintaining options available for the future
  - existence value: value of ethical feelings of existence of wildlife, cultural value, sacred or religious value.

Whichever system may be considered, all these values carry different weights and, importantly, these will vary according to the respective interests of the stakeholders involved. Virtual values, such as ethical value – although important – are not as powerful in terms of providing justifications for conserving wildlife when compared with pragmatic ones, such as economic values. Financial profitability, economic yield and environmental sustainability are often dominant values for high-level decision-makers as well as for grass-root level individuals who live in close proximity to wildlife (Chardonnet *et al.*, 2002).

In addition to the diversity of values, it is common to encounter interactions, either synergistic or antagonistic, between different types of values that further complicate their consideration, especially when making quantitative analyses. For example, aesthetic, humanistic and moralistic principles may be at the root of naturalistic, non-consumptive values that contribute to conservation and also to the generation of income based on tourism. Use through hunting generates business and provides utilitarian products, and also material for scientific research, but may be less compatible with the attraction of visitors that seek truly wild areas and feel uncomfortable with the idea of hunting. Non-consumptive values themselves may appear the most difficult to quantify, but it has been shown for instance that they significantly influence the price of land (e.g. Campos *et al.*, 2014).

2.2 Demonstrating value: Non-consumptive and consumptive use

2.2.1 Values and uses of wildlife in contemporary societies

Native and introduced ungulate species carry all kinds of values as outlined above. These values have changed over time and their actual weights can only be evaluated on the basis of ecological, historical, societal and traditional elements (Apollonio *et al.*, 2010a; Chapter 12). The most important elements shaping hunting and game management in a given country are historical and cultural traditions together with economic and legislative constraints. These kinds of cultural differences and similarities allow the classification of various European countries on the basis of the perspective of both hunters and the general public in relation to hunting, game management and related issues (Putman *et al.*, 2011). To evaluate the importance and the value of ungulates in different European countries it is necessary to have this perspective and to understand that current legislation, attitudes to ungulates, as well as the relative importance of different direct or indirect values are permanently changing as the society and the economics of a country change. The cornerstones of management are closely connected to the traditions of hunting, with sustainability and management actions being strongly determined by this framework (Putman *et al.*, 2011).

The wildlife values listed above include categories that either do not enter a market (or national) economy or cannot readily be estimated in comparative market terms. However, information about economic values is more and more important for managers and others working with wildlife resources (Gilbert and Dodds, 1992; Decker *et al.*, 2001). When looking for examples of wildlife valuation, the North American systems seem to be the best developed and most supported by quantitative information (Filion *et al.*, 1988; US Fish and Wildlife Service [US FWS], 2007, 2012). In these countries the direct and indirect values of wildlife were already identified in the 1940s and the efforts to measure cultural and intangible values connected to wildlife have an especially long history (Leopold, 1943; Bailey, 1984).

Today, outdoor recreation allows people to participate in activities that enrich their lives through direct and indirect interactions with wildlife. These opportunities give various environmental, social, educational, spiritual, health and cultural benefits (Kellert and Smith, 2000; Outdoor Foundation, 2012). Meeting with large mammals, like ungulates in the wild, increases the quality of the experience. The importance of this is shown in statistics of the US Fish and Wildlife Service (2002, 2007, 2012). Recreational hunting of, and watching of, large mammals are a significant pursuit of many Americans and a substantial contributor to the national economy.

In 2011, for example, 13.7 million people, 6% of the US population 16 years of age and older, went hunting. Hunters in the United States spent an average of 21 days pursuing game species. Ungulates like wapiti, white-tailed deer and moose attracted 11.6 million hunters (85%) who spent 212 million days in the field. Hunters spent 34.0 billion US dollars[1] on trips, equipment, licences and other items to support their hunting activities in 2011. The average expenditure per hunter was $2484. Total trip-related expenditures comprised 31% of all spending at $10.4 billion. Other expenditures, such as licences, stamps, land leasing and ownership, and plantings totalled $9.6 billion, 28% of all spending. Spending on equipment such as guns, camping equipment and four-wheel drive vehicles comprised 41% of spending, with $14.0 billion. Overall hunting participation increased by 9% from 2006 to 2011 in the United States (US FWS, 2002, 2012).

Wildlife watching is also a popular pastime for millions of people in the United States. Nearly 71.8 million people, aged 16 years or older, fed, photographed and observed wildlife in 2011. They spent $55.0 billion on their activities. Wildlife recreation is not only an important leisure activity but it is also a catalyst for economic growth. Hunters, anglers and wildlife watchers spent $145.0 billion on wildlife-related recreation in 2011. This spending contributed to local economies throughout the country, which added to employment, raised economic output and generated tax revenue (US FWS, 2012).

---

[1]In this and subsequent figures, values are in US billion; i.e. 1,000 million.

Compared to the wealth of information for North America, studies evaluating the consumptive and non-consumptive values/use of wildlife in countries in Europe (including ungulates) are scattered and use various approaches, data and methods (Chardonnet *et al.*, 2002; ESUSG, 2004; MacMillan and Phillip, 2008; Putman and Watson, 2010; Vajai, 2012). During the last few decades, a number of studies have been published but these have tended to focus upon particular elements of the whole: the contribution of European hunters to economies (Pinet, 1995), an economic analysis of ungulate management in Hungary (Vajai, 2012), the economic valuation of hunting in the Czech Republic (Feuereisel, 2012), a review of the use of wild living resources in the UK (Murray and Simcox, 2003; ESUSG, 2004) and a more specific assessment of consumptive and non-consumptive values of wild mammals in Britain (MacMillan and Phillip, 2008), an assessment of the contribution of deer shooting to the UK economy (BASC, 1997; Olstead, 2006; PACEC, 2006a, b) and an evaluation of the economic benefits and costs of wild deer and their management specifically in Scotland (Putman and Watson, 2010).

In the majority of European countries, only elementary, crude and incomplete data are available for the evaluation of any wildlife values in general, or ungulates as a specific group. Some measures can be extracted on the basis of game management statistics (particularly number of animals harvested according to species, supported in some cases with data reflecting their utilitarian/commercial value, such as body mass or trophy quality), number of hunters, price information related to shooting (license cost, shooting fees, etc.), income from animals sold (number or weight by species) and related business activities (e.g. ammunition sold). However, this information needs to be harmonized in order to develop consolidated databases and in any case addresses only consumptive parts of wildlife utilisation.

2.3 Non-consumptive use values of ungulates within a European context

2.3.1 Aesthetic and cultural value (sociocultural value)

Ungulates are well-known, sometimes iconic, species and people acknowledge their presence. The importance of ungulates becomes evident in place names, and family, regional, national and state symbols, as well as in regular media coverage. The cultural value can sometimes be called a *sacral value*, as in the case of Hungarians. According to the Hungarian legend of ancestry, the two forefathers of the Magyars and the Huns were chasing a 'wonder deer' when it lead them to a new mother country (Jankovics, 2004). For further exploration of the importance of this cultural value throughout history, see Chapter 12.

2.3.2 Recreational value (non-hunting)

Non-consumptive use of ungulates is mostly based on the aesthetic value of wildlife. Wildlife becomes the support of the recreation tourism industry and ecotourism, and wildlife viewing or photography is the typical expression of this. The economic contribution of these activities is almost always underestimated because non-consumptive uses and related employment are under-represented and because many small-scale uses and user groups are not adequately documented (ESUSG, 2004).

In Hungary, small game parks are typical cases where red deer, fallow deer, roe deer, mouflon and wild boar are presented for visitors. As a new kind of experience, small 'viewing areas' (of the order of 5–20 ha) are established within areas more generally used for hunting where visitors can watch wild animals from high seats, hides and other facilities. The purpose of these parks is to reduce the number of visitors in the open areas where they can disturb wild animals. In Slovenia, recreational use of ungulates is almost neglected and non-documented; no information exists on the overall contribution of hunting tourism to the national economy. But in a similar way one hunting ground (managed by the Slovenia Forest Service) now offers the possibility for deer watching and photography (the price for up to three persons is 150 euros), but there has been almost no interest from visitors over the last 3 years. On the other hand, on the same hunting ground, interest in watching brown bears (*Ursus arctos*) has been increasing, and in 2013 managers organised this activity for a total of 47 groups (Marinčič, Head of Hunting Ground with Special Purpose Jelen Snežnik, *pers. comm.*, 29 January 2014).

In Spain, wild ungulates constitute an important attraction for many natural areas. In some natural parks that attract many visitors, people expect to have the experience of watching wild ungulates in their natural habitats. As an example, the largest Natural Park of Spain is the Sierra de Cazorla Segura y Las Villas (Jaen province, in Andalusia), which covers more than 200,000 hectares and includes native Iberian ibex, red deer and wild boar, and introduced fallow deer and mouflon. This park receives 400,000 visitors per year, and accommodation capacity is over 11,000 beds in hotels and rural houses, and there are 247 small companies in the area with activities relating to ecotourism (Sánchez-Morales, 2008).

A recent report commissioned by Scottish Natural Heritage (Bryden *et al.*, 2010) concluded that the total visitor spend attributable to nature based tourism was £1.4 billion with an associated income from employment of £0.8 billion.[2] Of that total, 'wildlife watching' generated an economic contribution of £138 million (approximately 166 million euros) and supported 3943 FTEs (Full-time Employment Equivalents). By comparison the totals for consumptive field sports were £147.3 million (177 million euros) and 4209 FTEs. A separate report by the Deer Commission for Scotland ('*Challenges and opportunities of deer watching as a commercial activity: a critical review*') which looked specifically at 'deer watching' as the sole focus of an activity or included as part of a wider 'package' of wildlife watching, could only identify 40 potential 'deer watching' providers. Of the 14 respondents who offered dedicated deer-watching activities, 43% described the demand for these activities as high, 29% described as medium demand and 29% felt the demand was low. Participants were asked to provide information about the

---

[2]Equivalent to 1.7 billion euros and 1 billion euros, respectively.

numbers of people taking part in deer watching activity throughout the year and about the average spend per head. Using these figures Putman and Watson (2010) estimated that revenue generated by these enterprises is approximately £107,000–£113,000 annually (equivalent to some 130–135 thousand euros).

One element of tourist activities revolving around ungulates is the observation of, or even only listening to, red deer roaring during the rutting season in September and October. In an increasing number of places throughout Europe, there are accommodation sites and ecotourism companies that offer this seasonal nature experience. Unfortunately there are no studies that quantify the economy surrounding this natural service, but it is clearly increasing (see e.g. Drábková, 2013, for Cansiglio forest, Italy).

2.3.3 Scientific value, biodiversity value or biological value

Ungulates are important elements of ecosystems, and their effects on ecosystem processes are fundamental – with impacts often disproportional to actual abundance (McShea *et al.*, 1997; Reimoser and Putman, 2011; Smit and Putman, 2011). These effects include direct impacts on vegetation systems through grazing, browsing and rooting by wild boar, as well as seed dispersal and deposition, increasing soil productivity, nutrient transport and recycling (for review, see Reimoser and Putman, 2011). Ungulates are also very important prey species for large carnivores, particularly wolf (*Canis lupus*) and lynx (*Lynx lynx*), for which they represent predominant foodstuff of animal origin (e.g. Jedrzejewska and Jedrzejewski, 1998). This contribution to ecosystem processes is referred to as the biological value, and the values to humans are the services and products derived from the ecosystems.

While damage to agriculture or forests caused by free-ranging ungulates have been discussed and debated for many years, the effects on ecological processes are a rather more recent area of scientific study (Ripple and Beschta, 2012; Muhly *et al.*, 2013). Effects on vegetational systems and the underlying nutrient dynamics have been explored by for example Putman (1986) or Reimoser and Putman (2011). Understanding the cascading effects of large ungulates is especially important for the conservation of the ecosystems that these species inhabit (Milner *et al.*, 2013; Mathisen *et al.*, 2014). Although (or probably because) large predators can reduce ungulate populations and their effects on the environment (e.g. Beschta and Ripple, 2009; Hebblewhite and Smith, 2010), many European hunters oppose the comeback of wolf or lynx as they could reduce hunting opportunities as well as incomes related to ungulates (Szabó *et al.*, 2000). Nevertheless, although the importance of free-ranging ungulates for the existence of large carnivores is well-recognized and clearly stated also in strategic documents aimed towards conservation of these endangered species (e.g. for Slovenia: *Action Plan for Wolf Conservation*; Vlada RS, 2013), no single study has been made so far which attempts to quantify the economic value of these ecosystem services provided by ungulates.

As a part of their *biological values*, ungulates also carry *genetic values*. Our most important domestic animals are descendants of close relatives of the ungulate species living in Europe (Bartosiewicz, 2006; Roots, 2007). Wild ungulates are valuable genetic reserves for biodiversity and saving (the protection) of their genetic diversity is vital, as they represent an important resource for future generations. In this regard, there are potential risks associated with intensive ungulate management. In Chapter 6, Atle Mysterud explores the possible genetic consequences of selective harvesting (by age or sex), whether such selectivity is deliberate or inadvertent. But in addition to the consequences of selective harvesting, there have been widespread attempts over the years to alter the genetic status of native wildlife populations in order to 'improve' trophy quality.

In red deer farms and wild boar farms (or hunting parks) the breeding and genetic manipulation of animals is commonplace (Miller, 2012). The purpose of ungulate farming is, in some cases, to produce game meat and other products (e.g. velvet antler of deer), although in other cases it is to provide live animals to be released in hunting estates. Different breeding programmes can be used to improve the product quality or the productivity, and the breeding stock can be selected and/or hybridised with animals belonging to various subspecies/species (Carranza, 2010). Where animals escape from such parks or are deliberately released into wild populations, such manipulations may have a significant impact on the genetics of the native wild stock and may contaminate the local gene pool (genetic introgression, 'swamping'), leading to hidden extinctions of native lineages (Carranza *et al.*, 2003).

For example, red deer from Northern and Eastern Europe have been introduced into Spain, to be crossed with native Iberian red deer in farms or under controlled conditions in hunting estates, to produce hybrid, big antlered stags for the hunting market (Carranza *et al.*, 2003). Feral sika deer have hybridised with native red deer in the Czech Republic and in Scotland, affecting its appearance and the size of some measurable traits (e.g. Senn *et al.*, 2010). Such hybridisations may also have impacts on the population dynamics of ungulates. For example, hybridisation with domestic pigs with higher reproductive rate is supposed to be one of the main factors responsible for increasing litter size and a subsequent rapid increase in wild boar numbers in eastern Croatia (Šprem, 2009). Similarly, in south-western Slovenia, hybridisation of wild boar (*Sus s. scrofa*) with much smaller and presumably different subspecies escaping from an enclosure in northern Italy (near the Gulf of Trieste, in the close vicinity of the state border) is predicted to be one of the key reasons for extremely fast population growth during the last couple of years (Sila and Koren, 2011). The occurrence and effects of such genetic alterations of wild populations are increasingly being detected with modern methods (e.g. Lorenzini and Lovari, 2006; Royo *et al.*, 2007; Skog *et al.*, 2009; Valvo *et al.*, 2009; Scandura *et al.*, 2011; Fernández-García *et al.*, 2014). It is necessary to underline that in such cases not only the genetic value of a given species might be damaged but the conservation or diversity values are also put at risk.

Along with hybridisation, artificial selection of hybrids or even of native strains subjected to selective breeding programmes is also becoming a real, growing problem where these animals are being introduced into natural populations. Animals kept in captivity can be manipulated by traditional or modern breeding methods to produce animals which would otherwise not exist in nature (animals such as frankendeer; Kessler, 2012). Social issues related to confinement of ungulates for sporting purposes include cultural and legal issues of hunter's ethics, public perception of hunting, commercialisation, and illegal manipulation and fraud of wild animals or animals raised in breeding facilities (Jones, 2012), but the underlying issue we would wish to highlight here is the potential impact on intrinsic biological values. In order to protect the genetic purity of free-ranging ungulates (as well as the 'fair chase' principle), international wildlife organisations have made clear statements on this (CIC, 2006, 2011; TWS, 2009, 2012).

However, while biological arguments may advocate for more natural management principles, in practice it is difficult to achieve this simply through regulatory constraints. This problem is being faced in southern Spain under a new system of positive incentives. Genetic introgression is controlled by a genetic test that is applied to red deer trophies when they are submitted to the Spanish Trophy Commission where they are measured and catalogued according to the criteria of the International Council for Game and Wildlife Conservation (CIC). If genetic tests indicate that some trophies do not belong to pure Iberian red deer specimens, then the Spanish Commission will reject them as candidates to be included in the Spanish records (Carranza *et al.*, 2003). This measure has resulted in a decrease in the value of hybrids as trophies and is producing the desired effect i.e. a reduction in the introduction of foreign deer in Spain. At the same time, some public administrations have started using the genetic tests as a prerequisite for authorising translocations, and some owners ask for the genetic tests to be applied to samples of their live animals in order to implement programmes aimed at purging hybridisation from their populations.

In addition, a 'quality of game management system' is being implemented by regional administrations, aimed at promoting adequate types of management more compatible with conservation. Under this system, landowners can apply for a certificate of quality awarded by the Regional government after a formal evaluation. The main benefits for owners are the access to funds oriented towards conservation and the launch of their products on the market, together with a quality brand (Carranza and Vargas, 2007).

All those measures aimed at ensuring that the exploitation of game is compatible with conservation may contribute to maintaining the environmental value of ungulates, including non-consumptive value but also allowing sustainable consumptive use.

2.4 Consumptive use values of ungulates

Traditionally, hunting of ungulates for meat, hides or sport was the most important use of wild ungulate populations and this remains true today. Depending on the type of hunting system which applies in different European countries (see Putman, 2011), the lease of hunting areas, the sale of hunting licences, trophy fees and/ or the meat (and other body parts of ungulates) sold, can generate income for the landowner or the hunting area. In most instances, the actual value of these consumptive economic values can be measured more directly than estimates of non-consumptive values. Nevertheless, data on the economic values of ungulates are also rather lacking in Europe, and only in those countries where up-to-date hunting statistics do exist can the economic outcome of ungulate management be assessed on the basis of known culling quotas, amount of venison produced, trophy qualities and so forth.

For example, in Slovenia (where the economic issue of game management is of rather marginal importance due to a specific hunting system; for review, see Adamič and Jerina, 2010), 47,000 ungulates (31,773 roe deer, 4,458 red deer, 168 fallow deer, 2,270 chamois, 473 mouflons, and 7,858 wild boar) were harvested annually in the period 2009–11 (Oslis, 2014), which provided 2.8 million euros of income to hunters directly through the sale of the venison. However, much higher incomes were earned by meat producers and dealers; these incomes could be increased by a significant order of magnitude if there were an established tradition associated with the preparation of game meat as a delicacy. Apart from meat production, trophies may represent another important consumptive value of ungulates, although their economic value is almost completely neglected in the majority of Slovene hunting grounds, and hunters are allowed (as a means of compensation for their voluntary work) to shoot male deer free of charge regardless of the trophy quality. Nevertheless, it should be mentioned that in the last 3 years (2011–13) on these hunting grounds, managed by non-professional hunters, 1,042 harvested roe bucks exceeded the CIC bronze, silver and gold medal thresholds (712, 262 and 68 antlers, respectively), with a maximum of 167.30 CIC points for the highest scoring trophy. In the case of red deer, 507 stags would have received bronze, silver or gold medals during the same period (377, 115 and 15, respectively) with a maximum of 240.00 CIC points for the highest scoring trophy (Lisjak, 2014). According to the price list of the *Slovene Hunting Association*, the trophies of these medal-quality red deer stags alone would cost a total of 1.36 million euros were they sold on the market; similarly, trophies of roe deer bucks would cost 0.91 million euros.

The Spanish National Association of Hunters (*Real Federación Española de Caza*) has carried out annual estimates of the number of animals hunted from all game species in Spain and the economic impact of hunting (Garrido-Martín, 2012). In the hunting period 2010–11, over 400,000 ungulates were harvested in Spain, including 222,658 wild boar, 130,759 red deer, 29,975 roe deer, 5,024 Iberian ibex, 1,161 chamois, 12,059 fallow deer, 9,251 mouflon and 697 Barbary sheep (arrui). By estimating the amount of money that hunters paid for the hunting rights for these animals plus the value of the meat obtained, in both cases with prices for 2012, the author calculated a total figure of 302.71 million euros.

During the last half century, the actual numbers and harvested numbers of the most important ungulates (e.g. red deer, roe deer, wild boar, moose and fallow deer) have increased in Europe, and as much as a 10-fold increase of the harvest of these species (in total) can be documented (reviewed in Apollonio *et al.*, 2010a; Deinet *et al.*, 2013). The larger population sizes and greater ranges have increased the big game hunting opportunities for hunters as these species become more readily available for shooting in terms of both numbers and distributional range. On the basis of such figures, the following points can be drawn:

- Ungulate hunting has become available to more European hunters and in larger geographical regions; therefore, big game hunting has become more popular.
- The larger populations might allow an increase in rental fees for hunting grounds and more investment in game management for a long-term return of investments.
- Increasing populations of ungulates have influenced the prices of shooting licenses, probably by increasing prices in better areas and reducing them in poorer hunting grounds.
- As a consequence of the increasing ungulate harvest, more game meat is available, which has reduced its price in the market; however, when compared with the past, income from the sale of venison could become more important than shooting and/or trophy fees.
- With small game populations having shown a declining pattern during the last half century, ungulate hunting has at least partially replaced small game hunting in many areas of Europe.
- Hunting tourism has become much more focused on big game (ungulate) hunting, but the market is limited and oversupplied.
- In the last decades the number of hunters has declined in many European countries (see also Chapter 12), and this might contribute (together with environmental changes and favourable management policy in many countries) to the continued increase of ungulate populations.

This kind of change is closely linked to the changes in small and big game harvests in many areas of Europe, where the most important small game species have been showing a continued decline in numbers. For example, in Hungary, brown hare, ring-necked pheasant, grey partridge, mallard and also waterfowl numbers have declined since the 1960s and 1970s, while populations of five ungulate species have continued to expand, with roe deer taking over the large agricultural plains during the 1960s and 1970s (Farkas and Csányi, 1990) and with wild boar and red deer also extending their range into agricultural regions in the 1980s (Csányi, 1991, 1995). These changes are reflected in the harvest data for small game and big game, when these are pooled by decade between 1960 and 2012 (Table 2.1).

While during the period 1960–69 the average annual harvest of small game was 686,000, the same average during the period 2000–12 was 626,000, meaning a 0.9-fold decrease in the number of small game harvested. During the same period, the number of hunters increased 2.8-fold. Overall, the average hunting bag of small game per hunter decreased 0.32-fold, from 35.6 harvested small game/hunter between 1960 and 1969 to 11.4 between 2000 and 2012. By contrast, the average annual harvest of big game between 1960 and 1969 was 20,500 head; the same average for the period 2000–12 was 221,700, meaning a 10.3-fold increase. Between these two periods, the average culling bag per hunter increased 3.65-fold, from 1.1 harvested big game/hunter between 1960 and 1969 to 4.1 between 2000 and 2012. A similar situation can be depicted for other countries.

Table 2.1 Changes of small game and big game harvest in Hungary.

Decade	Mean harvest	Change compared to 1960–1969	Mean number of hunters	Change compared to 1960–1969	Harvest/ hunter	Change compared to 1960–1969
Big game harvest						
1960–1969	21,492		19,284		1.11	
1970–1979	80,452	3.74	25,318	1.31	3.18	2.85
1980–1989	101,995	4.75	33,124	1.72	3.08	2.76
1990–1999	116,391	5.42	45,275	2.35	2.57	2.31
2000–2012	221,666	10.31	54,811	2.83	4.06	3.65
Small game harvest						
1960–1969	686,214		19,284		35.58	
1970–1979	1,341,326	1.95	25,318	1.31	52.98	1.49
1980–1989	1,244,280	1.81	33,124	1.72	37.56	1.06
1990–1999	845,233	1.23	45,275	2.35	18.67	0.52
2000–2012	626,478	0.91	54,811	2.84	11.43	0.32

These changes are not only numerical but they also resulted in a complete change in the values and management attitudes of game managers and hunting. The Hungarian landscape was historically dominated by agricultural lands where small game species were extraordinarily abundant and available for hunters (Csőre, 2000). Big game (ungulate) management was focused on red deer, and ungulates were available only for a limited number of hunters. Trophies of ungulates received great recognition and a protective hunting culture determined their management (Csányi and Lehoczki, 2010; Rivrud *et al.*, 2013); big game shooting was a rather privileged activity and restricted to a few Hungarian and paying foreign hunters (Csányi and Lehoczki, 2010). However, as populations of small game have declined and ungulate populations have increased, management of and attitudes towards the hunting of these species has been altered significantly. Changes in numbers of small game species and the rapid expansion in numbers and distribution of ungulates documented above meant that by the early 1990s a large part of the traditionally small game management areas lost their small game populations and turned their efforts towards big game management (Csányi and Lehoczki, 2010). Changes in attitudes were also motivated by the anticipated better income opportunities from big game, as trophy animals also provide a significant amount of venison.

On the basis of the management statistics, it is possible to follow how the relative contribution of these different elements to the overall income resulting from game management changes over time (Table 2.2). For example, before the political and economic changes in Hungary in 1989–90, the most important source of game management income was the payments of foreign trophy hunters. At that time >50% of the income came from western European hunters visiting Hungary to shoot either red deer stags or roe deer bucks. After 1990, this gradually changed and the proportion of income deriving from foreign hunters decreased from 49.9% (1994) to 22.6% (2012). At the same time, payments by domestic hunters increased, as did their share of the market value, which increased from 3.0% (1994) to 19.0% (2009). The relationships between ungulate densities and the financial items (income/cost) can also be presented on the basis of management statistics (Table 2.3 and Figure 2.1). Although some of the values show high variation among counties, several important trends can be identified:

- In 2012, the income deriving from fees paid by foreign hunters showed a reasonable correlation with the density of ungulates per km^2 ($r^2 = 0.426$). The best red deer and wild boar areas of North West Hungary are receiving the highest incomes from these hunters.
- In 2012, the income originating from the payments of the domestic hunters showed a weak correlation with the density of ungulates per km^2 ($r^2 = 0.289$). This may be a result of the fact that paying domestic hunters are normally shooting younger, smaller trophy males or females and young ungulates (Rivrud *et al.*, 2013).

Table 2.2 Incomes of game management and hunting in Hungary between 1994 and 2012 (units: 1000 Hungarian forint).

Names			1994		1998		2002		2008		2010		2012	
Incomes	Foreign	Shooting fees	2,581,456	49.9%	4,165,579	43.0%	5,980,815	41.7%	4,678,605	32.7%	4,246,435	25.9%	4,579,856	22.6%
		Services	391,743	7.6%	650,659	6.7%	567,313	4.0%	403,303	2.8%	481,312	2.9%	629,170	3.1%
	Domestic	Shooting fees	153,424	3.0%	854,331	8.8%	1,622,539	11.3%	2,359,886	16.5%	2,993,971	18.3%	3,068,374	15.1%
		Services	108,223	2.1%	247,168	2.6%	453,061	3.2%	410,415	2.9%	575,486	3.5%	776,895	3.8%
	Income from	Live game sold	326,715	6.3%	577,129	6.0%	662,562	4.6%	594,805	4.2%	470,130	2.9%	796,083	3.9%
		Shot game sold	761,473	14.7%	1,609,740	16.6%	2,632,691	18.4%	2,337,119	16.3%	3,090,821	18.9%	5,265,569	26.0%
		Others	849,524	16.4%	1,574,276	16.3%	2,412,607	16.8%	2,688,175	18.8%	3,868,270	23.6%	4,272,713	21.1%
		Subsidies							836,502	5.8%	643,550	3.9%	894,878	4.4%
	Total		*5,172,558*	*100.0%*	*9,678,882*	*100.0%*	*14,331,588*	*100.0%*	*14,308,810*	*100.0%*	*16,369,975*	*100.0%*	*20,283,538*	*100.0%*
Spending	Labour cost		821,941	17.4%	1,700,622	18.9%	2,871,823	20.5%	3,273,619	22.9%	3,819,813	23.9%	4,457,278	23.6%
	Game management cost		2,048,559	43.4%	4,217,793	46.8%	6,200,147	44.2%	6,138,660	43.0%	6,389,493	40.0%	7,749,057	41.0%
	Compensation	Aricultural damages	389,114	8.2%	730,805	8.1%	1,618,354	11.5%	1,412,025	9.9%	2,141,876	13.4%	2,511,674	13.3%
		Forest damages	109,410	2.3%	78,135	0.9%	181,653	1.3%	199,282	1.4%	141,059	0.9%	136,898	0.7%
	Others		1,354,735	28.7%	2,276,999	25.3%	3,166,265	22.6%	3,267,221	22.9%	3,480,205	21.8%	4,053,517	21.4%
	Total		*4,723,759*	*100.0%*	*9,004,354*	*100.0%*	*14,038,242*	*100.0%*	*14,290,807*	*100.0%*	*15,972,446*	*100.0%*	*18,908,424*	*100.0%*
Balance			448,799		674,528		293,346		18,003		397,529		1,375,114	

Table 2.3 Ungulate (big game) harvest and incomes/spending of game management in Hungary at the level of the 19 counties in 2012 (financial unit: euro).

Code	County	Harvest data					Total big game harvested
		Red deer	Fallow deer	Roe deer	Mouflon	Wild boar	
		Number of animals harvested					
14	Somogy	8,168	2,783	4,273	53	15,037	**30,314**
5	Borsod-Abaúj-Zemplén	2,186	24	5,517	668	15,605	**24,000**
19	Veszprém	5,063	628	4,917	518	12,553	**23,679**
20	Zala	5,186	110	4,873	40	12,701	**22,910**
13	Pest	1,918	996	6,036	461	12,357	**21,768**
8	Győr-Moson-Sopron	4,344	291	7,370	54	8,833	**20,892**
2	Baranya	4,477	396	3,242	27	12,307	**20,449**
18	Vas	4,239	185	4,282	13	10,886	**19,605**
17	Tolna	4,067	2,797	3,402	0	8,827	**19,093**
3	Bács-Kiskun	1,734	864	8,378	0	5,584	**16,560**
12	Nógrád	2,128	236	1,966	346	11,437	**16,113**
10	Heves	2,097	27	3,398	1,164	8,440	**15,126**
15	Szabolcs-Szatmár-Bereg	274	385	7,668	53	4,757	**13,137**
11	Komárom-Esztergom	2,364	408	1,866	408	7,186	**12,232**
7	Fejér	1,329	467	3,461	197	6,154	**11,608**
4	Békés	71	1,204	7,478	0	1,115	**9,868**
9	Hajdú-Bihar	82	522	5,140	0	3,305	**9,049**
6	Csongrád	31	139	7,063	0	1,106	**8,339**
16	Jász-Nagykun-Szolnok	0	10	6,244	0	1,067	**7,321**
	Total	**49,758**	**12,472**	**96,574**	**4,002**	**159,257**	**322,063**

Code	County	Income						
		Foreign hunters		Domestic hunters		Live game	Income from sold	Total income
		Shooting fees	Services	Shooting fees	Services		Shot game	
14	Somogy	1,984,524	466,124	1,250,200	599,928	1,638	2,377,552	8,661,010
5	Borsod-Abaúj-Zemplén	161,241	8,493	942,948	194,362	33,238	635,548	3,158,503
19	Veszprém	1,627,241	164,734	367,948	180,003	78,348	1,586,179	6,088,890
20	Zala	1,045,607	155,662	1,436,652	186,938	0	1,697,821	5,765,890
13	Pest	891,455	112,752	719,528	177,534	11,438	842,797	3,820,107
8	Győr-Moson-Sopron	1,444,776	131,569	435,090	86,883	1,724	1,289,655	4,205,417
2	Baranya	1,111,983	94,041	311,269	59,976	15,369	1,518,776	3,820,534
18	Vas	1,703,724	530,803	128,655	42,386	46,793	1,415,586	4,383,603
17	Tolna	1,322,859	142,445	732,855	59,572	22,900	1,359,210	4,745,324
3	Bács-Kiskun	467,593	38,900	673,014	116,869	46,059	811,448	3,033,483
12	Nógrád	159,852	7,769	240,786	144,283	63,872	758,390	2,179,607
10	Heves	161,690	120,355	520,124	427,283	14,586	630,379	2,488,521
15	Szabolcs-Szatmár-Bereg	400,803	8,083	257,638	30,217	20,455	633,848	2,054,266
11	Komárom-Esztergom	373,917	17,859	884,776	88,028	21,393	649,721	2,502,345
7	Fejér	701,421	61,179	695,110	46,190	22,028	611,245	2,798,166
4	Békés	413,545	16,690	316,586	29,103	1,212,145	387,531	3,273,948
9	Hajdú-Bihar	754,338	42,414	252,376	31,307	97,790	360,141	2,102,897
6	Csongrád	493,072	9,069	171,879	124,600	569,993	334,162	2,295,269
16	Jász-Nagykun-Szolnok	572,966	40,610	243,166	53,486	465,345	257,145	2,565,455
	Total	**15,792,607**	**2,169,552**	**10,580,600**	**2,678,948**	**2,745,114**	**18,157,134**	**69,943,234**

(Continued)

Table 2.3 (*Continued*)

Code	County	Spending				Total	Balance of income and spending
		Labour cost	Game management	Compensation paid to			
				Aricultural damages	Forest damages		
14	Somogy	1,379,448	3,584,500	1,878,059	31,793	8,546,655	114,355
5	Borsod-Abaúj-Zemplén	906,234	871,962	279,552	51,214	2,878,262	280,241
19	Veszprém	911,917	2,100,759	678,997	24,014	5,068,548	1,020,341
20	Zala	1,013,138	1,582,300	1,443,562	39,500	5,331,707	434,183
13	Pest	816,721	1,741,748	200,155	15,583	3,451,552	368,555
8	Győr-Moson-Sopron	980,041	1,340,445	448,269	31,666	3,714,193	491,224
2	Baranya	862,831	1,456,852	778,452	20,983	3,689,910	130,624
18	Vas	788,014	1,453,497	1,226,714	67,297	4,173,897	209,707
17	Tolna	861,541	2,086,907	483,907	9,414	3,968,945	776,379
3	Bács-Kiskun	907,386	1,216,983	122,369	11,069	2,850,007	183,476
12	Nógrád	409,024	695,683	287,003	79,862	1,969,824	209,783
10	Heves	667,345	843,417	58,317	345	2,084,828	403,693
15	Szabolcs-Szatmár-Bereg	668,924	710,341	84,734	121	1,928,866	125,400
11	Komárom-Esztergom	451,966	729,766	367,128	12,793	2,186,586	315,759
7	Fejér	719,969	1,549,524	125,307	53,855	3,032,669	-234,503
4	Békés	870,183	1,532,514	36,659	6,993	3,058,045	215,903
9	Hajdú-Bihar	597,910	949,441	89,914	0	2,102,317	579
6	Csongrád	605,900	1,132,031	45,469	15,562	2,270,531	24,738
16	Jász-Nagykun-Szolnok	951,431	1,142,217	26,379	0	2,894,121	-328,666
	Total	**15,369,924**	**26,720,886**	**8,660,945**	**472,062**	**65,201,462**	**4,741,772**

Big game counties: 2, 5, 8, 13, 14, 18, 19, 20; Small game counties: 4, 6, 7, 9, 16; others show intermediate characteristics

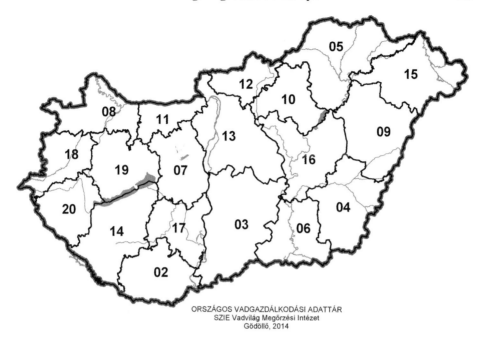

Figure 2.1 Borders of the 19 counties of Hungary; numbering as in Table 2.3.

- A very strong correlation was found between the income from the shot game sold (carcass and meat) and the density of ungulates harvested per km^2 ($r^2 = 0.868$). This relationship shows that the venison which is sold has become an essential income for the counties which have small game and roe deer. This relationship is especially important in the big game areas of South West Hungary (counties 2, 14, 17, 18, 19, 20; see figure 2.1).
- Similar to the total income, the total cost of game management shows a strong correlation with the big game harvest density ($r^2 = 0.696$) meaning that larger incomes allow more payments for game management.

We noted a strong correlation between income from the shot game sold (carcass and meat) and the density of ungulates harvested per km^2, suggesting that the sale of game meat has become an important additional source of income (and in this example, particularly so in the big game areas of South West Hungary). Game meat is consumed worldwide but in most regions it contributes only a small part to the overall meat and food supply (Chardonnet *et al.*, 2002). Despite differences in game species, ante mortem conditions (free-range or fenced; wild or semi-domesticated), hunting or harvesting procedures and further handling of the carcass, there are common requirements with regards to meat safety and quality. Indeed, European legislation sets high standards of public health and animal health problems relating to harvesting game species and placing venison on the market (EU, 1992). This means that ungulates shot and processed must fulfil strict rules right from the site of the kill through to the consumer (Paulsen *et al.*, 2011).

The importance of wild ungulates as an alternative source of animal protein production is well documented (e.g. Hoffman and Cawthorn, 2012; Bleier *et al.*, 2013). While historically much of the venison produced was consumed directly by the hunters themselves and comparatively little was sold, as ungulate populations/harvests have been increasing, the larger quantities of venison now available mean that game meat is more available for customers (Sonkoly *et al.*, 2013). Moreover, venison is gaining a good reputation and consumers are ready to pay more for such meat either in the market or in restaurants. Changes in game meat production in Hungary can again be used as an example of the consequences of increasing ungulate numbers and declining small game populations (*ibid.*). Traditionally, the meat of game ungulates was exported to western European markets and the domestic market was not developed; domestic consumption does not absorb the 10,000 tonnes of venison that is now harvested each year. In the 1960s, the total amount of game meat was far below 2,000 tonnes, but production had reached 4,000 tonnes by the mid-1980s. A further rapid increase in the total game meat production started in 1998 and this is still continuing today. When considering also the proportions of the meat of ungulate species, marked changes can be identified. In the 1970s, >50% of venison was the result of red deer shooting and 20–30% was wild boar meat; roe deer contributed approximately 30% of the total game meat production. Recently, the share of red deer venison has decreased to around 30% and wild boar increased to >50%; the contribution from roe deer venison is between 15–20%. As a comparison in Slovenia, around 1,085 tonnes of free-ranging ungulate venison has been produced annually over the last decade, and the total amount is composed of the following: 39% roe deer, 29% wild boar, 28% red deer, 4% chamois, and <1% mouflon and fallow deer, together.

Altogether, the several uses of ungulates in Europe are believed to generate billions of euros every year (Apollonio *et al.*, 2010b). Recreation related to them is not only a widespread leisure activity but can also be a catalyst for economic growth. This contributes to local economies through employment, increased economic output (money circulation) and tax revenue generation, all of which can act as motivating forces for conservation initiatives (ESUSG, 2004). Further, in many EU countries national strategies for wildlife management are dependent not only on, for example, farmers and other landowners, but importantly also on recreational hunting and part-time hunters. Thus, the whole 'ethos' of current and future management of ungulate populations is entirely dependent on 'recreational' hunting, which is seen as a fundamental mechanism for the overall management of free-ranging ungulates and their impacts (Dickson *et al.*, 2009).

The value of efforts carried out on the ground by hunters and wildlife managers to create new breeding sites/resting places, to control predators or pest species and to manage the balance of wildlife populations, still needs to be estimated (Pinet, 1993, 1995; BASC, 1997; Marshall and McCormick, 2006). For example, in Slovenia in the period 2007–11 around 60 professional hunters, employed in 11 hunting grounds with special purposes (total area of 259,000 ha), spent >87,000 working hours on activities connected with conservation and monitoring of wild-life, and they improved >7,000 ha of wildlife habitats by mowing, preparation of pastures, maintaining of hedgerows and forest edges etc.; moreover, they prepared and maintained 2,219 mires and 109 large water sources (Pokorny *et al.*, 2012). However, these figures are of minor importance in comparison with efforts carried out by >20,600 non-professional Slovene hunters, who spend almost 400,000 gratis working hours a year on the conservation of wildlife and improvement of the carrying capacity of the environment (Lisjak, 2014). Although these activities are primarily orientated towards ungulates, importantly, they also improved the living conditions for other wildlife and hence represent an important but almost neglected value of ungulates and their management. Indeed, unlike, for example, agricultural products, it is rare for a market price to be placed on wildlife, therefore it remains an undervalued asset. This highlights the importance of developing non-market valuation techniques to measure the value of wildlife, especially when monetary values are implicitly embedded in other marketed goods and services (Marjainé Szerényi, 2007; Catlin *et al.*, 2013).

The majority of studies to date of the value of wild ungulates have been undertaken at national level and primarily on the consumptive use of wildlife, yet there has been no uniformity or standardisation amongst them concerning the boundaries and methodologies employed. Some studies measured the direct value whilst others included the multiplier effects to local and national economies (e.g. Filion *et al.*, 1988; US FWS, 2002, 2007, 2012; Feureisel, 2012). It is impossible therefore to collate and compare the resulting economic figures. There is a need to develop a common understanding about what to measure and how. Therefore, a standardised methodology is necessary to quantify adequately the value of wildlife in general (and ungulates in particular) across regions, which in turn can be used to extrapolate values at different scales.

2.5 Capturing value: Environmental-economic accounting in monetary terms

Social scientists have not quantified all the values of wildlife. The economic value of goods exchanged in the marketplace, subscription to magazines, or equipment related to hunting or wildlife watching can be estimated with reasonable accuracy where an appropriate system of information collection is set up. Economists have also developed methods to estimate values of benefits that lie outside the marketplaces (Decker *et al.*, 2001; Tisdell, 2005; Nagy and Kiss, 2011), such as wildlife recreational experience, but those are often considered less accurate than estimates of expenditures. However, to the degree that people are willing to pay for those values, they have an economic component (Decker *et al.*, 2001).

Together, consumptive and non-consumptive uses of wildlife provide an excellent example of a 'green economy': a sustainable system which produces increasing wealth while conserving the environment and systems on which we rely. Wildlife, when managed sustainably, is part of a functioning renewable system characterised by a symbiotic and/or competitive relationship between natural resources and their users resulting from historical human demand. However, as we have noted already, it is monetary measurements of costs and benefits which tend to govern managerial decisions, including the management of natural resources. Economic value is an important (and very often the only relevant) determinant for decision-makers when they are faced with choices. When seeking to conserve wildlife, for example, we are often faced with competing land uses (e.g. agronomy, forestry, development of infrastructure), many of which have a clear economic value. In addition, the general public is better able to perceive value in monetary terms (Bolen and Robinson, 1995).

There are, as we have noted, few published studies that attempt to measure the economic value of wildlife and such as do exist tend to be rather localised/country specific and do not employ consistent methodologies. In addition, research efforts have tended to concentrate on user expenditures associated with extractive recreational uses (BASC, 1997; MacMillan and Phillip, 2008; Feuereisel, 2012). Combining extractive and non-extractive use values for wildlife (and ungulates) through market valuations, either direct or embedded in other products, may provide much more holistic estimates of the total economic value of wildlife.

A recent project (RECAMAN: Campos and Caparrós, 2011) in the Andalusia region of Spain can provide an example of a monetary accounting approach. Andalusia is located in the South of Spain, covering 8.76 million hectares, of which 4.39 million hectares are forest areas (Mediterranean forest, shrublands, oak woodlands and grasslands) that can be regarded as suitable habitat for ungulates. Land-usage over the bulk of this land (3.10 million hectares) includes hunting. The RECAMAN project has focused on accounting income and capital of forest areas in Andalusia, following Hicks (1939) and Krutilla (1967), considering both manufactured and environmental incomes. RECAMAN measures the economic flows and stocks of forest ecosystems by focusing on those scarce goods and services for which a person and/or an institutional entity are willing to pay a sum of money in order to access its use. The total income of an ecosystem can be classified, based on the criterion of ownership, either public or private, and by the criterion of the market, whether commercial or non-commercial (the latter includes private and public products; Campos and Caparrós, 2006). Forests produce natural goods and services depending amongst other things on the circumstances surrounding demand, location and property rights. These goods and services have an economic value in some places and circumstances, whilst in others the same goods and services have no economic value or are free, e.g. when the owner does not find a person and/or institutional entity willing to pay a sum of money for its consumption and/or appropriation (see also Campos *et al.*, 2009).

The project has estimated that ungulate populations within forest areas of Andalucia number in excess of 650,000, with red deer (369,000) and wild boar (214,000) the most abundant species. Every year, more than 96,000 ungulates are hunted (approximately 48,000 red deer and 32,000 wild boar as the main quarry species). The environmental income (excluding manufactured income) from game exploitation of ungulates in 2010 was 7.19 euros per hectare of hunting properties in forest areas, which amounts a total of 22.29 million euros per year. Other environmental services which may be taken into account in formal economic evaluation include, for example, those services consumed by the land owners themselves and public recreation. These private (self-consumption of amenities) and public (recreation) environmental incomes in 2010 contribute to a total aggregated value of 577.10 million euros (Campos, Carranza *et al.*, unpublished data).

In a recent study, which took as a basis the current market and customer positions, the latest harvest data, as well as the pricing of game meat and shooting fees, the total market value of the Hungarian ungulates was estimated at 19,47 billion Hungarian forint (*c.* 66 million euros). If the full values for game meat sales and income from hunting fees were to enter the market then the total actual theoretical value of big game reaches 37,95 billion Hungarian forint (*c.* 128.6 million euros). These values indicate that (1) ungulates (especially red deer, roe deer and wild boar) represent a large part of the total economic value of game species in Hungary (>50%), (2) the efficiency of the use of these values is low, and can be increased in the future. Despite the second observation being rather theoretical, it shows that the financial efficiency of game management needs to be measured with the appropriate methods. In order to assess the values of game species (or of ungulates), it is necessary to use and understand the methods developed in economics either in general or in specific areas for valuation (Vajai, 2012).

2.6 Conclusions

Ungulates are widely recognised and clearly shown to be important values in natural ecosystems contributing significantly to ecosystem services. However, to have the influence that they deserve in terms of management decisions, it is necessary that they be valued in monetary terms. Research on monetary accounting of ungulates may focus on pricing and valuation of ungulates within ecosystem services and ecosystem assets and the possible augmentation of the standard economic accounts of the System of National Accounts (SNA), using these valuations (see SEEA, 2013).

For the future of European hunters and the conservation of ungulates as resources it is essential to improve databases on consumptive use of ungulates and to set up suitable systems in order to provide information about non-consumptive uses and their economic importance for the public. Such data, collected in a standardised form, will be essential if we are to truly assess the economic benefits of ungulates and other wildlife, to balance against known negative economic impacts, and to assess the cost-effectiveness of management options.

In the context of the Common Agricultural Policy (CAP) and the Rural Development directive, ungulates become a crucial biological tool to produce valuable viable landscapes in marginal European natural areas. This additional role for ungulates should be incorporated by policy makers into wildlife management policies as an additional source of commercial income for land owners when livestock is not a functional option, particularly in many marginal areas and where low-cost alternatives are needed for conserving working landscapes like, for instance, the Iberian dehesas (see Macaulay *et al.*, 2013).

We suggest that in future work, the first and most important step is to develop a systematic and detailed database of all free-ranging ungulate species, both native and introduced, across Europe. The Slovene database which is accessible on-line (https://apl.logos.si/LIS) and was developed by the *Slovene Hunting Association* may serve as a good starting point. It contains many essential data on any ungulate specimens harvested either by hunters or which have other causes of death. A similar long-term national database is available for Hungary: www.vmi.info.hu (Csányi *et al.*, 2013). This proposed Europe-wide database should establish a straightforward and standardised set of measures/parameters which can be collected for all European countries (harvest, prices, population size, body mass, trophy values, etc.) with this defined and standardised set acting as a kind of lowest common denominator. It will be necessary to develop measures within the database for both consumptive and non-consumptive uses of ungulates.

In order adequately to recognise and measure both costs and benefits associated with wild ungulates and their management, it will also be necessary to establish an equivalent system for a more formal recording of economic costs of management (paid stalkers, where management is not undertaken on a voluntary basis by recreational hunters; gamekeepers; costs of fencing, etc.), and costs associated with negative impacts of ungulates and their mitigation (e.g. costs of damage to agriculture, forestry, conservation habitats, comprehensive costs of ungulate – vehicle collisions; and costs of prevention measures implemented to reduce such damages/impacts). Collating data from existing studies and developing a standardised methodology for new studies on both national and pan-European level is essential in order to create reproducible and comparable studies in the future.

Whilst such analyses may *not* take account of less obvious values, such as biological values, cultural values or values conveyed through ecosystem services, a more formal assessment of those economic costs and benefits which *can* be directly measured should provide a means for scientists and policy makers to assess and quantify the monetary and non-monetary values of ungulates in Europe and other parts of the world. This is essential in designing measures to improve the conservation, strategic impact assessment and sustainable management of ungulates (as well as wildlife resources in a broader sense). The establishment of such a database needs a long-term commitment of founding organisations and the efforts of both the research and end-user communities to build up a useful and detailed body of appropriate information.

References

Adamič, M. and Jerina, K. (2010) Ungulates and their management in Slovenia. In M. Apollonio, R. Andersen and R. Putman (eds), *European Ungulates and their Management in the 21st Century*. Cambridge, UK: Cambridge University Press, pp. 507–526.

Apollonio, M., Andersen, R. and Putman, R. (2010a) Present status and future challenges for European ungulate management. In M. Apollonio, R. Andersen and R. Putman (eds), *European Ungulates and their Management in the 21st Century*. Cambridge, UK: Cambridge University Press, pp. 578–604.

Apollonio, M., Andersen, R. and Putman, R. (eds) (2010b) *European Ungulates and Their Management in the 21st Century*. Cambridge, UK: Cambridge University Press.

Bailey, J.A. (1984) *Principles of Wildlife Management*. New York: J. Wiley and Sons.

Barrios-Garcia, M.N. and Ballari, S.A. (2012) Impact of wild boar (*Sus scrofa*) in its introduced and native range: a review. *Biological Invasions* **14**, 2283–2300.

Bartosiewicz, L. (2006) *Régenvolt háziállatok. Bevezetés a régészeti állattanba.* [*Domestic Animals of the Past. An Introduction to Archeozoology.*] Budapest, Hungary: L'Hartmattan Kiadó. [In Hungarian.]

BASC (1997) *Deer, Deer Stalking and the Future: A Survey of Deer Stalking Members of BASC and Their Views on Their Sport and its Future.* Wrexham, UK: The British Association for Shooting and Conservation.

Békés, S. (2012) *Vadászetikai beszélgetések.* [*Conversations on Hunting Ethics.*] Budapest, Hungary: Immosensus Kft. [In Hungarian.]

Belházy, G. (1903) *A vadász tíz parancsolatja.* [Decalogue of the hunters]. Budapest, Hungary: Szerző kiadása, Bács-Doroszló. [In Hungarian.]

Beschta, R.L. and Ripple, W.J. (2009) Large predators and trophic cascades in terrestrial ecosystems of the western United States. *Biological Conservation* **142**, 2401–2414.

Bleier, N., Galló, J., Szabó, L., Balázs, B., Tóth, B., Biró, Zs., Heltai, M. and Szemethy, L. (2013) Comparison of the fat and protein content in the red deer hind meat from confined and open areas. In M. Beuković (ed.), *Modern Aspects of Sustainable Management of Game Populations.* 2nd International Symposium on Hunting, 17–20 October 2013. Novi Sad, Serbia: Faculty of Agriculture, pp. 258–261.

Bolen, E.G. and Robinson, W.L. (1995) *Wildlife Ecology and Management.* Upper Saddle River, NJ: Prentice Hall.

Borsy, A., Podani, J., Steger, V., *et al.* (2009) Identifying novel genes involved in both deer physiological and human pathological osteoporosis. *Molecular Genetics and Genomics* **281**, 301–313.

Bryden, D.M., Westbrook, S.R., Burns, B., Taylor, W.A. and Anderson, S. (2010) *Assessing the economic impacts of nature based tourism in Scotland.* Scottish Natural Heritage Commissioned Report No. 398, Scottish Natural Heritage, Inverness.

Bubenik, G.A. (1990) The antler as a model in biomedical research. In G.A. Bubenik and A.B. Bubenik (eds), *Horns, Pronghorns, and Antlers: Evolution, Morphology, Physiology, and Social Significance.* New York: Springer-Verlag, pp. 74–484.

Campos, P. and Caparrós, A. (2006) Social and private total Hicksian incomes of multiple use forests in Spain. *Ecological Economics* **57**, 545–557.

Campos, P. and Caparrós, A. (2011) RECAMAN Project: Mediterranean monte ecosystems total income green accounting. Presentation at the Expert Meeting on Ecosystem Accounting. Copenhagen, Denmark: European Environment Agency, http://unstats.un.org/unsd/envaccounting/seearev/meetingMay2011/lod.htm.

Campos, P., Oviedo, J.L., Caparrós, A., Huntsinger, L. and Coelho, I. (2009) Contingent valuation of private amenities from oak woodlands in Spain, Portugal, and California. *Rangeland Ecology and Management* **62**, 240–252.

Carranza, J. (2010) Ungulates and their management in Spain. In M. Apollonio, R. Andersen and R. Putman (eds), *European Ungulates and their Management in the 21st Century.* Cambridge, UK: Cambridge University Press, pp. 419–440.

Carranza, J. and Vargas, J.M. (eds) (2007) *Criterios para la Certificación de la Calidad Cine-gética en España.* Extramadura, Spain: Universidad de Extremadura.

Carranza, J., Martínez, J.G., Sánchez-Prieto, C., *et al.* (2003) Game species: extinctions hidden by census numbers. *Animal Biodiversity and Conservation* **26**, 81–84. [In Spanish.]

Catlin, J., Hughes, M., Jones, T. and Campbell, R. (2013) Valuing individual animals through tourism: Science or speculation? *Biological Conservation* **157**, 93–98.

Chardonnet, P., Clers, B.D., Fischer, J., Gerhold, R., Jori, F. and Lamarque, F. (2002) The value of wildlife. *Revue Scientifique et Technique (International Office of Epizootics)* **21**, 15–51.

CIC (2006) *Limassol declaration of the 53rd CIC General Assembly.* Limassol, Cyprus: International Council for Game and Wildlife Conservation.

CIC (2011) *CIC Recommendation: wildlife and commercially-bred formerly wild animals.* 58th General Assembly of the International Council for Game and Wildlife Conservation, St. Petersburg, Russia.

Csányi, S. (1991) Red deer population dynamics in Hungary: management statistics versus modelling. In R.D. Brown (ed.), *The Biology of Deer.* New York: Springer Verlag, pp. 37–42.

Csányi, S. (1995) Wild boar population dynamics and management in Hungary. *IBEX Journal of Mountain Ecology* **3**, 222–225.

Csányi, S. and Lehoczki, R. (2010) Ungulates and their management in Hungary. In M. Apollonio, R. Andersen and R. Putman (eds), *European Ungulates and their Management in the 21st Century.* Cambridge, UK: Cambridge University Press, pp. 291–318.

Csányi, S., Lehoczki, R. and Sonkoly, K. (2013) National game management database of Hungary: a 20 years stewardship of Hungarian wildlife resources. *Hungarian Agricultural Research* **22**(2), 19.

Csőre, P. (2000) A magyar vadászat története. [The history of Hunting in Hungary.] In J. Pechtol (ed.), *Vadászévköny 2000.* Budapest, Hungary: Országos Magyar Vadászkamara, pp. 33–48. [In Hungarian.]

de Ferrieres, H. (2006) Modus király vadászkönyve. [The hunting of King Modus.] *Zalaerd Zrt. Nagykanizsa* **134**, 1354–1377. [In Hungarian.]

Decker, D.J., Brown, T.L. and Siemer, W.F. (eds) (2001) *The Human Dimensions of Wildlife Management in North America.* Bethesda, MD: The Wildlife Society.

Deinet, S., Leronymidou, C., McRae, L., Burfield, I.J., Foppen, R.P., Collen, B. and Bohm, M. (2013) *Wildlife Comeback in Europe: the recovery of selected mammal and bird species.* Final report to Rewilding Europe. London, UK: ZSL, BirdLife International and the European Bird Census Council.

Dickson, B., Hutton, J. and Adams, W.M. (eds) (2009) *Recreational Hunting, Conservation and Rural Livelihoods.* London: Wiley-Blackwell.

Drábková A. (2013) Tourists in Cansiglio forest, Italy: case study about forest visitors and their opinions. *Human Geographies: Journal of Studies and Research in Human Geography* **7**(2), 35–43.

Edens, B. and Hein, L. (2013) Towards a consistent approach for ecosystem accounting. *Ecological Economics* **90**, 41–52.

ESUSG (2004) *Wealth from the Wild. A Review of the Use of Wild Living Resources in the United Kingdom.* UK Committee of IUCN, European Sustainable Use Specialists Group of IUCN, London, UK.

EU (1992) Council Directive 92/45/EEC of 16 June 1992 on public health and animal health problems relating to the killing of wild game and the placing on the market of wild-game meat. *Official Journal L* **268**, 35–53.

European Commission (2011) *Our Life Insurance, our Natural Capital: an EU Biodiversity Strategy to 2020.* Communication from the Commission to the European Parliament, the Council, the Economic and Social Committee and the Committee of the Regions. Brussels, Belgium: European Commission.

Farkas, D. and Csányi, S. (1990) Current problems of the roe deer (*Capreolus capreolus*) management in Hungary. *Folia Zoologica* **38**, 37–46.

Fernández-García, J.L., Carranza, J., Martínez, J.G. and Randi, E. (2014) Mitochondrial D-loop phylogeny signals two native Iberian red deer (*Cervus elaphus*) populations genetically different to western and eastern European red deer and infers human-mediated translocations. *Biodiversity and Conservation,* **23**, 537–554.

Ferroglio, E., Gortazar, C. and Vicente, J. (2011) Wild ungulate diseases and the risk for livestock and public health. In R. Putman, M. Apollonio and R. Andersen (eds), *Ungulate Management in Europe: Problems and Practices.* Cambridge, UK: Cambridge University Press, pp. 192–204.

Feuereisel, J. (2012) Economical approach on hunting. In M. Beukovic (ed.), *Modern Aspects of Sustainable Management of Game Populations.* 1st International Symposium on Hunting, 22–24 June 2012. Zemun-Belgrade, Serbia: Faculty of Agriculture, pp. 168–175.

Filion, F.L., Parker, S.A.D. and DuWors, E. (1988) *The Importance of Wildlife to Canadians. Demand for Wildlife to 2001.* Ottawa, Canada: Canadian Wildlife Service.

Garrido-Martín, J.L. (2012) *La Caza. Sector Económico. Valoración por subsectores.* Real Federación Española de Caza. http://www.fecaza.com/images/stories/CAZA_Sector_ economico.pdf [In Spanish.]

Gilbert, F.F. and Dodds, D.G. (1992) *The Philosophy and Practice of Wildlife Management* (2nd edition). Malabar, FL: Krieger Publishing Company.

Hanley, N., Shogren, J.F. and White, B. (2007) *Environmental Economics in Theory and Practice* (2nd edition). New York: Palgrave Macmillan.

Hebblewhite, M. and Smith, D. (2010) Wolf community ecology: ecosystem effects of recovering wolves in Banff and Yellowstone National Park. In M.P.C. Musiani and P.C. Paquet (eds), *The Wolves of the World: New Perspectives on Ecology, Behavior, and Policy.* Calgary, Alberta: University of Calgary Press, pp. 69–120.

Hicks, J.R. (1939) *Value and Capital.* Oxford: Clarendon Press.

Hoffman, L.C. and Cawthorn, D.M. (2012) What is the role and contribution of meat from wildlife in providing high quality protein for consumption? *Animal Frontiers* **2**, 40–53.

Hull, D.B. (1964) *Hounds and Hunting in Ancient Greece.* Chicago, IL: University of Chicago Press.

Jankovics, M. (2004) *A Szarvas könyve.* [*The Book of Deer.*] Debrecen, Hungary: Csokonai Kiadó. [In Hungarian.]

Jedrzejewska, B. and Jedrzejewski, W. (1998) *Predation in Vertebrate Communities: The Bialowieza Primeval Forest as a Case Study. Ecological Studies* **135**. Berlin: Springer.

Jones, T. (2012) Canned hunting in Europe and how to avoid getting scammed: answers from Dr Rolf D. Baldus of CIC. *The Hunting Report* **32**, 16–17.

Kawata, J. (2010) An economic analysis of wild animal management under non-consumptive use. *Journal of Social and Economic Development* **12**, 1–11.

Kellert, S. (1978) Attitudes and characteristics of hunters and antihunters. *Transactions of the forty-third North American Wildlife and Natural Resources Conference* **43**, 412–423.

Kellert, S.R. and Smith, C.P. (2000) Human values toward large mammals. In S. Demarais and P.R. Krausman (eds), *Ecology and Management of Large Mammals in North America.* Upper Saddle River, NJ: Prentice Hall, pp. 38–63.

Kenward, R. and Putman, R. (2011) Ungulate management in Europe: towards a sustainable future. In R. Putman, M. Apollonio and R. Andersen (eds), *Ungulate Management in Europe: Problems and Practices.* Cambridge, UK: Cambridge University Press, pp. 376–395.

Kessler, W.B. (2012) Frankendeer: The fear is real. *Fair Chase*, winter, 62.

Kowalsky, N. (ed.) (2010) *Hunting: Philosophy for Everyone: In Search of the Wild Life.* Oxford: Wiley-Blackwell.

Krutilla, J.V. (1967) Conservation reconsidered. *American Economic Review* **57**, 777–786.

Landete-Castillejos, T.S., Gallego, L. and Garcia, A. (2012) From broken antlers to human osteoporosis. *Deer* [The Journal of the British Deer Society] Summer 2012, 20–24.

Langbein, J., Putman, R. and Pokorny, B. (2011) Traffic collision involving deer and other ungulates in Europe and available measures for mitigation. In R. Putman, M. Apollonio and R. Andersen (eds) *Ungulate Management in Europe: Problems and Practices.* Cambridge, UK: Cambridge University Press, pp. 215–259.

Lecocq, Y. and Meine, K. (eds) (1995) *FACE Handbook of Hunting in Europe.* Brussels, Belgium: Fédération des Associations de Chasseurs de l'EU.

Leopold, A. (1943) Wildlife in American culture. *Journal of Wildlife Management* **7**, 1–6.

Lisjak (2014) *Slovenski lovski informacijski sistem.* [Slovene hunting information system.] Accessible at https://apl.logos.si/LIS. [In Slovene.]

Lorenzini, R. and Lovari, S. (2006) Genetic diversity and phylogeography of the European roe deer: the refuge area theory revisited. *Biological Journal of the Linnean Society* **88**, 85–100.

Macaulay, L.T., Starrs, P.F. and Carranza, J. (2013) Hunting in managed oak woodlands: contrasts among similarities. In P. Campos, L. Huntsinger, J.L. Oviedo, P.F. Starrs, M. Díaz, R.B. Standiford, G. Montero (eds), *Mediterranean Oak Woodland Working Landscapes.* Landscape Series 16. Amsterdam, Netherlands: Springer, pp. 311–350.

MacMillan, D.C. and Phillip, S. (2008) Consumptive and non-consumptive values of wild mammals in Britain. *Mammal Review* **38**, 189–204.

Marjainé Szerényi, Z. (ed.) (2007) *A természetvédelemben alkalmazható kozgazdasági értékelési módszerek. A Kornyezetvédelmi és Vízugyi Minisztérium Természetvédelmi Hivatalának tanulmánykotete.* [Economic methods for the use in nature conservation. Publication of the Nature Conservation Authority of the Ministry for Environment and Water Management.] Budapest, Hungary: Budapesti Corvinus Egyetem, Kornyezetgazdaságtani és Technológiai Tanszék. [In Hungarian.]

Marshall, P. and McCormick, A. (2006) *Deer Stalking Survey: A Detailed Account of Deer Stalking by Members of BASC.* Wrexham, UK: The British Association for Shooting and Conservation.

Mathisen, K.M., Milner, J.M., van Beest, F.M. and Skarpe, C. (2014) Long-term effects of supplementary feeding of moose on browsing impact at a landscape scale. *Forest Ecology and Management* **314**, 104–111.

McShea, W.J., Underwood, H.B. and Rappole, J.H. (eds) (1997) *The Science of Overabundance: Deer Ecology and Population Management.* Washington DC: Smithsonian Institution Press.

Miller, J.E. (2012) A growing threat: how deer breeding could put public trust wildlife at risk. *Wildlife Professional* **6**, 22–27.

Milner, J.M., Beest, F.M. and Storaas, T. (2013) Boom and bust of a moose population: a call for integrated forest management. *European Journal of Forest Research* **132**, 959–967.

Muhly, T.B., Hebblewhite, M., Paton, D., Pitt, J.A., Boyce, M.S. and Musiani, M. (2013) Humans strengthen bottom-up effects and weaken trophic cascades in a terrestrial food web. *PLoS ONE* **8**, e64311.

Murray, M. and Simcox, H. (2003) *Use of Wild Living Resources in the United Kingdom: A Review.* Flintshire, UK: UK Committee for IUCN.

Murray, B.D., Webster, C.R. and Bump, J.K. (2013) Broadening the ecological context of ungulate-ecosystem interactions: the importance of space, seasonality, and nitrogen. *Ecology* **94**, 1317–1326.

Myrberget, S. (1990) *Wildlife Management in Europe Outside the Soviet Union.* Trondheim, Norway: Norsk institutt for naturforskning (NINA).

Nagy, G.G. and Kiss, V. (eds) (2011) *Borrowing Services from Nature: Methodologies to Evaluate Ecosystem Services Focusing on Hungarian Case Studies*. Budapest, Hungary: EEweb for Biodiversity.

Newsome, D., Dowling, R.K. and Moore, S.A. (2005) *Wildlife Tourism*. Clevedon, UK: Channel View Publications.

Olstead, J. (2006) *Shooting sports: findings of an economic and environmental survey*. PACEC Report. Stamford, UK: BPG (Bourne) Ltd.

Oslis, 2014. Osrednji slovenski lovski informacijski sistem. [Central hunting information system of Slovenia]. https://www.google.si/#q=oslis+gozdis. [In Slovene]

Outdoor Foundation (2012) *Outdoor recreation participation report 2012*. www.outdoorfoundation.org.

PACEC (2006a) *The Contribution of Deer Management to the Scottish Economy*. Cambridge: Public and Corporate Economic Consultants.

PACEC (2006b) *The Economic and Environmental Impact of Sporting Shooting*. A report prepared by PACEC on behalf of The British Association for Shooting and Conservation, Country Land and Business Association, and Countryside Alliance and in association with Game Conservancy Trust. Cambridge and London: PACEC.

Paulsen, P., Bauer, A., Vodnansky, M., Winkelmayer, R. and Smulders, F.J.M. (eds) (2011) *Game Meat Hygiene in Focus: Microbiology, Epidemiology, Risk Analysis and Quality Assurance*. Wageningen, Netherlands: Wageningen Academic Publishers.

Pinet, J.M. (1993) *Les Chasseurs de France*. Paris: Union Nationale des Fédérations Départmentales des Chasseurs, Institut National Agronomique Paris, Grignon Laboratoire de la Faune Sauvage. [In French.]

Pinet, J.M. (1995) The hunter in Europe. In Y. Lecocq and K. Meine (eds) *FACE Handbook of Hunting in Europe*. Brussels, Belgium: Fédération des Associations de Chasseurs de l'EU, pp. 1–12.

Pokorny, B. and Jelenko, I. (2013) Ecological importance and impacts of wild boar (*Sus scrofa* L.). *Zlatorogov Zbornik* **2**, 2–30. [In Slovene with an English summary.]

Pokorny, B., Jelenko, I., Mazej Grudnik, Z., Kos, I. and Jerina, K. (2012) *Strokovno mnenje o upravičenosti delovanja lovišč s posebnim namenom v Republiki Sloveniji. [Expert opinion on the justification of the existence of hunting grounds with special purposes in Slovenia.]* Final report for the Ministry of Agriculture and the Environment of the Republic of Slovenia. Velenje, Slovenia: Institute ERICo. [In Slovene.]

Putman, R.J. (1986) *Grazing in Temperate Ecosystems; Large Herbivores and their Effects on the Ecology of the New Forest*. Beckenham, UK: Croom Helm.

Putman, R. (2011) A review of the legal and administrative systems governing management of large herbivores in Europe. In R. Putman, M. Apollonio and R. Andersen (eds) *Ungulate Management in Europe: Problems and Practices*. Cambridge, UK: Cambridge University Press, pp. 54–79.

Putman, R. and Watson, P. (2010) *Scoping the Economic Benefits and Costs of Wild Deer and their Management in Scotland*. Report for Scottish Natural Heritage, Inverness, UK.

Putman, R., Andersen, R. and Apollonio, M. (2011) Introduction. In R. Putman, M. Apollonio and R. Andersen (eds), *Ungulate Management in Europe: Problems and Practices*. Cambridge, UK: Cambridge University Press, pp. 1–11.

Reimoser, F. and Putman, R. (2011) Impacts of wild ungulates on vegetation: costs and benefits. In: R. Putman, M. Apollonio and R. Andersen (eds), *Ungulate Management in Europe: Problems and Practices*. Cambridge, UK: Cambridge University Press, pp. 144–191.

Ripple, W.J. and Beschta, R.L. (2012) Large predators limit herbivore densities in northern forest ecosystems. *European Journal of Wildlife Research* **58**, 733–742.

Rivrud, I.M., Sonkoly, K., Lehoczki, R., *et al.* (2013) Hunter selection and long-term trend (1881–2008) of red deer trophy sizes in Hungary. *Journal of Applied Ecology* **50**, 168–180.

Roots, C. (2007) *Domestication.* Westport, CT: Greenwood Press.

Royo, L.J., Pajares, G., Alvarez, I., Fernández, I. and Goyache, F. (2007) Genetic variability and differentiation in Spanish roe deer (*Capreolus capreolus*): a phylogeographic reassessment within the European framework. *Molecular Phylogenetics and Evolution* **42**, 47–61.

Sánchez-Morales, J.D.S. (2008) *Uso turístico en el Parque Natural Sierras de Cazorla, Segura y Las Villas.* CONAMA8. Segura y Las Villas, Spain: Parques Naturales de Despeñaperros y Sierras de Cazorla. [In Spanish.]

Scandura, M., Iacolina, L. and Apollonio, M. (2011) Genetic diversity in the European wild boar *Sus scrofa*: phylogeography, population structure and wild × domestic hybridization. *Mammal Review* **41**, 125–137.

SEEA (2013) *System of Environmental-Economic Accounting 2012. Experimental Ecosystem Accounting.* http://unstats.un.org/unsd/envaccounting/pubs.asp.

Senn, H.V., Swanson, G.M, Goodman, S.J., Barton, N.H. and Pemberton, J.M. (2010) Phenotypic correlates of hybridisation between red and sika deer (genus *Cervus*). *Journal of Animal Ecology* **79**, 414–425.

Sila, A. and Koren, I. (2011) Divji prašič ob zahodni državni meji – problemi populacije in upravljanja z njo. [Wild boar near the western state border – problems and management with the population] In H. Poličnik and B. Pokorny (eds) *Divji prašič: zbornik prispevkov. [Wild boar: Proceedings of the 2nd International Conference on Game Management]* Velenje, Slovenia: Institute ERICo, pp. 72–76. [In Slovene.]

Skog, A., Zachos, F.E., Rueness, E.K., Feulner, P.G.D., *et al.* (2009) Phylogeography of red deer (*Cervus elaphus*) in Europe. *Journal of Biogeography* **36**, 66–77.

Smit, C. and Putman, R. (2011) Large herbivores as 'environmental engineers'. In R. Putman, M. Apollonio and R. Andersen (eds), *Ungulate Management in Europe: Problems and Practices.* Cambridge: Cambridge University Press, pp. 260–283.

Sonkoly, K., Bleier, N., Heltai, M., *et al.* (2013) Big game meat production in Hungary: a special product of a niche market. *Hungarian Agricultural Research* **22**, 12–16.

Šprem, N. (2009). *Morfološke i genetske osobine divljih svinja (Sus scrofa L.) u Republici Hrvatskoj. [Morphological and genetic characteristics of the wild boar (Sus scrofa L.) in the Republic of Croatia.]* PhD dissertation, University of J. J. Strossmayer, Osijek, Croatia. [In Croation.]

Stéger, V. (2011) *Antler development and coupled osteoporosis in the skeleton of red deer Cervus elaphus: expression dynamics for regulatory and effector genes.* PhD thesis, Eötvös Lóránt University, Budapest, Hungary.

Szabó, Á., Szemethy, L., Firmánszky, G. and Heltai, M. (2000) Visszatelepülő nagyragadozók természetvédelmi és vadgazdálkodási problémái. [Nature conservation and wildlife management problems associated with the returning large carnivores.] *A Vadgazdálkodás Időszerű Tudományos Kérdései* **1**, 62–72. [In Hungarian.]

TEEB (2010) *The Economics of Ecosystems and Biodiversity: Mainstreaming the Economics of Nature: A synthesis of the approach, conclusions and recommendations of TEEB.* Valletta, Malta: Progress Press.

Tisdell, C.A. (2005) *Economics of Environmental Conservation* (2nd edition). Cheltenham, UK: Edward Elgar Publishing.

TWS (The Wildlife Society of America) (2009) *Final Position Statement Confinement of Wild Ungulates within High Fences.* Bethesda, MD: The Wildlife Society.

TWS (The Wildlife Society of America) (2012) *Captive cervid breeding: fact sheet and position statement.* Bethesda, MD: The Wildlife Society.

US FWS (US Fish and Wildlife Service, 2002) *2001 National Survey of Fishing, Hunting, and Wildlife-Associated Recreation*. Washington DC: US Department of the Interior, Fish and Wildlife Service and US Department of Commerce, US Census Bureau.

US FWS (US Fish and Wildlife Service, 2007) *2006 National Survey of Fishing, Hunting, and Wildlife-Associated Recreation*. Washington DC: US Department of the Interior, Fish and Wildlife Service and US Department of Commerce, US Census Bureau.

US FWS (US Fish and Wildlife Service, 2012) *2011 National Survey of Fishing, Hunting, and Wildlife-Associated Recreation: National Overview*. Washington DC: US Fish and Wildlife Service.

Vajai, P. (2012) *A vadgazdálkodás okonómiai hatékonyságának elemzése nagyvadas teruleteken*. [*An analysis of economic efficiency of big game management*.] Post-graduate Thesis, University of West Hungary, Sopron, Hungary. [In Hungarian.]

Valvo, G., Sturaro, E., Maretto, F. and Ramanzin, M. (2009) Genetic analysis reveals roe deer (*Capreolus capreolus*) population structure in North-Eastern Italian Alps. *Italian Journal of Animal Science* **8** (Suppl. 3), 104–108.

Vlada RS (2013) *Akcijski načrt za upravljanje populacije volka (*Canis lupus*) v Sloveniji za obdobje 2013–2017*. [*Action plan for management of population of wolf (*Canis lupus*) in Slovenia for the period 2013–2017*.] Vlada RS, Document 00728-7/2013/4. Ljubljana, Slovenia. [In Slovene.]

WFSA (2010) *Ecologic and Economic Benefits of Hunting*. Proceedings of the Symposium on Hunting Activities, 14–17 September 2009. World Forum on Shooting Activities, Windhoek, Namibia.

Chapter 3

Reintroductions as a Management Tool for European Ungulates

Marco Apollonio, Massimo Scandura and Nikica Šprem

At the turn of the 20th century, populations of a number of species of large mammals in Europe were severely compromised (see also Chapter 12). Most species of large carnivores and large herbivores had greatly restricted distributional ranges and were still declining in numbers in many regions (see Breitenmoser, 1998; Apollonio *et al.*, 2010a). Overhunting and the loss of suitable habitats were the main causes of this decline. This trend was reversed around the middle of the century, when populations began to recover and communities of large mammals were progressively restored to large parts of mainland Europe (Apollonio *et al.*, 2010a).

For wild ungulates, the recolonisation of former range resulted from two processes: a natural expansion of remnant populations due to legal protection and/or increased availability of suitable habitats, and direct human intervention by reintroduction or restocking. The balance of importance of these two processes varied both with region and species. In some countries the bulk of the recovery was through natural processes; in others the recovery was effected primarily by human activities. In the same way, some species recovered mostly naturally, some others would have not recovered without active human intervention.

In the current chapter, we examine the role that reintroductions have had in shaping the current distribution of European species and explore some of the perils of such introductions, where the negative effects may have outweighed the positive ones, or compromised the intended objective. Finally, after exploring possible factors determining the success of reintroductions, we provide some remarks on the possible role that such tool could play in the future management of European wild ungulates.

3.1 Reintroductions as a tool to effect the recovery of nearly extinct taxa

Within Europe as a whole, no ungulate species has declined so severely in recent centuries as to risk real extinction, apart from two: the Alpine ibex and the European bison. After over-hunting had driven the species to extinction in the rest of its former range, the alpine ibex reached a population size of as few as 100 individuals during the 19th century, restricted to a small area of the north-western Italian Alps, the royal hunting reserve Gran Paradiso (meaning 'great heaven'). It was really from this 'great heaven' that this mountain species came back to repopulate the Alps, as the result of a significant management effort involving a great number of translocations. This remnant population, occupying the present Gran Paradiso National Park, was the source for captive breeding programmes that started from 1906 in two Swiss zoos (in St Gallen and Interlaken). By 1942, 109 ibex caught in Gran Paradiso National Park had been brought into these parks. From 1911 to 1938 the first reintroductions were made within the Swiss Alps and the first reintroduced Italian colony was established in 1921. A number of further reintroductions followed in Switzerland, Italy, France (from 1959), Austria (from 1924), Germany (from 1938) and Slovenia (from 1953). Nowadays there are about 160–170 populations in the Alps, amounting to around 47,000 individuals (Tosi *et al.*, 2009). In the last decade, the number of reintroductions has declined because of the reduced need of conservation efforts, but also because of the lack of hunting opportunities in some countries (e.g. Italy and France) which, in consequence, reduced the interest of important stakeholders (the hunters) toward these kinds of interventions.

The second species to be rescued in this way was the European bison. After having been exterminated through most of its historical range, by the beginning of the 20th century no wild bison inhabited the lowland forests of Europe any more and only a few individuals survived in the mountains of Caucasus. This state of affairs was announced in 1923 during an International Congress of Nature Protection in Paris, and in the same year an international society was founded with the aim of launching and coordinating a joint effort to recover the species. The first goal of the society was to compile a register of all bison living in captivity (the European Bison Pedigree Book); the second was to start an international captive-breeding programme, which involved a number of zoological gardens, parks and reserves in different countries. In the meantime, the last wild Caucasian bison was shot in 1927.

This concerted conservation effort brought the European bison back to the wild in 1952, in exactly the place where the last wild lowland bison had died in 1919, i.e. the Białowieża Forest in Poland. This initiative was followed by a number of successful reintroductions in several sites of Central and Eastern Europe. However, although in 2004 the world population of free-living bison reached around 2000 individuals scattered in six countries, one third of them inhabited the Polish and Belarusian sides of the Białowieża Forest and only five populations numbered more than 100 individuals (Krasinska and Krasinski, 2007). Nowadays, the European bison is listed as 'vulnerable' by the IUCN, but with an increasing trend (Olech, 2008).

These two examples present some similarities and some essential differences. In both cases, human intervention was decisive for the recovery of the species, and the role of reintroductions was far more significant in the establishment of new populations than natural dispersal. On the other hand, the recovery of the two species was achieved in rather different ways. In bison, it was the product of a wide programme of cooperation, which has seen the establishment of an international society, the creation of a pedigree book and a large number of coordinated exchanges of breeding stocks between breeding centres in several countries. By contrast, the restoration of ibex was managed in the main by single states (or local administrations within states). For instance, even as early as 1875, the first law on hunting agreed by the Swiss Confederation included an expressed intention to reintroduce Alpine ibex to the whole country. This led to the situation that the founding group of the first captive nucleus in Switzerland was in fact illegally transferred from the Gran Paradiso population (Imesch-Bebiè *et al.*, 2010), as the Italian king refused to sell his ibex to a foreign country.

Although no other full species of ungulate faced extinction as such, a number of populations with unique characteristics (often classified as separate subspecies) have dramatically declined over recent years, raising concerns for their conservation. There are three cases that may be considered paradigmatic in this sense: the Eurasian forest reindeer (*Rangifer tarandus fennicus*), the Apennine chamois (*Rupicapra pyrenaica ornata*) and the Corsican red deer (*Cervus elaphus corsicanus*).

Forest reindeer were once quite widespread in Finland but extensive hunting reduced their number to extinction in the country. Some animals recolonised northeastern regions by natural immigration from Russian Karelia, but in the period 1979–84, a number of reindeer were translocated from this founder population to the western part of the country in order to speed up the process of re-establishment. Interestingly, the reintroduced population is now doing consistently better than the source population, with a population growth rate exceeding 1.2 (Ruusila and Kojola, 2010). In the case of Apennine chamois the only remnant population in Italy was in the Abruzzo National Park, which was established in 1922 with the deliberate intention of saving this taxon from extinction (together with the Abruzzo brown bear, *Ursus arctos marsicanus*) (Sipari, 1926). It was obvious that a subspecies represented by only a single population was at significant risk, so a number of reintroductions were planned and executed from 1990 to 2013 in the massifs of Gran Sasso, Majella, Sibillini and Velino Sirente, in order to build up a number of daughter populations in case of some catastrophic event within the source population.

A similar conservation success was achieved for the Corsican red deer (see also Linnell and Zachos, 2011). Natural or pre-historically introduced populations of this red deer subspecies were present in both Sardinia and Corsica (see Hmwe *et al.*, 2006). Until the end of the 19th century it occurred almost everywhere on the two islands; it became extinct in Corsica in the early 1970s and at that time also, the remaining red deer in Sardinia were represented by three small populations in the south of the island reaching less than 100 individuals (Jenkins, 1972). A strong effort to preserve the species in those three refuge areas, together with protection in a number of enclosures scattered across the island (Beccu, 1993) permitted populations to rise to sufficient levels that animals could be reintroduced to Corsica in 1985 (Kidjo *et al.*, 2007) and in North Sardinia in 2003 and 2010 (Apollonio, pers. comm.).

These three reintroductions had the common aim of expanding relict nuclei of vulnerable/threatened subspecies, in order to establish metapopulations able to withstand catastrophic events which might affect any one single nucleus: in all the three cases the subspecies concerned are now less threatened than before by actual extinction and overall population size is much increased.

3.2 Reintroductions as a tool to promote (artificial) range expansion

While, in the cases of bison and ibex and the various subspecies considered above, reintroductions were useful in achieving genuine conservation goals, they have also been widely used to restore wildlife populations for recreational interest. Wild ungulates represent valuable game species and a great number of initiatives of reintroduction, mostly after World War II, were aimed at reintroducing species like red deer, roe deer, chamois and wild boar in areas of their former range, as quarry species for hunting. It is interesting to note that only some species, not surprisingly the most common ones in Europe like roe deer, red deer and wild boar, were the primary focus of such reintroductions while others were very rarely involved in these operations and relied for their recovery mainly on the natural expansion of residual populations, as happened in the case of moose throughout Fennoscandia (Andersen *et al.*, 2010; Liberg *et al.*, 2010; Ruusila and Koijola, 2010).

An analysis conducted on the 28 countries included in a review of the management of European ungulates (Apollonio *et al.*, 2010a) reveals the extent of such intervention (Table 3.1).

The red deer is the most manipulated species out of the 11 species in total that are native to the European continent. In all 27 countries where it occurs (absent in Finland), some reintroductions/ restocking has taken place at some point (see Table 3.1). In at least 12 countries the importance of these operations was substantial, as it gave rise to the majority of present populations if not to all of them. And in most of these countries, reintroductions are recorded already during the 19th century: for instance, in Italy a reintroduction of deer from Bohemia took place in the Casentinesi Forest, Tuscany, in 1835 (Crudele, 1988); in England, during the Victorian age, red deer from parks, often of continental or even North American origin, were used to repopulate the southern part of the country (Putman, 2010); in Slovakia red deer was almost extinct in the middle of the 19th century and was then widely reintroduced starting from enclosures where deer of at least three different subspecies were bred (Findo and Skuban, 2010).

One of the most clear examples of the effects of such operations comes from Italy: the original populations living on the mainland were totally exterminated,

Table 3.1 Data on releases of the four most common ungulate species in Europe. Only the countries that were considered in the volume edited by Apollonio *et al.* (2010a) are listed. Documented cases of reintroductions are in bold. Approximate period of release is mentioned if known.

Country	Red deer *Cervus elaphus*	Roe deer *Capreolus capreolus*	Wild boar *Sus scrofa*	Alpine chamois *Rupicapra rupicapra*	References
Norway	1900–03 releases from Central Europe to Is. Otterøya and translocations from Otterøya to continental Norway	No	–	–	Andersen *et al.* 2010, Haanes *et al.* 2010, 2013
Sweden	1950–today several transfers from Scania, Denmark and other countries	No	1979–90 non-native animals escaped from enclosures	–	Liberg *et al.* 2010, Höglund *et al.* 2013
Denmark	1970–today escapes and releases in Jutland and Sjaelland	No	1926 **reintroduction** in Jutland from Germany	–	Niethammer 1963, Andersen and Holthe 2010
Finland	–	by 1941 translocations to Aland archipelago and SW Finland	No	–	Ruusila and Kojola 2010
Baltic States	17th century–1987 **reintroductions** from Germany, Poland and Russia	19th century translocation from Siberia to Estonia	No	–	Andersone-Lilley *et al.* 2010

Great Britain	19th century–today releases of Central European deer and North American wapiti and translocations from English parks	18th–19th century in England **reintroductions** from Scotland, Germany, Austria and other sources, 20th century introduction of Siberian roe deer	1990–today escapes and releases of hybrids in Dorset, Kent and Sussex	—	Danilkin 1996, Putman 2010, Baker and Hoelzel 2012, Perez-Espona *et al.* 2013
Ireland	19th century–1949 **reintroductions** from Scotland and England and translocations from other areas in Ireland	—	2000–today feral pigs escaped from captivity	—	Carden *et al.* 2012, McDevitt *et al.* 2013
The Netherlands	By 1940 translocations from Scotland, Germany, Carpathians, 1992 **reintroduction** in Oostvaardersplassen	No	Early 20th century **reintroductions** from Germany to Veluwe	—	Niethammer 1963, van Wieren *et al.* 2010
Belgium	19th century **reintroductions** in several places	19th century **reintroductions**, 20th century introductions of Siberian roe deer	No	—	Libois 1992, Danilkin 1996, Casaer and Licoppe 2010
Germany	19th–20th century **reintroductions** in Bavarian Forest, Palatinate Forest and north of Hamburg from different sources in Central Europe, translocations from several areas in the country and introductions of exotic deer	1935 **reintroduction** in Fehmarn island from Denmark, 20th century introductions in Frisian islands and translocations from Central and Eastern Europe, introductions of Siberian roe deer	19th–20th century translocations from Poland	1937–39 introductions in Saxon Switzerland from Bavarian Alps, 1935–40 **reintroduction** to Black Forest from Alps	Niethammer 1963, Danilkin 1996, Wotschikowsky 2010, Linnell and Zachos 2011

(Continued)

Apollonio, Scandura and Šprem

Table 3.1 (Continued)

Country	Red deer Cervus elaphus	Roe deer Capreolus capreolus	Wild boar Sus scrofa	Alpine chamois Rupicapra rupicapra	References
Poland	19th–20th century **reintroductions** from Germany and Czech Rep. and translocations from other regions of Poland	Late 19th century translocation from Siberia to Bialowieza Forest	No	No	Karcov 1903, Wawrzyniak et al. 2010, Niedziałkowska et al. 2012
Czech Republic	19th–20th century **reintroductions** from deer parks (various sources) and introduction of exotic deer species	Introductions of Siberian roe deer	No	20th century introductions in North Bohemia and Altvatergebirge from several areas in the Alps	Niethammer 1963, Bartos et al. 2010
Slovakia	19th–20th century **reintroductions** in Javorina and other locations from Czech Rep. Russia, introductions of North American wapiti	20th century introductions of Siberian roe deer	No	1969–76 **reintroductions** from Tatra mt. to Low Tatra, 1956–62 introductions of Alpine chamois from Czech Rep. in 2 areas	Niethammer 1963, Danilkin 1996, Findo and Skuban 2010
Hungary	End 19th century–1945 **reintroductions** and restocking	No	No	–	Csanyi and Lehoczki 2010
Romania	19th–20th century **reintroductions** from Austria and Germany and translocations from other areas	No	No	No	Micu et al. 2010

Austria	translocations from other areas in the country and from Hungary, introduction of exotic deer species	Local releases	Local releases	Local releases	Niethammer 1963, Reimoser and Reimoser 2010
Switzerland	19th–20th century releases of deer from Carpathians and other sources	19th century releases from unknown sources in the south	No	1950–62 **reintroduction** from Swiss Alps in Yura Mt.	Imesch-Bebié *et al.* 2010
Portugal	1970–today **reintroductions** from enclosures containing deer of various origin (Spain, Scotland, Hungary)	1990–today **reintroductions** south of the Douro river with roe from France and Northern Portugal	No	–	Carvalho *et al.* 2008, Vingada *et al.* 2010
Spain	1950–today **reintroductions** in Aragon with deer from other Iberian populations and translocations from Iberia and other countries	1993 **reintroduction** in Montnegre massif, Catalunya, from NE France	No	–	Rosell *et al.* 1996, Carranza 2010, Vingada *et al.* 2010, Gonzalez *et al.* 2013, Fernandez-Garcia *et al.* 2014
France	1950–70 **reintroductions** from Central and NE France, Germany, Austria and Hungary, 1985–87 **reintroductions** in Corsica from Sardinia	1952–79 releases, 1970–today **reintroductions** and restocking from Chizé and Trois Fontaines	1980s releases from parks and translocations	1956 **reintroduction** in Vosges, 1978 **reintroduction** in Massif Central	Niethammer 1963, Danilkin 1996, Kidjo *et al.* 2007, Maillard *et al.* 2010

(Continued)

Table 3.1 *(Continued)*

Country	Red deer *Cervus elaphus*	Roe deer *Capreolus capreolus*	Wild boar *Sus scrofa*	Alpine chamois *Rupicapra rupicapra*	References
Italy	1960–2003 **reintroductions** from France, Scotland, Switzerland, Germany and other countries, translocations from Eastern Alps	1959–today **reintroductions** in several regions from Alps and other countries	1950–today **reintroductions** and restocking with boar from Italy and from Central and Eastern Europe	1955–today translocations from Western Alps to Central and Eastern Alps	Mattioli *et al.* 2001, Crestanello *et al.* 2009, Apollonio *et al.* 2010b
Slovenia	1888–99 **reintroductions** in 2 areas from Carpathians, Austria, Germany, Hungary and Poland, restocking in 1962 with deer from Croatia	No	1913 escape from a private enclosure keeping animals from Germany	1927–59 translocations of chamois from Karawanke Alps	Adamic and Jerina 2010
Croatia	No	1960s introduction in Mt. Biokovo (failed)	No	1964–1978 **reintroductions** of Balkan and Alpine chamois in Velebit and Balkan chamois in Mt. Biokovo	Frković, 2008, Kusak and Krapinec 2010, present study
Greece	19th–20th century releases of deer from Germany, Denmark and Balkans in Mt. Parnitha, and **reintroduction** from here to Koziakas	No	1980–today releases from unknown sources in several areas of the country, **reintroduction** in Peloponnese	No	Papaioannou 2010

Serbia	1950–70 **reintroductions** from Croatia to mountains, 2009 **reintroduction** from Hungary to Fruška Gora NP	No	No	1963 **reintroductions** in NE, Central and SW Serbia, 1983 translocation from NE to E Serbia, 2007 **reintroduction** in Stara Planina NP from Bulgaria	Paunovic *et al.* 2010, Ristić *et al.*, 2010
Macedonia	After 1945 **reintroductions** from Slovenia, Croatia and Serbia	No	No	No	Stojanov *et al.* 2010

Notes: 'No' means that no release was reported, '–' means that the species is currently absent in the country.

with the single exception of a small nucleus inhabiting a costal forest in the Po delta area (Lorenzini *et al.*, 1998), and the present populations on Western Alps and Apennines are totally derived from a number of successful reintroductions following World War II (Mattioli *et al.,* 2001; see Figure 21.2 in Apollonio *et al.*, 2010b). Even in Norway, which seems to have been immune from the many reintroductions that characterised red deer management elsewhere in Europe, 17 animals of German–Hungarian origin were introduced into the much reduced stock on the island Otterøya over one century ago (Haanes *et al.*, 2010). Genetic analyses show that the present population in Otterøya consists of an admixture of Norwegian and Hungarian red deer (Haanes *et al.*, 2010).

Perhaps because they were already fairly widespread, roe deer seem to have been somewhat less affected by this type of manipulation: in at least 11 countries roe deer seem not to have been subject to any reintroduction (Apollonio *et al.*, 2010a). Yet, in other cases, populations have been affected by introductions, whether historic, as in the case of England where roe deer was reintroduced from Scotland, Germany, Austria and other countries (Whitehead, 1964; Chapman *et al.*, 1985; Prior, 1995; Baker and Hoelzel, 2013), or more recent, as in the case of Gardunha Mountains in Portugal (Carvalho *et al.*, 2008) or Petit Luberon in France (Maillard *et al.*, 2002; Calenge *et al.*, 2005). Where local native subspecies were still present, reintroductions inevitably have compromised the genetic integrity of local ecotypes, as in Central Italy where introduction of roe deer from Alps, Balkans and other regions threatened the conservation of native gene pools (see below).

Reintroductions had a limited impact also on alpine chamois. Actually, this species is present in 14 countries and in only five of them have no reintroductions taken place. However chamois was not subject to large scale operations related to the recovery of its full historical range, but rather to local management efforts aiming to restore original populations on isolated mountain chains: two good examples are the reintroduction of alpine chamois in France on the Massif Central and the Vosges (Maillard *et al.*, 2010) and in Croatia on Mt Biokovo (see Box 3.1). Only in the case of Czech Republic and in Germany were reintroductions directed towards the full re-establishment of the species once extinct (Bartos *et al.*, 2010; Wotschikowsky, 2010). Even more negligible was the role of human management on the present distribution of moose. Only one case of reintroduction is reported for the examined countries and was aimed at recovering the pristine range of moose in Central Poland (Wawrzyniak *et al.*, 2010).

The different extent of human intervention in affecting the distribution of these various species clearly reflects the cultural value attributed to them. Red deer have historically represented a species of major hunting interest; therefore it is not surprising that human interference was so extensive that we can claim the present distribution and status of red deer in Europe as largely the outcome of direct action of man (see also Linnell and Zachos, 2011). In some cases, translocations were clearly devoted to improve trophy quality and this played an important role in the choice of source populations (Lowe and Gardiner, 1974). In other species, too, greater efforts

Box 3.1 Fifty years since the successful reintroduction of the Balkan chamois to Mt Biokovo, Croatia.

The Dinarides Mountain Range represents a unique habitat in south-east Europe, extending from Slovenia through to Albania in the western part of the Balkan Peninsula (one of the 35 world biodiversity hotspots; Zachos and Habel, 2011). Mt Biokovo is part of the Dinarides Mountain Range and extends in the direction north-west–south-east in the middle of the Dalmatia region, at the point where Dinarides most closely approach the Adriatic Sea. The highest elevation is peak Sveti Jure (1762 m a.s.l.).

Historical data and archaeological research confirm a large abundance of chamois in these sites in former times, however the local population went extinct in the early 20th century (Miracle and Sturdy, 1991). The main reasons for its disappearance have been attributed to intensive livestock keeping and to over-hunting. After the disappearance of the chamois from Mt Biokovo, no other ungulate species inhabited this area until around 50 years ago. In the 1960s, due to a decline in the interest for livestock keeping and the gradual abandonment of Mt Biokovo, hunters of the Makarska region decided to try to re-establish wild ungulate populations in the mountain. The assistance of the Institute for Forestry and Hunting Research from Zagreb was requested; it compiled a project on the introduction of Balkan chamois (*R. rupicapra balcanica*), mouflon and roe deer on Mt Biokovo, edited by Dr Oto Röhr and Dr Zvonko Car (Šabić and Lalić, 2005).

The assessment of the local environmental conditions indicated that Mt Biokovo was a favourable habitat, with the possibility of supporting autumn populations of 6–10 chamois per 100 ha. Reintroduction in the original habitats took place in the period from 1964 to 1969, when 48 Balkan chamois (36 females, 12 males) were relocated here from Mts Čvrsnica and Prenj in the neighbouring Bosnia and Herzegovina. The first chamois were transported and released on 31 October 1964 above the village Veliko Brdo near Makarska, when seven individuals (three males, four females) were released. Twelve years later, in 1976, when the population was assessed at about 500 individuals, hunting was permitted for the first time; between 1976 and 1994 a total of 946 chamois were culled (657 males and 289 females; Šabić and Lalić, 2005).

In addition to chamois, mouflon and roe deer were also released in the area in the same period, though the results of such introductions were significantly poorer. The current mouflon population is assessed at approximately 80 individuals, while roe deer have completely disappeared from Mt Biokovo. The reason of the failure of roe deer establishment lies in the fact that local habitat and climatic conditions are not very suitable for this species. The lack of natural water sources is the main problem of this habitat, and therefore numerous pools in karst valleys were built and stone water troughs installed. Currently, some 20 pools are maintained, with fresh water regularly transported by trucks in the summer months (Šabić and Lalić, 2005).

The wolf (*Canis lupus*) has begun to appear in these areas in the past decade, and nowadays it is a constant presence in the area, with two resident packs (12–20 individuals in total, B. Šabić; pers. comm.). Hunters have also observed the tracks of two lynx (*Lynx lynx*) in the area (R. Bekavac, pers. comm.), and the presence of

Continued

Box 3.1   Continued

brown bear (*Ursus arctos*) is increasing to almost constant levels in recent years (B. Šabić, pers. comm.). The reintroduction of chamois is one of the main reasons for the reappearance of these large predators, which has ultimately had a positive consequence in restoring a natural community of large mammals, increasing the biodiversity of the area.

Mt Biokovo was declared Natural Park in 1981 and annually more than 20,000 vehicles and up to 80,000 visitors visit the park. The increased tourism pressure, particularly outside the marked hiking trails (collecting of medicinal herbs, berries and mushrooms, or playing extreme sports such as free climbing, trekking, mountain biking, motocross, etc.), has directly or indirectly impacted the park and brought disturbances to the habitats (Zwijacz-Kozica *et al.*, 2012). Due to the isolation of the mountain area, the current and future anthropogenic impacts (construction of motorways, tunnels, cable cars, urbanisation and mass tourism), and the appearance of large predators (wolf, lynx and bear), it is becoming increasingly important to undertake research studies on the biology of the population in order to ensure a sustainable management and a long-term conservation (Šprem *et al.*, 2011). According to the most recent data of the Ministry of Agriculture, which reports on a total number of about 1300 chamois in Croatia, the population on Mt Biokovo represents an important stronghold for the species, with about 450 individuals. As the current range of the Balkan chamois in the country does not correspond to the actual habitat suitability, and it is listed as 'regionally extinct' in the Red Book of Mammals of Croatia (Tvrtković and Grubešić, 2006), the feasibility could be evaluated to use the Mt Biokovo population (possibly supplemented by other stocks) as a source for further reintroductions.

Balkan chamois in Mt. Biokovo (photo: Igor Ilic-Serval).

to restore declining populations were observed when the link with hunting interests was similarly strong. It is not surprising, for instance, that wild boar reintroductions were more common in countries like Italy, where a long tradition of wild boar hunting with dogs was present. Conversely, in the case of moose, the lack of a strong hunting tradition in previous centuries acted against active management policies, even if nowadays, thanks to its natural recovery, the species represents a very popular quarry species in Fennoscandia. Apart from in Italy and Portugal, reintroductions seem to have played a minor impact on roe deer, possibly, because this species, among ungulates, was affected by a higher percentage of failures (Niethammer, 1963).

Most of the reintroductions in which ungulate distribution was manipulated in order to provide recreational opportunities, took place during the second half of the 20th century and were aimed at filling the distributional gaps generated between the middle of the 19th century and the end of World War II. Such management choice was the consequence of a combination of low numbers and scattered distribution of ungulate species, a greater demand by hunters and the improved economic conditions of many European countries after World War II. This brought consequences, such as a general tendency not to plan such introductions in any coordinated way, since they were conceived mostly as a pure wildlife management action undertaken by local administrations and/or by groups of hunters.

3.3 Genetic aspects: lessons from inappropriate reintroductions

The history of wild ungulates and their reintroduction in Europe is full of examples in which human activities have deliberately or inadvertently modified the original gene pool of populations. Translocation of animals out of their native range, release of captive-reared individuals, co-occurrence of related domestic animals, or the introduction of related exotic species (offering potential for hybridisation) all represent possible sources of genetic changes in wild ungulate populations (see Linnell and Zachos, 2011 and Chapter 4 in this volume). International authorities acknowledged the relevance of genetic biodiversity in conservation and stressed its significance in relation to reintroductions in the late '90s (IUCN, 1998), already recognising the potential problems arising from genetically inappropriate translocation. More recently IUCN has also remarked that reintroduced animals and plants should reflect, as far as possible, the patterns of genetic variation occurring in the original or (if extinct) in the extant neighbouring, populations, or in populations inhabiting similar habitats (IUCN/SSC, 2013). Modern genetic methods offer the potential to assess the genetic characteristics of possible source populations, as well as to define important parameters (like genetic variation, relatedness and number of founders, number of source populations) for reintroduction projects, thus maximising the chance of success (Allendorf and Luikart, 2007).

All these aspects have been partially or completely neglected in many reintroductions performed during the last century, in part because of lack of understanding, in part because so many of these interventions were never coordinated as such and often carried out by individuals or individual organisations without wider consultation. Unfortunately, genetic tools have become popular in wildlife management and conservation biology only in the last 30 years, and thus they were not available when most of the reintroductions were performed. Thus, historically, and even in more recent times, source population and individuals to release have most often been selected on the basis of other criteria, like physical traits (e.g. trophy size), local abundance or logistics (e.g. personal contacts, ease of the bureaucratic procedure, costs, etc.).

Ultimately, neglecting genetic aspects in wildlife reintroductions has three main potential consequences: (i) the establishment of a population of inbred or maladapted individuals (Groombridge *et al.*, 2012), (ii) the breakdown of the wider natural phylogeographic structure of the species (Olden *et al.*, 2004; Sommer *et al.*, 2009). (iii) the contact and genetic admixture with neighbouring native populations, with loss of local adaptation or even genetic extinction of endangered taxa (subspecies) (Moritz, 1999; Storfer, 1999).

3.3.1 Inbreeding and maladaptation

Using an insufficient number of founders or a sufficient number of inadequately selected founders (e.g. related individuals or from a genetically depleted source population) can produce inbreeding depression or low adaptability, due to a restricted genetic variation (Keller *et al.*, 2012). Both are likely to cause a reduced viability of the new population. Similarly, transferring animals from a wrongly selected source can imply a mismatch between the environmental conditions in the reintroduction area and the genetic background of founders, which makes the newly established population more susceptible to ecological stresses and stochastic events. The latter risk can be even higher if the released animals come from captivity, where selective forces are relaxed and founders can be vectors of maladapted genetic variants (Lynch and O'Hely, 2001; Frankham *et al.*, 2004).

Maximising the genetic variation and minimising inbreeding in the founder group are of utmost importance, as these individuals will represent the starting gene pool of the new population. In international conservation programmes where a reintroduction follows a captive-breeding program, this aspect usually receives great attention (Frankham *et al.*, 2004). Conversely, in reintroductions undertaken for recreational purposes by local administrations or by hunters, animals are usually chosen on other grounds (i.e. age, sex, phenotype and especially trophy quality).

The alpine ibex represents a paradigmatic case of ungulate species, endemic to the European Alps, whose present genetic status was strongly influenced by past reintroductions. All current ibex populations have their ancestry in the remnant nucleus of around 100 ibex which survived in the 19th century in the Gran Paradiso area (see 3.1). Animals from this population were transferred to Swiss zoos and parks, which represented the main source, together with the Gran Paradiso National Park, for subsequent reintroductions in the wild. Further translocations brought ibex into a number of additional areas in all the alpine countries. In most cases such transferred stocks had multiple origins and were composed of fewer than 50 individuals (Biebach and Keller, 2009).

Accordingly, all present populations have undergone from one (Gran Paradiso) to four or more successive bottlenecks (Fig. 3.1), being exposed to the negative effects of inbreeding and genetic drift.

In practice, although the general outcome of ibex reintroduction programmes was positive, the species was saved from extinction and a large (yet fragmented) trans-national population was reconstituted, genetic data highlighted possible drawbacks of such species management. Biebach and Keller (2009, 2010, 2012) analysed historical data and genetic variation in around 40 reintroduced ibex populations in Switzerland. They observed that, although Swiss nuclei were very close to representing the genetic diversity of the ancestral Italian population, many single populations had only a small fraction of it. This was associated to a loss of rare alleles and to a reduction of heterozygosity, both influenced by the founder group size and the number of bottlenecks experienced. While allelic loss was an immediate consequence of a founder event, heterozygosity diminished more gradually with time (Biebach and Keller, 2009). An admixed ancestry was able to counter-balance a low number of founders, as a higher genetic diversity was, not surprisingly, found in populations receiving a genetic contribution from multiple sources (Biebach and Keller, 2012). High levels of inbreeding were detected in several populations, especially those that had been founded by captive-bred individuals, and it appeared to depend on the number of released individuals and on the historical effective population size since founding (Biebach and Keller, 2010).

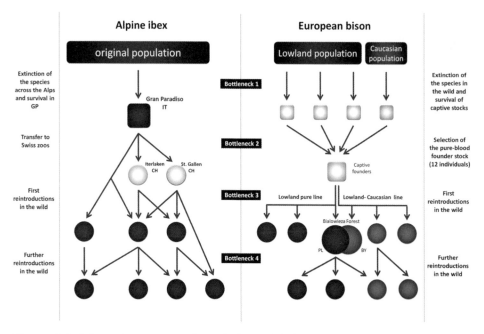

Figure 3.1 Schematic representation of the serial founder events (and bottlenecks), which accompanied the restoration of alpine ibex (*Capra ibex*) and European bison (*Bison bonasus*) populations in Europe. Grey tones are used to distinguish captive (lighter) from wild (darker) populations.

The negative effect of a high level of inbreeding was suggested by correlation with slower rates of post-release population growth (Biebach, 2009), but other variables could have played a role and conclusive data are lacking. Interestingly, it was estimated that on average only about half of the animals released in Switzerland actually contributed to the present-day genetic variation of the reintroduced populations (i.e. were 'genetic founders', Biebach and Keller, 2012). This highlights one of the most important factors in reintroductions: released individuals do not contribute equally to the gene pool of the new population. The difference between the founder group size and the estimated number of genetic founders can be due to the presence of relatives in the initial group or to a difference in individual fitness. Here, the skewed male reproductive success of ibex (Willisch *et al.*, 2012) might have played a role, warning about the importance of taking biological information into account in planning reintroductions.

Reintroductions of game species are often carried out using animals derived from captive populations, which ensure a reduced effort if compared to the use of live-captured individuals. However, relevant genetic problems are linked to the use of animals that have been raised in captivity (Frankham *et al.*, 2004). Under captive conditions selective forces are different from the wild: tameness and ability to breed are often selected for, while the ability to search for food and escape predators are no longer favoured. Also selection for disease and parasite resistance is usually weaker due to hygiene and veterinary care. Furthermore, if populations are small, inbreeding depression, loss of genetic diversity or mutational accumulation due to drift can be substantial.

On the other hand, captive populations have sometimes a high genetic variation because of deliberate crossings between individuals from multiple sources or between the wild species and the corresponding domestic form. In such cases, the major risk associated to the reintroduction of admixed captive animals is the establishment of a population with anomalous morphological and behavioural traits, which can become invasive, impacting natural and agriculture systems and threatening native populations. This is the case in many wild boar reintroductions or illegal releases, which were realised across Europe with individuals originating from captive populations that were often introgressed by domestic pigs (Scandura *et al.*, 2011). Most Italian wild boar populations north of the Po river have this origin (Monaco *et al.*, 2007), as well as free-ranging populations in England (Frantz *et al.*, 2012), in north-western Europe (Goedbloed *et al.*, 2013) and probably other regions across the continent.

Therefore, when captive founders are going to be used for a reintroduction, a genetic evaluation is strongly recommended beforehand in order to limit the detrimental effects of captivity, whether this may happen through prejudicing the viability of the reintroduced population or by impacting the local environment.

3.3.2 Breakdown of species phylogeography

At a wide geographical scale, empirical patterns of genetic variation often show regional differences across a species' range. Arrays of populations form intraspecific genetic clusters or phylogroups, which may be recognised or not as 'evolutionarily significant units' (Avise, 2000). Most of the time, a genetic legacy of retreats into Pleistocenic glacial refugia and subsequent post-glacial recolonisation routes are invoked to explain such phylogeographic discontinuities (Taberlet *et al.*, 1998; Sommer and Zachos, 2009). This often implies the occurrence of barriers which limited gene flow at suture zones and the establishment of mutation-drift equilibria which would maintain differentiation over time. Simulation studies, however, revealed that genetic breaks can arise in continuously distributed species even in absence of historical barriers to gene flow (Irwin, 2002).

Because of overhunting and habitat disruption, European ungulate populations were exposed to perceptible demographic and range fluctuations in the past (Sommer and Zachos, 2009), which temporarily interrupted and reshaped the natural phylogeographic structure. Reintroductions contributed to the restoration of a geographic, but often not genetic, continuity in the species' range. In fact, the local replacement of extinct native gene pools with alien genotypes introduces artificial spikes in the genetic divergence among populations, which are incoherent with any model based on spatial distance, landscape resistance to gene flow or local adaptation (e.g. Frantz *et al.*, 2006; Sommer *et al.*, 2009; Linnell and Zachos, 2011). As a consequence, in European ungulates (as in other game species) phylogeographic patterns which are otherwise difficult to explain were usually justified with undocumented translocations that would have been carried out for hunting purposes (Randi *et al.*, 2004; Crestanello *et al.*, 2009).

This was clearly the case in the number of inconsistencies observed in the genetic composition of red deer populations, which were subject to a number of reintroductions (see above). Two main lineages are present in European red deer, an eastern and a western lineage, plus an additional Mediterranean lineage occurring in Sardinia, Corsica and North Africa (Zachos and Hartl, 2011). Since the western lineage is largely dominant in Europe, spreading from Iberia to north-eastern Europe, including Scandinavia and the British Isles, most of the translocations have occurred across regions occupied by this clade and little disruption to phylogeographic patterns is thus apparent. However, some exceptions exist. Corsican red deer mitochondrial DNA (mtDNA) haplotypes were identified in Scottish islands (Nussey *et al.*, 2006) and in Ireland (Carden *et al.*, 2012), testifying to the use of animals from very distant populations for past (probably ancient) introductions. Translocations of deer from western Europe, linked to reintroductions or restocking operations, were considered the probable explanation for the wide suture zone between the two lineages occurring in Poland, Belarus and the Baltic Countries (Niedziałkowska *et al.*, 2011).

A similar situation is reported for roe deer. As consequence of introductions from eastern countries, populations of north-western Italy show a clear phylogeographic mismatch, carrying mtDNA belonging to the eastern *Capreolus capreolus* lineage (Vernesi *et al.*, 2002). Other reintroductions artificially implanted Balkan and Central European haplotypes into central Italy (Randi *et al.*, 2004; and see below). Among chamois, *Rupicapra pyrenaica pyrenaica* mtDNA sequences were found in *R. r. rupicapra* populations of the Italian Alps (Crestanello *et al.* 2009; Rodriguez *et al.*, 2009). By discarding the alternative hypothesis of ancient hybridisation between the two species, Crestanello and colleagues attributed this phylogeographic mismatch to a more recent translocation of Pyrenean chamois to the Italian Alps, although this interpretation has been questioned (Corlatti *et al.*, 2011).

As a comment on these examples, we should remark that, as a general rule, the genetic impact of translocations increases with the distance over which animals are moved, as it often correlates with genetic distance (the so-called 'isolation by distance' pattern). This also means that non-native genotypes are more likely to be identified if they come from distant populations, whereas genetic signals of relocations are more difficult to track if animals are moved over medium or short distances. Yet, local adaptations and habitat fragmentation can lead to a strong genetic differentiation even between relatively close populations (Orsini *et al.*, 2013). In practice, there is no absolute distance threshold that can make an introduction detectable, because it depends on the species' phylogeography and dispersal ability, as long with landscape features. Furthermore, as climate represents one of the most important selective forces, translocations over long latitudinal ranges are expected to cause a potential loss of adaptive ability. Evolutionary consequences of such artefacts are mostly unexplored and can perhaps be investigated in the future by modern genomic approaches.

Overall, phylogeographic studies on several species in Europe have revealed that, although translocations may have muffled genetic diversity at a regional scale, the overall phylogeography of ungulate species across the continent appears to have not been impacted that much and, for most species, the signature of past (peri-glacial) demographic and spatial dynamics is still detectable (Lorenzini and Lovari, 2006; Scandura *et al.*, 2008; Skog *et al.*, 2009; Vilaça *et al.*, 2014).

3.3.3 Genetic admixture and loss of endemic variation

Under natural conditions, the genetic identity of a population or species arises from adaptation to external and internal modifications and from the effect of chance. As a consequence, genetic diversity, even at a local or regional scale, is geographically structured, with either clinal or abrupt differentiation between adjoining areas, depending on local environmental features and on the presence of physical barriers to gene flow (e.g. Coulon *et al.*, 2004; Pérez-Espona *et al.*, 2008). Stochastic events may induce local extinctions, breaking the species' continuity, and such gaps are naturally sutured by dispersal from neighbouring demes. In case of human intervention, the release of translocated or captive-bred individuals can introduce alien genes, which are incoherent with the previous population structuring. With the spread of the new population, genetic variants that had been selected in different environmental conditions may come into contact and mix up, producing hybrid genotypes which are able to replace the original ones (Laikre *et al.*, 2010). This leads to a depletion of native variation that can ultimately drive to genetic extinction of local taxa (Rhymer and Simberloff, 1996) and prejudice the future adaptive potential of the species (Storfer, 1999). Notably, introgressive hybridisation can induce an augmentation of fitness in admixed individuals (the so called 'hybrid vigour'), thus hiding the important, and mostly irreversible, negative effects.

This risk was ignored, or at least not foreseen, in most translocations of wild ungulates performed across Europe to supplement existing populations or to reintroduce species where they had died out (Carranza *et al.*, 2003; Crestanello *et al.*, 2009; Sommer *et al.*, 2009; Zachos and Hartl, 2011). In Italy, for instance, the recovery of red deer, roe deer and wild boar populations from previously low levels was largely obtained through a number of reintroductions and restocking activities (Apollonio *et al.*, 2010b; and see earlier in this chapter). In the case of roe deer, after World War II, non-native stocks, mostly originating in the Alps or Eastern Europe, were released in many areas of central and southern Italy. Once the scientific community acknowledged the existence of isolated remnant populations of an endemic subspecies (*Capreolus c. italicus* Festa 1925), the risk of genetic admixture with introduced conspecifics and loss of endemic diversity by the native taxon became a major conservation issue (Focardi *et al.*, 2009). Unfortunately, especially in its northern range, the endemic subspecies had already come into contact with the expanding reintroduced populations (e.g. in Tuscany) and hybrid assemblages had resulted and are currently present (Gentile *et al.*, 2009; Mucci *et al.*, 2012).

Admixed populations may also have arisen in Portugal, where the natural spread of native roe deer populations in the north of the country was reinforced with reintroductions of animals from France (Vingada *et al.*, 2010). Subsequent molecular investigations suggested however that the native population may belong to an endemic Iberian subspecies (*C. c. decorus*) shared with north-western Spain (Royo *et al.*, 2007), questioning the appropriateness of introducing additional animals from the genetically distinct populations in France.

A similar situation threatens the genetic integrity of the Tatra chamois (*Rupicapra r. tatrica*), an endemic subspecies surviving in the Tatra Mountains at the border between Slovakia and Poland. Between 1950 and 1970, when the original population had declined to around 200 individuals, reintroductions were planned in surrounding massifs in Slovakia. However, while in one case (Low Tatra) the new population was established by translocating *R. r. tatrica* individuals, the other reintroductions were realised with alpine chamois *R. r. rupicapra* (Findo and Skuban, 2010). Hence, current populations, occurring within a radius of <100 km, host two divergent taxa, which can come into contact and hybridise, and genetic data suggest an incipient hybridisation (Crestanello *et al.*, 2009). This represents a serious risk to the genetic integrity of the native subspecies.

3.3.4 All in one: the case of the rescued European bison

The wisent, or European bison, provides a comprehensive example of all the different abovementioned issues, highlighting the importance of genetic management in reintroduction plans of severely bottlenecked species. After having disappeared in the wild in 1927 and the already mentioned great international effort to recover the species, the European bison was reintroduced into at least 35 different areas across Eastern Europe. Nonetheless, the restoration of this species in the wild raised a number of genetic issues, previously examined by Linnell and Zachos (2011), that can sum up, in one species, the three points mentioned above.

First, all existing bison populations descend from 12 individuals surviving in the 1920s, which belonged to two different subspecies: one male being the only surviving mountain (or Caucasian) bison, *B. b. caucasicus*, and 11 individuals of the Lowland subspecies *B. b. bonasus* (Krasinska and Krasinski, 2007). Not surprisingly, the species' genetic variability is very low (Tokarska *et al.*, 2011), due to the initial dramatic bottleneck and to additional bottlenecks associated with any new reintroduction (Figure 3.1). Also, high levels of inbreeding were observed in bison populations (Olech, 1987), and an impairment of the immune response was a possible sign of inbreeding depression (Radwan *et al.*, 2006; Wolk and Krasinska, 2004). Second, due to the nature of the surviving group, two different lines were constituted: a pure Lowland and an admixed Lowland–Caucasian line. Given the use of this admixed Lowland–Caucasian line to reintroduce bison into the former range of *B. b. bonasus* (Tokarska *et al.*, 2011), the recovery of the species in the wild implied a mismatch with its historical phylogeography. In fact, most pure Lowland populations are concentrated in Northern Poland, Belarus and Lithuania, while the admixed line was reintroduced in lowland forests of Ukraine and Russia (Krasinska and Krasinski, 2007). The present phylogeographic pattern is thus basically the consequence of the choice of animals for reintroductions, which was constrained by the limited availability of source populations. Finally, at the start of the breeding programme, a number of hybridisation events occurred in the captive populations between European × American bison (*Bison bison*) and European bison × cattle (*Bos taurus*), and hybrids were often in mixed groups with pure-blood individuals. One of the key actions in the rescue of the wisent was the creation of the European Bison Pedigree Book, which recorded all living pure-blood specimens, enabling the removal of hybrids from the breeding herds and the establishment of pure lines. In the Polish side of the Białowieża Forest, a concerted effort was made to ensure a pure Lowland bison population over a period of time, although an admixed reintroduced population was present on the Belarusian side of the forest (Krasinska and Krasinski, 2007). Nonetheless, hybrids with cattle and with North American bison were sometimes used in reintroductions (e.g. in the Caucasus region; Olech, 2008).

Lessons from past reintroductions highlight that, in planning any new reintroduction, the following three genetic aspects must be seriously taken into account: (1) genetic identity of the former extinct population (by using ancient DNA) or of living populations that lie in proximity of the target area, (2) genetic identity of the source population, (3) genetic variation of the reintroduced stock of individuals. A growing literature on the genetic make-up of European populations offers a solid background for most species of ungulates.

Post-release genetic monitoring of reintroduced populations is highly recommended. As census size may be fairly different from the effective population size, especially in polgynous species (Miller *et al.*, 2009), genetic analyses should be performed at different time intervals, in order to verify the possible rise of inbreeding or loss of genetic variation due to genetic drift in the population.

Finally, as the number of reintroductions is declining, it could be wise to put a major effort in evaluating the genetic consequences of past reintroductions, with a special focus on the risk of genetic pollution for indigenous taxa.

3.4 Factors affecting the success of reintroductions in European ungulates

What are the determinants of a successful reintroduction of ungulates? Review papers on the main factors responsible for the success of a reintroduction (Griffith *et al.*, 1989; Wolf *et al.*, 1996, 1998; Fischer and Lindenmayer 2000) stress the importance of the number and origin of individuals released and habitat quality at the release site. Some more ungulate-specific papers insist on the need to use large numbers of founders in order to reach successful outcomes. In particular, in the review by Komers *et al.* (1999), it was shown that reintroduction success increases substantially with the size of the founder population but that the function becomes asymptotic at about 20 animals, as demographic and environmental stochasticity affect smaller populations more than larger ones (Caughley and Sinclair, 1994).

It is therefore interesting to note that this primary rule was not generally taken into account in ungulate reintroductions in Europe; on the contrary, there is wide evidence of the use of very limited sets of founders, especially in earlier times. If we analyse the number of red deer used for reintroductions realised in the Italian Alps and Apennines before 2000 (Mattioli *et al.*, 2001), we can see that out of 22 reintroductions on the Alps, only in 6 cases were 20 deer or more released; on the Apennines out of 11 reintroductions, 7 were done with fewer than 20 founders. It is however to be noted that all these reintroductions were successful and some quite surprisingly so: in Val di Susa, Piedmont, starting with 10 founders in 1962–64, further supplemented by five deer in 1986, a population of thousands of deer is now present; in the Acquerino Forest, Tuscany, starting from seven deer released in 1958 and 1965 a similar result was achieved; and in Poland, the only well documented reintroduction of moose in Europe was done in the Kampinos area with five individuals (Dzieciolowski and Pielowski, 1993).

In contrast, in some well organised reintroductions the number of founders was clearly much more substantial and could have exceeded 100 individuals, either released in groups or in small units across a large area. Mattioli *et al.* (2001) noted that in Italy, red deer reintroduction in Abruzzo National Park in 1972–75, involved the release of 64 deer in 3 years while more than 100, also in 3 years, were released in the Velino Sirente Regional Park (in the Abruzzo region as well) in 1990–93. A different approach implies the use of many small groups of individuals released across a large area in order to reach an homogeneous coverage: this was, for instance, the case of wild boar restoration within the Italian province of Arezzo, Tuscany. There, wild boar was absent from the 19th century and the aim was to bring back the species in the whole mountain and hilly areas of the province, more than 150,000 ha. In the 3 years from 1972 to 1974 groups of 2–6 boars were released in up to 20 different localities up to an overall number of 130 boar; the annual cull of wild boar is now about 20,000 and the species covers more than 250,000 of the 320,000 ha of the province (Arezzo Provincial Government, 1980).

One of the main reasons that should induce the use of large numbers of individuals in a reintroduction is to maximise genetic variation, which should confer a greater adaptive potential and reduce the risk of inbreeding depression (see above). However, the same result can be achieved by using founders from multiple sources (Biebach and Keller, 2012). Actually, the choice of the source population may represent a critical factor: many founders from one environmentally similar area should ensure a quick adaptation and higher initial growth rate to the reintroduced population; on the other hand, multiple sources can prevent the risk of inbreeding depression and ensure a higher long-term resistance to ecological perturbations. Again, in the history of ungulate reintroductions in Europe both situations are found. In red deer, many releases included animals from different populations or intentional crosses undertaken for trophy improvement (e.g., Haanes *et al.*, 2010; Niedziałkowska *et al.*, 2012). In most cases, releases from different sources were performed at different times. Out of 40 reintroduced ibex populations in Switzerland, 18 were established with the contribution of multiple sources and such admixture strongly influenced their current genetic diversity (Biebach and Keller, 2012). Unfortunately, the origin of animals used for many ungulate reintroductions is unknown and no clear evidence is available on the impact of one versus multiple sources on the final outcome of reintroductions.

Another important factor to promote success of reintroduction is considered the habitat quality at release site (Griffith *et al.*, 1989; Wolf *et al.*, 1996, 1998). As in the case of the number of founders there are contradictory reports in the record of European reintroductions; earlier attempts were obviously less accurate in defining habitat suitability with proper predictive models. With respect to this even the reintroduction of Alpine ibex in Switzerland, a high valuable species from a conservation viewpoint, posed some questions. Even as early as 1966, Nievergelt, looking at population dynamics of single established nuclei, remarked that some locations were unsuitable for the species and this ended up with a reduced viability of the reintroduced population itself.

It is interesting to note that the low success rate commonly attributed to animal reintroduction (Griffith *et al.*, 1989; Wolf *et al.*, 1996) seems not to be typical for the reintroduction of European ungulates. Ungulate reintroductions out of Europe already proved to be quite successful (Stanley Price, 1989; Matson *et al.*, 2004; Wronski, 2010); Komers *et al.* (1999) showed a success of 81% out of 31 reintroductions/introductions of European and non-European ungulates. In Europe other data confirmed this tendency: almost no reintroduction attempt of European bison failed (Krazinska and Krazinski, 2007), out of 36 reintroductions of alpine ibex in Italy, 72% were successful and 14% could not be evaluated at time of the review (Tosi *et al.*, 2009). No information of obvious failures is available over the scores of reintroductions of red deer in the same country (Mattioli *et al.*, 2001; Apollonio, pers. comm.). As a reference, we may consider that out of 116 re-introductions of bird and mammals worldwide, 26% were classified as successful, 27% as failures, while 47% were classified as unknown at the stage of publication (Fisher and Lindenmayer, 2000).

A first remark is that reintroductions of herbivores are in general more successful than those of carnivores (Griffith *et al.*, 1989). Specifically for ungulates, a possible explanation for their enhanced success in comparison to other species can be given by a higher tolerance of most (but not all) species to stress, that is an important factor in determining the outcome of a reintroduction attempt (Teixeira *et al.*, 2007). Some species are highly tolerant to the presence of humans to the point where they have been subjected to some form of domestication in the past (e.g. red deer in Norway, Andersen *et al.*, 2010; mouflon in Sardinia, Vigne, 1992; wild boar, Albarella *et al.*, 2011). Furthermore, in many cases these reintroductions were conducted using soft release: even if the question of the pros and cons of soft versus hard release is still open (Parker *et al.*, 2012), in the case of ungulates, it seems that soft release confers some advantages (Ryckman *et al.*, 2010). A further point is the extreme environmental plasticity of the most common European species (red deer, roe deer, wild boar) that are able to live from the seashore to the upper mountains and extend their distribution from Mediterranean areas to the far north in Fennoscandia/Baltic countries.

3.5 Conclusions

It is pretty clear that in Europe there is no more room for further reintroduction of non-endangered species, except in the few countries where ungulates have not yet reached high numbers, like Greece and Portugal (Papaioannou *et al.*, 2010; Vingada *et al.*, 2010). In general in Europe the priority is now constituted by a proper management of the rich resource represented by almost 20 million ungulates rather than by their increase (both in number and distribution). But there should remain two priorities, both related to the need to preserve the local genetic identities of the native ungulate populations with special reference to endemic subspecies:

• To restore viable populations (or metapopulations) of native, endangered subspecies within their range and to protect them from the risk of crossbreeding with exotic subspecies;
• To amend, as much as possible, genetically depleted local populations of even common species and subspecies that are now living in sympatry or even crossbreeding with introduced stocks of different origin.

Acknowledgements

We are grateful to T. Podgorski, R. Putman and F. Zachos for useful comments and information on reintroduction events.

References

Adamic, M. and Jerina, K. (2010) Ungulates and their management in Slovenia. In M. Apollonio, R. Andersen and R.J. Putman (eds), *European Ungulates and their Management in the 21st century*. Cambridge, UK: Cambridge University Press, pp. 507–526.

Albarella, U., Dobney, K. and Rowley-Conwy, P. (2011) The domestication of the pig (*Sus scrofa*): new challenges and approaches. In M.A. Zeder, D.G. Bradley, E. Emshwiller and B.D. Smith (eds), *Documenting Domestication: New Genetic and Archeological Paradigms*. Berkeley, CA: University of California Press, pp. 209–227.

Allendorf, F.W. and Luikart, G.H. (2007) *Conservation and Genetics of Populations*. Oxford: Blackwell Publishing.

Andersen, R. and Holthe, V. (2010) Ungulates and their management in Denmark. In M. Apollonio, R. Andersen and R. Putman (eds), *European Ungulates and Their Management in the 21st Century*. Cambridge, UK: Cambridge University Press, pp. 71–85.

Andersen, R., Lund, H., Solberg E.J. and Sæther, B-E. (2010) Ungulates and their management in Norway. In M. Apollonio, R. Andersen and R. Putman (eds), *European Ungulates and Their Management in the 21st Century*. Cambridge, UK: Cambridge University Press, pp. 14–36.

Andersone-Lilley, Ž. Balčiauskas, L., Ozolinš, J., Randveer, T. and Tõnisson J. (2010) Ungulates and their management in the Baltics (Estonia, Latvia and Lithuania). In M. Apollonio, R. Andersen and R. Putman (eds), *European Ungulates and Their Management in the 21st Century*. Cambridge, UK: Cambridge University Press, pp. 103–128.

Apollonio, M., Andersen, R. and Putman, R. (2010a) *European Ungulates and Their Management in the 21st Century*. Cambridge, UK: Cambridge University Press.

Apollonio, M., Ciuti, S., Pedrotti, L. and Banti, P. (2010b) Ungulates and their management in Italy In M. Apollonio, R. Andersen and R. Putman (eds), *European Ungulates and Their Management in the 21st Century*. Cambridge, UK: Cambridge University Press, pp. 475–506.

Arezzo Provincial Government (1980) *Cinghiali, reintroduzione ed ambientamento di questa specie nel territorio aretino*. Arezzo, Italy: Poligrafico Aretino.

Avise, J.C. (2000) *Phylogeography: The History and Formation of Species*. Cambridge, MA: Harvard University Press.

Baker, K.H. and Hoelzel A.R. (2013) Evolution of population genetic structure of the British roe deer by natural and anthropogenic processes (*Capreolus capreolus*). *Ecology and Evolution*, **3**, 89–102.

Bartos, L., Kotrba, R. and Pintir, J. (2010) Ungulates and their management in Czech Republic. In M. Apollonio, R. Andersen and R. Putman (eds), *European Ungulates and Their Management in the 21st Century*. Cambridge, UK: Cambridge University Press, pp. 243–261.

Beccu, E. (1993) Consistenza e prospettive di salvaguardia della popolazione di *Cervus elaphus corsicanus* presente in Sardegna. *Supplementi alle Ricerche di Biologia della Selvaggina* **21**, 277–287.

Biebach, I. (2009) *Genetic structure, genetic diversity and inbreeding in reintroduced Alpine ibex (Capra ibex ibex) populations*. PhD Dissertation, University of Zurich.

Biebach, I. and Keller, L.F. (2009) A strong genetic footprint of the re-introduction history of Alpine ibex (*Capra ibex ibex*). *Molecular Ecology* **18**, 5046–5058.

Biebach, I. and Keller, L.F. (2010) Inbreeding in reintroduced populations: the effects of early reintroduction history and contemporary processes. *Conservation Genetics* **11**, 527–538.

Biebach, I. and Keller, L.F. (2012) Genetic variation depends more on admixture than number of founders in reintroduced Alpine ibex populations. *Biological Conservation* **147**, 197–203.

Breitenmoser, U. (1998) Large predators in the Alps: fall and rise of man's competitors. *Biological Conservation* **83**, 279–289.

Calenge, C., Maillard, D., Invernia N. and Gaudin, J-C. (2005) Reintroduction of roe deer *Capreolus capreolus* into a Mediterranean habitat: female mortality and dispersion. *Wildlife Biology* **11**, 153–161.

Carden, R.F., McDevitt, A. D., Zachos, F. E., Woodman, P. C., O'Toole, P., Rose, H., Monaghan, N. T., Campana, M. G., Bradley, D. G. and Edwards, C. J. (2012) Phylogeographic, ancient DNA, fossil and morphometric analyses reveal ancient and modern human introductions of a large mammal: the complex case of red deer (*Cervus elaphus*) in Ireland. *Quaternary Science Reviews* **42**, 74–84.

Carranza, J. (2010) Ungulates and their management in Spain. In M. Apollonio, R. Andersen and R. Putman (eds), *European Ungulates and Their Management in the 21st Century.* Cambridge, UK: Cambridge University Press, pp. 419–440.

Carranza, J., Martínez, J.G., Sánchez–Prieto, C.B. *et al.* (2003) Game species: extinction hidden by census numbers. *Animal Biodiversity and Conservation* **26**, 81–84.

Carvalho, P., Nogueira, A.J.A., Soares, A.M.V.M. and Fonseca C. (2008) Ranging behaviour of translocated roe deer in a Mediterranean habitat: seasonal and altitudinal influences on home range size and patterns of range use. *Mammalia* **72**, 89–94.

Casaer, J. and Licoppe, A. (2010) Ungulates and their management in Belgium. In M. Apollonio, R. Andersen and R. Putman (eds), *European Ungulates and Their Management in the 21st Century.* Cambridge, UK: Cambridge University Press, pp. 184–200.

Caughley, G. and Sinclair, A.R.E. (1994) *Wildlife Ecology and Management.* Boston, MA: Blackwell Scientific Publications.

Chapman, N.G., Claydon, K., Claydon, M. and Harris, S. (1985) Distribution and habitat selection by muntjac and other species of deer in a coniferous forest. *Acta Theriologica* **30**, 287–303.

Corlatti, L., Lorenzini, R. and Lovari, S. (2011) The conservation of the chamois *Rupicapra* spp. *Mammal Review* **41**, 163–174.

Coulon, A., Cosson, J.F., Angibault M. *et al.* (2004) Landscape connectivity influences gene flow in a roe deer population inhabiting a fragmented landscape: an individual–based approach. *Molecular Ecology* **13**, 2841–2850.

Crestanello, B., Pecchioli, E., Vernesi, C. *et al.* (2009) The genetic impact of translocations and habitat fragmentation in chamois (*Rupicapra* spp). *Journal of Heredity* **100**, 691–708.

Crudele, G. (1988) La fauna. In M. Padula and G.Crudele (eds), *Le Foreste Campigna-Lama nell'Appennino tosco-romagnolo.* Bologna, Italy: Regione Emilia-Romagna Editore, pp. 4–11.

Csanyi, S. and Lehoczki, R. (2010) Ungulates and their management in Hungary. In M. Apollonio, R. Andersen and R. Putman (eds), *European Ungulates and Their Management in the 21st Century.* Cambridge, UK: Cambridge University Press, pp. 291–318.

Danilkin, A. (1996) *Behavioural Ecology of Siberian and European Roe Deer.* London: Chapman and Hall.

Dzięciołowski, R. and Pielowski, Z. (1993) *Łoś Warszawa, Poland.* (In Polish.) Warsaw: Anton-5.

Fernández-García, J.L., Carranza, J., Martínez, J.G. and Randi, E. (2014) Mitochondrial D-loop phylogeny signals two native Iberian red deer (*Cervus elaphus*) lineages genetically different to Western and Eastern European red deer and infers human-mediated translocations. *Biodiversity and Conservation* **23**, 537–554.

Findo, S. and Skuban, M. (2010) Ungulates and their management in Slovakia. In M. Apollonio, R. Andersen and R. Putman (eds), *European Ungulates and Their Management in the 21st Century.* Cambridge, UK: Cambridge University Press, pp. 262–290.

Fischer, J. and Lindenmayer, D.B. (2000) An assessment of the published results of animal relocations. *Biological Conservation* **96**, 1–11.

Focardi, S., Montanaro, P., La Morgia, V. and Riga, F. (eds), (2009) Piano d'azione nazionale per il capriolo italico (*Capreolus capreolus italicus*). *Quaderni Conservazione della Natura* **31**, Ministero Ambiente, ISPRA, Italy.

Frankham, R., Ballou, J.D. and Briscoe, D.A. (2004) *A Primer of Conservation Genetics.* Cambridge: Cambridge University Press.

Frantz, A.C., Tigel Pourtois, J., Heuertz, M., Schley, L., Flamand, M.C., Krier, A., Bertouille, S., Chaumont, F. and Burke, T. (2006) Genetic structure and assignment tests demonstrate illegal translocation of red deer (*Cervus elaphus*) into a continuous population. *Molecular Ecology* **15**, 3191–3203.

Frantz, A.C., Massei, G. and Burke, T. (2012) Genetic evidence for past hybridization between domestic pigs and English wild boars. *Conservation Genetics* **13**, 1355–1364.

Frković, A. (2008) Reintroduction of chamois in Northern Velebit. *Stručni Članci* **11–12**, 543–550.

Gentile, G., Vernesi, C., Vicario, S., Pecchioli, E., Caccone, A., Bertorelle, G. and Sbordoni, V. (2009) Mitochondrial DNA variation in roe deer (*Capreolus capreolus*) from Italy: evidence of admixture in one of the last *C. c. italicus* pure populations from central-southern Italy. *Italian Journal of Zoology* **76**, 16–27.

Goedbloed, D.J., van Hooft, P., Megens, H.-J., Langenbeck, K., Lutz, W., Crooijmans, R.P.M.A., van Wieren, S.E., Ydenberg, R.C and Prins, H.H.T. (2013) Reintroductions and genetic introgression from domestic pigs have shaped the genetic population structure of Northwest European wild boar. *BMC Genetics* **14**, 43.

González, J., Herrero, J., Prada, C. and Marco, J. (2013) Changes in wild ungulate populations in Aragon, Spain between 2001 and 2010. *Galemys* **25**, 51–57.

Griffith, B., Scott J.M., Carpenter G.W. and Reed, C. (1989) Translocations as a species conservation tool. *Science* **245**, 477–480.

Groombridge, J.J, Raisin, C., Bristol, R. and Richardson, D.S. (2012) Genetic consequences of reintroductions and insights from population history. In J.G. Ewen, D.P. Armstrong, K.A. Parker and P.J. Seddon (eds), *Reintroduction Biology: Integrating Science and Management.* Oxford: Blackwell Publishing, pp. 395–440.

Haanes, H., Røed, K.H., Mysterud, A., Langvatn, R. and Rosef, O. (2010) Consequences for genetic diversity and population performance of introducing continental red deer into the northern distribution range. *Conservation Genetics* **11**, 163–1665.

Haanes, H., Rosvold, J. and Røed K.H. (2013) Non-indigenous introgression into the Norwegian red deer population. *Conservation Genetics* **14**, 237–242.

Hmwe, S.S., Zachos, F.E., Eckert, I., Lorenzini, R., Fico, R. and Hartl, G.B. (2006) Conservation genetics of the endangered red deer from Sardinia and Mesola with further remarks on the phylogeography of *Cervus elaphus corsicanus*. *Biological Journal of the Linnean Society* **88**, 691–701.

Höglund, J., Cortazar-Chinarro, M., Jarnemo, A. and Thulin, C-G. (2013) Genetic variation and structure in Scandinavian red deer (*Cervus elaphus*): influence of ancestry, past hunting, and restoration management. *Biological Journal of the Linnean Society* **109**, 43–53.

Imesch-Bebiè, N., Gander, H. and Schnidring-Petrig, R. (2010) Ungulates and their management in Switzerland. In M. Apollonio, R. Andersen and R. Putman (eds), *European Ungulates and Their Management in the 21st Century.* Cambridge, UK: Cambridge University Press, pp. 357–391.

Irwin, D.E. (2002) Phylogeographic breaks without geographic barriers to gene flow. *Evolution* **56**, 2383–2394.

I.U.C.N. (1998) *Guidelines for Re-introductions*. Prepared by the IUCN/SSC Re-introduction Specialist Group. Gland, Switzerland and Cambridge: IUCN.

I.U.C.N./S.S.C. (2013) *Guidelines for Reintroductions and Other Conservation Translocations. Version 1.0*. Gland, Switzerland: IUCN Species Survival Commission, pp. viii.

Jenkins, D. (1972) The status of red deer (*Cervus elaphus corsicanus*) in Sardinia. In AA.VV. *Una vita per la natura*. Camerino, Italy: WWF.

Karcov, G. (1903) *Białowieża Primeval Forest. Historical description, current hunting management, and royal hunts in the forest*. A. Marks, Saint Petersburg, Russia. [In Russian.]

Keller, L.F., Biebach, I., Ewing, S.R. and Hoeck, P.E.A. (2012) The genetics of reintroductions: inbreeding and genetic drift. In J.G. Ewen, D.P. Armstrong, K.A. Parker and P.J. Seddon (eds), *Reintroduction Biology: Integrating Science and Management*. Oxford: Blackwell Publishing, pp. 360–375.

Kidjo, N., Feracci, G., Bideau, E. *et al.* (2007) Extirpation and reintroduction of the Corsican red deer *Cervus elaphus corsicanus* in Corsica. *Oryx* **41**, 488–494.

Komers, P.E., Curman, G.P., Birgersson, B. and Ekwall, K.(1999) The success of ungulate reintroductions: effects of age and sex composition. In L.M. Darling (ed.), *Proceedings of a Conference on the Biology and Management of Species and Habitats at Risk*, Kamloops, BC, Canada, 179–188.

Krasinska, M. and Krasinski, Z. (2007) *European Bison: The Nature Monograph*. Białowieża, Poland: Mammal Research Institute, Polish Academy of Sciences.

Kusak, J. and Krapinec, K. (2010) Ungulates and their management in Croatia. In M. Apollonio, R. Andersen and R. Putman (eds), *European Ungulates and Their Management in the 21st Century*. Cambridge, UK: Cambridge University Press, pp. 527–539.

Laikre, L., Schwartz, M.K., Waples, R.S., Ryman, N. and The GeM Working Group (2010) Compromising genetic diversity in the wild: unmonitored large-scale release of plants and animals. *Trends in Ecology and Evolution* **25**, 520–529.

Liberg, O., Bergström, R., Kindberg, J. and von Essen, H. (2010) Ungulates and their management in Sweden. In M. Apollonio, R. Andersen and R. Putman (eds), *European Ungulates and Their Management in the 21st Century*. Cambridge, UK: Cambridge University Press, pp. 37–70.

Libois R.M. (1992) Introductions et reintroductions de Mammiferes en Belgique: bilan et reflexions. Actes du XIVème Colloque Francophone de Mammalogie. *Annales biologiques du Centre* **4**, 17–28.

Linnell, J.D.C. and Zachos, F.E. (2011) Status and distribution patterns of European ungulates: genetics, population history and conservation. In R. Putman, M. Apollonio and R. Andersen (eds), *Ungulate Management in Europe: Problems and Practices*. Cambridge, UK: Cambridge University Press, pp. 12–53.

Lorenzini R and Lovari S. (2006) Genetic diversity and phylogeography of the European roe deer: the refuge area theory revisited. *Biological Journal of Linnean Society* **88**, 85–100.

Lorenzini, R., Mattioli, S. and Fico, R. (1998) Allozyme variation in native red deer *Cervus elaphus* of Mesola wood, Northern Italy: implications for conservation. *Acta Theriologica* suppl. 5, 63–74.

Lowe, V.P.M. and Gardiner, A.S. (1974) A re-examination of the sub-species of red deer (*Cervus elaphus*) with particular reference to the stocks in Britain. *Journal of Zoology, London* **174**, 185–201.

Lynch, M. and O'Hely, M. (2001) Captive breeding and the genetic fitness of natural populations. *Conservation Genetics* **2**, 363–378.

Maillard, D., Calenge, C., Invernia, N. and Gaudin, J.C. (2002) Home range size and reproduction of female roe deer re-introduced into a Mediterranean habitat. *Zeitschrift für Jagdwissenschaft* **48**, 194–200.

Maillard, D., Gaillard, J.-M., Hewison, M., Ballon, P., Duncan, P., Loison, A., Togo, C., Baubet, E., Bonnefant, C., Garel, M. and Saint-Andrieux, C. (2010) Ungulates and their management in France. In M. Apollonio, R. Andersen and R. Putman (eds), *European Ungulates and Their Management in the 21st Century*. Cambridge, UK: Cambridge University Press, pp. 441–474.

Matson, T. K., Goldizen, A.W. and Jarman, P. J. (2004) Factors affecting the success of translocations of the black-faced impala in Namibia. *Biological Conservation* **116**, 359–365.

Mattioli, S., Meneguz, P.G., Brugnoli, A. and Nicoloso, S. (2001) Red deer in Italy: recent changes in range and numbers. *Hystrix: Italian Journal of Mammalogy* **12**, 27–35.

McDevitt, A.D., Carden, R.F., Coscia, I. and Frantz A.C. (2013) Are wild boars roaming Ireland once more? *European Journal of Wildlife Research* **59**, 761–764.

Micu, I., Náhlik, A., Neguş, S., Mihalache, I. and Szabó, I. (2010) Ungulates and their management in Romania. In M. Apollonio, R. Andersen and R. Putman (eds), *European Ungulates and Their Management in the 21st Century*. Cambridge, UK: Cambridge University Press, pp. 319–337.

Miller, K.A., Nelson, N.J., Smith, H.G. and Moore, J.A. (2009) How do reproductive skew and founder group size affect genetic diversity in reintroduced populations. *Molecular Ecology* **18**, 3792–3802.

Miracle, P. and Sturdy, D. (1991) Chamois and the karst of Herzegovina. *Journal of Archeological Science*, **18**, 89–108.

Monaco, A., Carnevali, L., Riga, F. and Toso, S. (2007) Il Cinghiale sull'arco alpino: status e gestione delle popolazioni. In H.C. Hauffe, B. Crestanello. and A. Monaco (eds), *Il Cinghiale sull'arco alpino: status e gestione*. Trento, Italy: Centro di Ecologia Alpina, Report nr. 38.

Moritz C. (1999) Conservation units and translocations: strategies for conserving evolutionary processes. *Hereditas* **120**, 217–228.

Mucci, N., Mattucci, F. and Randi, E. (2012) Conservation of threatened local gene pools: Landscape genetics of the Italian roe deer (*Capreolus c. italicus*) populations. *Evolutionary Ecology Research* **14**, 897–920.

Niedziałkowska, M., Jędrzejewska, B., Honnen, A.-C. *et al.* (2011) Molecular biogeography of red deer *Cervus elaphus* from eastern Europe: insights from mitochondrial DNA sequences. *Acta Theriologica* **56**, 1–12.

Niedziałkowska, N., Jędrzejewska, B., Wojcik, J.M. and Goodman, S.J. (2012) Genetic structure of red deer population in North-eastern Poland in relation to the history of human interventions. *Journal of Wildlife Management* **76**, 1264–1276.

Niethammer, G. (1963) *Die Einbürgerung von Säugetieren und Vögeln in Europa*. Berlin, Germany: Paul Parey.

Nievergelt, B. (1966) *Der Alpensteinbock (*Capra ibex *L.) in seinen lebensraum*. Berlin, Germany: Paul Parey.

Nussey, D.H., Pemberton, J., Donald, A. and Kruuk, L.E. (2006) Genetic consequences of human management in an introduced island population of red deer (*Cervus elaphus*). *Heredity* **97**, 56–65.

Olden, J.D., Podd, N.L., Douglas, M.R., Douglas, M.E. and Fausch, K.D. (2004) Ecological and evolutionary consequences of biotic homogenization. *Trends in Ecology and Evolution* **19**, 18–24.

Olech,W. (1987) Analysis of inbreeding in European bison. *Acta Theriologica* **32**, 373–387.

Olech, W. - IUCN SSC Bison Specialist Group (2008) *Bison bonasus*. In IUCN 2013 *IUCN Red List of Threatened Species*. Version 2013.2.

Orsini, L., Vanoverbeke, J., Swillen, I., Mergeay, J. and De Meester, L. (2013) Drivers of population genetic differentiation in the wild: isolation by dispersal limitation, isolation by adaptation and isolation by colonization. *Molecular Ecology* **22**, 5983–5999.

Papaioannou, H. (2010) Ungulates and their management in Greece. In M. Apollonio, R. Andersen and R. Putman (eds), *European Ungulates and Their Management in the 21st Century.* Cambridge, UK: Cambridge University Press, pp. 540–562.

Parker, K.A., Dickens, M.J., Clarke, R.H. and Lovegrove, T.G. (2012) The theory and practice of catching, holding, moving and releasing animals. In J.G. Ewen, D.P. Armstrong, K.A. Parker and P.J. Seddon (eds), *Reintroduction Biology.* Oxford: Blackwell Publishing, pp. 105–137.

Paunovic, M., Ćirović, D. and Linnell, J.D.C. (2010) Ungulates and their management in Serbia. In M. Apollonio, R. Andersen and R. Putman (eds), *European Ungulates and Their Management in the 21st Century.* Cambridge, UK: Cambridge University Press, pp. 563–571.

Pérez-Espona, S., Pérez-Barbería, F.J., Mcleod, J.E., Jiggins, C.D., Gordon, I. J. and Pemberton, J.M. (2008) Landscape features affect gene flow of Scottish Highland red deer (*Cervus elaphus*). *Molecular Ecology* 7, 981–996.

Pérez-Espona, S., Hall, R.J., Pérez-Barbería, F.J., Glass, B.C., Ward, J.F. and Pemberton, J.M. (2013) The impact of past introductions on an iconic and economically important species, the red deer of Scotland. *Journal of Heredity* **104**, 14–22.

Prior, R. (1995) *The Roe Deer: Conservation of a Native Species.* Shrewsbury, UK: Swan Hill Press.

Putman, R. (2010) Ungulates and their management in Great Britain and Ireland. In M. Apollonio, R. Andersen and R. Putman (eds), *European Ungulates and Their Management in the 21st Century.* Cambridge, UK: Cambridge University Press, pp. 129–164.

Radwan, J., Kawalko, A., Wojcik, J.M. and Babik, W. (2006) MHC-DRB3 variation in a free-living population of the European bison, *Bison bonasus. Molecular Ecology* **16**, 531–540.

Randi, E., Alves, P.C., Carranza, J. *et al.* (2004) Phylogeography of roe deer (*Capreolus. capreolus*) populations: the effects of historical genetic subdivisions and recent nonequilibrium dynamics. *Molecular Ecology* **13**, 3071–3083.

Reimoser, F. and Reimoser, S. (2010) Ungulates and their management in Austria. In M. Apollonio, R. Andersen and R. Putman (eds), *European Ungulates and Their Management in the 21st Century.* Cambridge, UK: Cambridge University Press, pp. 338–356.

Rhymer, J.M. and Simberloff, D.S. (1996) Extinction by hybridization and introgression. *Annual Review of Ecology and Systematics* **27**, 83–109.

Ristić, Z., Marković, V., Barović, V. and Ristanović, B. (2010) Application of GIS in re-introduction of red deer in National Park Fruška Gora (Vojvodina, Serbia). *Geographia Technica* **9**, 58–66.

Rodríguez, F., Hammer, S., Pérez, T., Suchentrunk, F., Lorenzini, R., Michallet, J., Martínková, N., Albornoz, J. and Domínguez, A. (2009) Cytochrome b phylogeography of chamois (*Rupicapra* spp.). Population contractions, expansions and hybridizations governed the diversification of the genus. *Journal of Heredity* **100**, 47–55.

Rosell, C., Carretero, M.A., Cahill, S. and Pasquina, A. (1996) Seguimiento de una reintroducción de corzo (*Capreolus capreolus*) en ambiente mediterráneo. Dispersión y área de campeo. *Doñana, Acta Vertebrata* **23**, 109–122.

Royo, L., Pajares, G., Álvarez, I., Fernandez, I. and Goyache, F. (2007) Genetic variability and differentiation in Spanish roe deer (*Capreolus capreolus*): a phylogeographic reassessment within the European framework. *Molecular Phylogenetics and Evolution* **42**, 47–61.

Ruusila, V. and Kojola, I. (2010) Ungulates and their management in Finland. In M. Apollonio, R. Andersen and R. Putman (eds), *European Ungulates and Their Management in the 21st Century.*Cambridge, UK: Cambridge University Press, pp. 86–102

Ryckman, M.J., Rosatte, R.C., McIntosh, T., Hamr, T. and Jenkins, D. (2010) Postrelease dispersal of reintroduced elk (*Cervus elaphus*) in Ontario, Canada. *Restoration Ecology* **18**, 173–180.

Šabić, B. and Lalić, N. (2005) *Game management plan for state hunting ground no. XVII/1 "Biokovo" for the period from 01 April 2005 to 31 March 2015.* Makarska, Croatia: Hrvatskešumed.o.o. Zagreb. [In Croatian.]

Scandura, M., Iacolina, L., Crestanello, B. *et al.* (2008) Ancient vs. recent processes as factors shaping genetic variation of the Europian wild boar: are the effects of the last glaciation still detectable? *Molecular Ecology* **17**, 1745–1762.

Scandura, M., Iacolina, L. and Apollonio, M. (2011) Genetic diversity in the European wild boar *Sus scrofa*: phylogeography, population structure and wild × domestic hybridization. *Mammal Review* **41**, 125–137.

Sipari, E. (1926) *Relazione del Presidente del Direttorio provvisorio dell'Ente autonomo del Parco nazionale d'Abruzzo alla Commissione amministratrice dell'Ente stesso, nominata con Regio Decreto 25 marzo 1923.* Tivoli, Italy: Tipografia Maiella.

Skog, A., Zachos, F.E., Rueness, E.K. *et al.* (2009) Phylogeography of red deer (*Cervus elaphus*) in Europe. *Journal of Biogeography* **36**, 66–77.

Sommer, R.S. and Zachos, F.E. (2009) Fossil evidence and phylogeography of temperate species: 'glacial refugia' and post-glacial recolonization. *Journal of Biogeography* **36**, 2013–2020.

Sommer, R.S., Fahlke, J.M., Schmölcke, U., Benecke, N. and Zachos, F.E. (2009) Quaternary history of the European roe deer *Capreolus capreolus. Mammal Review* **39**, 1–16.

Šprem, N. (2011) *Monitoring of the chamois (Rupicapra rupicapra) population in Nature Park Biokovo - II.Report,* University of Zagreb, Faculty of Agriculture, Zagreb, Croatia. [In Croatian.]

Stanley Price, M.R. (1989) *Animal re-introductions: the Arabian oryx in Oman.* Cambridge, UK: Cambridge University Press.

Stojanov, A., Melovski, D., Ivanov, G. and Linnell, J.D.C. (2010) Ungulates and their management in Macedonia. In M. Apollonio, R. Andersen and R. Putman (eds), *European Ungulates and Their Management in the 21st Century.* Cambridge, UK: Cambridge University Press, pp. 572–577.

Storfer, A. (1999) Gene flow and endangered species translocations: a topic revisited. *Biological Conservation* **87**, 173–180.

Taberlet, P., Fumagalli, L., Wust-Saucy, A.G. and Cosson, J.F. (1998) Comparative phylogeography and postglacial colonization routes in Europe. *Molecular Ecology* **7**, 453–464.

Teixeira, C.P., Schetini de Azevedo, C., Mendl, M., Cipreste, C.F. and Joung, R.J. (2007) Revisiting translocation and reintroduction programmes: the importance of considering stress. *Animal Behaviour* **73**, 1–13.

Tokarska, M., Pertoldi, C., Kowalczyk, R. and Perzanowski, K. (2011) Genetic status of the European bison *Bison bonasus* after extinction in the wild and subsequent recovery. *Mammal Review* **41**, 151–162.

Tosi, G., Apollonio, M., Giacometti, M. *et al.* (2009) *Piano di conservazione, diffusione e gestione dello stambecco sull'arco alpino italiano.* Agricultural and Environmental Resources, Provincia di Sondrio, Italy.

Tvrtković, N., Grubešić, M. (2006) Chamois. In N. Tvrtković (ed.), *Red Book of Mammals of Croatia.* Zagreb, Croatia: Ministry of Culture and State Institute for Nature Protection, pp. 42–43.

Van Wieren, S.E. and Groot Bruinderink, G.W.T.A. (2010) Ungulates and their management in the Netherlands. In M. Apollonio, R. Andersen and R. Putman (eds), *European Ungulates and Their Management in the 21st Century.* Cambridge, UK: Cambridge University Press, pp. 165–183.

Vernesi, C., Pecchioli, E., Caramelli, D. *et al.* (2002) The genetic structure of natural and reintroduced roe deer (*Capreolus capreolus*) populations in the Alps and central Italy, with reference to the mitochondrial DNA phylogeography of Europe. *Molecular Ecology* **11**, 1285–1297.

Vigne, J.-D. (1992) Zooachaeology and the biogeographical history of the mammals of Corsica and Sardinia since the last ice age. *Mammal Review* **22**, 87–96.

Vilaça, S.T., Biosa, D., Zachos, F.E. *et al.* (2014) Mitochondrial phylogeography of the European wild boar: the effect of climate on genetic diversity and spatial lineage sorting across Europe. *Journal of Biogeography* **41**, 987–998.

Vingada, J., Fonseca, C., Cancela, J., Ferreira, J. and Eira C. (2010) Ungulates and their management in Slovakia. In M. Apollonio, R. Andersen and R. Putman (eds), *European Ungulates and Their Management in the 21st Century.* Cambridge, UK: Cambridge University Press, pp. 392–418.

Wawrzyniak, P., Jędrzejewski, W., Jędrzejewska, B., Borowik, T. (2010) Ungulates and their management in Poland. In M. Apollonio, R. Andersen and R. Putman (eds), *European Ungulates and Their Management in the 21st Century.* Cambridge, UK: Cambridge University Press, pp. 223–242.

Whitehead, G.K. (1964) *The Deer of Great Britain and Ireland.* London: Routledge and Kegan Paul.

Willisch, C.S., Biebach, I., Koller, U., Bucher, T., Marreros, N., Ryser-Degiorgis, M.-P., Keller, L. and Neuhaus, P. (2012) Male reproductive pattern in a polygynous ungulate with a slow life-history: the role of age, social status and alternative mating tactics. *Evolutionary Ecology* **26**, 187–206.

Wolf, C.M., Griffith, B., Reed, C. and Temple, S.A. (1996) Avian and mammalian translocations: update and reanalysis of 1987 survey data. *Conservation Biology* **10**, 1142–1154.

Wolf, C. M., Garland, T. and Griffith, B. (1998) Predictors of avian and mammalian translocation success: reanalysis with phylogenetically independent contrasts. *Biological Conservation* **86**, 243–255.

Wolk, E. and Krasinska, M. (2004) Has the condition of European bison deteriorated over last twenty years? *Acta Theriologica* **49**, 405–418.

Wotschikowsky U. (2010) Ungulates and their management in Germany. In M. Apollonio, R. Andersen and R. Putman (eds), *European Ungulates and Their Management in the 21st Century.* Cambridge, UK: Cambridge University Press, pp. 201–222.

Wronski, T. (2010) Factors affecting population dynamics of re-introduced mountain gazelles (*Gazella gazella*) in the Ibex Reserve, central Saudi Arabia. *Journal of Arid Environments* **74**, 1427–1434.

Zachos, F.E. and Habel, J.C. (eds), (2011) *Biodiversity Hotspots: Distribution and Protection of Conservation Priority Areas.* Berlin: Springer-Verlag.

Zachos, F.E. and Hartl, G.B. (2011) Phylogeography, population genetics and conservation of the European red deer *Cervus elaphus. Mammal Review* **41**, 138–150.

Zwijacz-Kozica, T., Selva, N., Barja, I. *et al.* (2012) Concentration of fecal cortisol metabolites in chamois in relation to tourist pressures in Tatra National Park (South Poland). *Acta Theriologica,* **58**, 215–222.

Chapter 4

Introducing Aliens: Problems Associated with Invasive Exotics

Francesco Ferretti and Sandro Lovari

In this chapter an 'alien' taxon (allochtonous, non-native, non-indigenous, exotic) is taken as a species or subspecies which occurs outside its natural range (past or present) and dispersal potential (i.e. outside the range it occupies naturally), as a result of intentional or accidental introduction by man (IUCN, 2000; CBD Secretariat, 2002). An alien taxon is considered 'invasive' when it becomes established in natural or semi-natural ecosystems or habitats, is an agent of change, and threatens native biological diversity (IUCN, 2000). Biological invasions caused by alien species are considered one of the main threats to native biological diversity (IUCN, 2000), with significant negative effects also on ecosystem services to man (Vilà *et al.*, 2010; McLaughlan *et al.*, in press).

Movements of animal taxa by humans, including introductions, were a common practice in past centuries and such deliberate translocations determined the establishment of most non-native species currently living in Europe. These exotic introductions were part of a deliberate policy popular in the 19th century to enrich the fauna of European countries and, in the case of ungulates, were also commonly motivated by a desire to increase the number of species available for hunting. Such introductions were conducted by well organised 'acclimatisation societies' such as the French *La Societé Zoologique d'Acclimatation* founded in 1854 (see Linnell and Zachos, 2011). Movements of animals were not regulated by official guidelines or international agreements until as late as 1987, when the International Union for the Conservation of Nature issued the '*IUCN position statement on translocation of living organisms: introductions, re-introductions and re-stocking*' (IUCN, 1987; edited by P.N. Munton). Because these species have been introduced into novel environments with which they have not co-evolved, however, they may have strong negative impacts on those environments and on native species (see for example, Spear and Chown, 2009). Increasing concerns worldwide about invasive non-native species have resulted in international and national commitments to assess the status and impact of non-native species, and to devise action plans for their control. There are initiatives at both the global (Boyle, 1996) and national levels aiming to achieve these goals. For example, in England, the Invasive Non-Native Species Secretariat was established in response to the statutory obligations of signatory member states. It has set out the UK Government's approach to the sustainable management of wild deer in England in the Wild Deer Action Plan (DEFRA, 2004).

Ungulates play key roles in ecosystems. Their populations interact with those of carnivores, generating predator–prey dynamics which may have a significant impact on populations of both predators and prey (e.g. McLaren and Peterson, 1994; Post *et al.*, 1999; Sinclair *et al.*, 2003; White and Garrott, 2005; Jędrzejewski *et al.*, 2011). Grazing or browsing can determine top-down effects on vegetation and, in turn, on other animals which depend on it (see, for example, Reimoser and Putman, 2011; Smit and Putman, 2011, for reviews). Ecological processes driven by ungulates can bring important implications for management, when threatened taxa, priority habitats or human activities are influenced (e.g. Putman and Moore, 1998; Côté *et al.*, 2004; Gordon *et al.*, 2004; Putman *et al.*, 2011). Because of the increase of protected areas, abandonment of countryside and mountain ecosystems by humans and other land-use changes, as well as local removal of predators, populations of ungulates, including those of introduced species, have been increasing in number and distribution in Europe, with no obvious reason why this trend should not continue (e.g. Fuller and Gill, 2001; Ward, 2005; Apollonio *et al.*, 2010; Putman *et al.*, 2011). It is thus most likely that the impact of alien invasive species of ungulates on native ecological communities, as well as on ecosystem services, will increase (Vilà *et al.*, 2010).

Alien, invasive ungulates can represent a serious threat to native biodiversity. Their action can produce changes in vegetation composition and shifts in the distribution of dominant plants, which could result in effects from the population to the community and ecosystem level (see, for example, Nugent *et al.*, 2001; Vazquez, 2002; Courchamp *et al.*, 2003; Parker *et al.*, 2006a; Nunez *et al.*, 2010, for reviews). The presence of invasive ungulates can also determine severe impacts on human activities (e.g. through damage to agriculture and forestry, vehicle collisions or through problems associated with transmission of disease to humans or their domestic livestock). Within the European context, health issues and other effects of ungulates on human activities (damage to crops and forestry; vehicle collisions) have been treated elsewhere (e.g. Ferroglio *et al.*, 2011; Reimoser and Putman, 2011; Langbein *et al.*, 2011). In this chapter, we have concentrated on the ecological effects of alien ungulates on other components of ecosystems (both fauna and flora). Firstly, we have collated available information on distribution of alien ungulates in Europe (Section 4.1). Secondly, we summarise the effects of non-native ungulate species on vegetation (Section 4.2) and available information on interspecific competition between non-native and native species (Section 4.3); finally we consider problems arising from hybridisation with native ungulate species (Section 4.4).

4.1 Alien ungulates in Europe

According to the IUCN Red List, 2.6% of extant land mammals are invasive, with Artiodactyla containing the highest proportion of successful invaders (Clout and Russell, 2008). Because of their aesthetic and economic appeal, as exotic species, as well as their perceived value as game animals, large ungulates (especially deer) have been introduced many times worldwide and comprise the largest proportion of all invasive mammal species (Clout and Russell, 2008). But not all introductions are successful; those species which have been successfully established upon intro-duction are but a small subset of all those whose introduction has been attempted. Fautley *et al.* (2012) have assessed the various demographic and ecological factors associated with the success of establishment of different deer species following introduction and, separately, those associated with their subsequent spread (inva-sive success). They suggest that there is a distinction between those species which make successful colonists (in terms of success of the initial establishment) and those which do not, related to species-specific weaning age, age at sexual matu-rity and native range size, whereas weaning body mass was predictive of the rate of subsequent spread. Within species, different introductions also show varying degrees of success. Fautley *et al.* (2012) suggest that establishment success is pri-marily determined by the number of introduced individuals, whereas breadth of habitat and diet determine which of the established populations will spread.

There are, in Europe, ten species of alien ungulates: the aoudad or Barbary sheep, the American bison, the Mediterranean mouflon, as well as the fallow deer, the spotted deer or chital, the wapiti, the sika deer, the Chinese muntjac, the Chinese water deer and the white-tailed deer (Linnell and Zachos, 2011; http://www.europe-aliens.org; www.iucnredlist.org; http://www.cabi.org/isc/). Four ungu-lates native to Europe have been introduced to European countries where they are non-native: the muskox, the reindeer, the ibex and the Alpine chamois (Linnell and Zachos, 2011; http://www.europe-aliens.org; www.iucnredlist.org; http://www.cabi.org/isc/). Almost all European countries hold at least one species of alien ungu-late; indeed, except for Andorra, Liechtenstein, San Marino Republic and Vatican, all the countries of western Europe hold at least one alien ungulate species. Russia is the country with the greatest number of alien ungulates ($n = 7$ species), followed by Germany ($n = 6$), Croatia, the Czech Republic and UK ($n = 5$, Figure 4.1). Six countries hold four alien species of ungulates, ten countries have three alien species, six countries keep two alien species and three countries have one species (Figure 4.1). Thus, alien species of wild ungulates are present in 32 European countries. These figures confirm that the presence and the impact of alien ungulates on ecosys-tems are a potential, widespread problem at the continental scale.

4.2 Impacts on vegetation and habitats

Ungulates can modify vegetation, through browsing, grazing but also trampling (see, for example, Putman, 1986a, 1996a; Milchunas and Lauenroth, 1993; Hobbs,

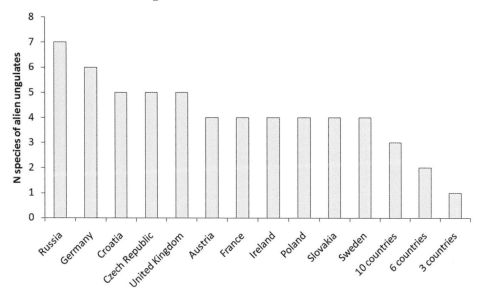

Figure 4.1 Number of species of alien ungulates present in European countries.
Sources: Delivering Alien Invasive Species Inventories for Europe: http://www.europe-aliens.org; *IUCN Red List*: www.iucnredlist.org; *Invasive Species Compendium*: http://www.cabi.org/isc/.

1996; Côté *et al.*, 2004; Gordon *et al.*, 2004; Gill and Morgan, 2010; Reimoser and Putman, 2011; Smit and Putman, 2011). Usually, some plant species are browsed or grazed more frequently than others (Gill, 1992) which, in turn, would lead to changes of species composition, if pressure of herbivory is high (Gill, 1992; Putman, 1996b; Gill and Beardall, 2001; Morecroft *et al.*, 2001). Furthermore, ungulates can determine effects on ground flora, through grazing or trampling (e.g. Kirby, 2001; Morecroft *et al.,* 2001). When the rate at which grasses or browse are consumed exceeds the rate of regrowth or that of production of new shoots, then the feeding activity of herbivores affects the regeneration of vegetation. In turn, grazing, browsing and trampling can induce habitat modifications, which could be detrimental, for example, for the stability of soil, the viability of other organisms and, *sensu lato*, for the conservation of habitats and biodiversity (see, for example, Smit and Putman, 2011, for a review).

There is increasing concern about the negative role of alien ungulates in affecting vegetation dynamics and habitats (e.g. Morecroft *et al.*, 2001; Nugent *et al.*, 2001; Vazquez, 2002; Cooke, 2005; Parker *et al.*, 2006a, b; Ricciardi and Ward, 2006; Dolman and Wäber, 2008; Spear and Chown, 2009; Nunez *et al.*, 2010; McLaughlan *et al.*, in press). It has been suggested that the natural history of grazing/browsing in an area influences the local impact of herbivores on vegetation (e.g. Milchunas *et al.*, 1988; Hobbs and Huenneke, 1992). The impact of alien ungulates would be greater in systems evolved in absence of herbivores than in those with a long evolutionary history of herbivory (Milchunas *et al.*, 1988). In turn, island ecosystems are greatly vulnerable to the action of invasive ungulates, as they have often evolved in absence of large herbivores (e.g. Bowden and van

Vuren, 1997; Nugent *et al.*, 2001; Courchamp *et al.*, 2003). Furthermore, there is evidence that exotic herbivores can facilitate abundance and species richness of exotic plants, through the reduction of native ones, which strongly suggests that plants are especially susceptible to the introduction of novel herbivores, which they have not been selected to resist (Parker *et al.*, 2006a). Thus, one could also expect that significant changes in the herbivory regime will affect vegetation (Hobbs and Huenneke, 1992). In turn, the imposition of new or different herbivores on a system could determine a disturbance to vegetation, especially if the new herbivores can reach greater densities than native ones (Hobbs and Huenneke, 1992; Vazquez, 2002; Parker *et al.*, 2006a).

Perhaps the classic illustration of the wide potential impacts of introduced ungulates on native flora and fauna comes from analysis of the various impacts of those species introduced to New Zealand (e.g. Challies, 1985; Tanentzap *et al.*, 2009). In New Zealand, seven introduced deer species have established viable populations (red deer, wapiti, white-tailed deer, sambar deer, rusa deer, sika deer and fallow deer; e.g. Fraser *et al.*, 2000; Nugent *et al.*, 2001; Forsyth *et al.*, 2010). Prior to introduction of deer, most forests in New Zealand had a dense understorey, with a great abundance of palatable plants (Nugent *et al.*, 2001, and references therein). Then, the action of deer determined heavy changes in vegetation composition and structure, with a strong reduction of understorey, a decrease of palatable plant species and a concurrent increase of browse-tolerant ones (e.g. Nugent *et al.*, 2001, for a summary). These changes do not seem to be totally reversible, even when control operations of deer have been conducted (e.g. Nugent *et al.*, 2001; Coomes *et al.*, 2003; Forsyth *et al.*, 2010, for reviews; see also Tanentzap *et al.*, 2009). Heavy browsing pressure, even at low deer densities, occupation by uneaten plants of ecological niches vacated by palatable ones, extinction of seed sources and alterations to succession and ecosystem processes are all factors preventing a recovery of forests in New Zealand (Coomes *et al.*, 2003).

Within Europe, sika deer, fallow deer and Reeves' muntjac seem to be the exotic species which are the most involved in damage to vegetation and habitats. It is notable that sika deer are one of the only three terrestrial vertebrate species included within a list of the ten most invasive species in Europe, considered to have the highest number of different impact types on ecosystem services (Vilà *et al.*, 2010; McLaughlan *et al.*, in press). Often these species do not occur alone, but in sympatry with other ungulates, which makes disentangling the contribution of different species to observed impacts on vegetation (e.g. Putman *et al.*, 2011) quite difficult. Some studies in allopatry sites or field experiments have allowed the assessment of single-species effects (e.g. Putman, 1996a; Cooke, 2005, 2006; Diaz *et al.*, 2006). In the next section, we present a summary of the main effects on vegetation reported for exotic deer (Reeves' muntjac, sika deer and fallow deer) in the British Isles. We present it as a case study, not because these are the only places where these deer species have been introduced, but because the situation in the UK has been the best-studied one.

4.2.1 Fallow deer, sika deer and Reeves' or Chinese muntjac, with particular reference to UK

Fallow deer are native to Asia Minor and, most likely, their present distribution in Europe has been determined by human intervention, after the Neolithic (Chapman and Chapman, 1980; Masseti, 1996; see also Chapter 12). The fallow deer is one of the most widespread ungulates in Europe and it has been introduced to over 50% of European countries (Linnell and Zachos, 2011; http://www.europe-aliens. org). In the UK, in the last 3–4 decades these deer seem to have increased their distribution range at a slow rate (Ward *et al.*, 2005; Ward and Etherington, 2008; see also Carnevali *et al.*, 2009; Liberg *et al.*, 2010, for other European countries). Fallow deer have relatively high rates of recruitment, which seem apparently comparatively insensitive to density dependence (Putman *et al.*, 1996). Because of the slow dispersal from source, populations may reach high local densities, which may in turn impose significant impacts on natural vegetation and wilder habitats (e.g. Smale *et al.*, 1995), as well as on human activities (e.g. damage to forestry, vehicle collisions, disease transmission: Ferroglio *et al.*, 2011; Langbein *et al.*, 2011; Putman *et al.*, 2011).

In the UK, regeneration in semi-natural woodlands can be kept down as the result of heavy browsing by fallow deer (e.g. Harris, 1981; Ward *et al.*, 1994). Intensive selective browsing of seedlings and coppice can also affect the regeneration of preferred species, resulting in changes of species composition and species dominance among woody plants (Putman, 1986a, b). Impacts of fallow deer at high density have been also suggested on ground flora (e.g. on oxlips: Rackham, 1975, 2003; Tabor, 1993, 1999; Taylor and Woodell, 2008). There is some indication that fallow deer may reach greater densities than those of sympatric native deer (e.g. Putman, 1996a; Gill and Morgan, 2010; see e.g. Garcia-Gonzalez and Cuartas, 1992; Mattioli *et al.*, 2003; Focardi *et al.*, 2006; Ferretti *et al.*, 2011a, for other European study areas), which suggests that the magnitude of their impacts on vegetation and habitats may be greater than that of native species of European deer.

Sika deer are native to Eastern Asia (Japan, Taiwan, China and Far Eastern Russia; McCullough *et al.*, 2009b). Especially between the 1890s and 1930s, they were widely introduced to other countries and populations are now established in New Zealand, North America and Europe (Austria, Czech Republic, Denmark, France, Germany Ireland, Poland, UK; McCullough *et al.*, 2009b; Feldhamer and Demarais, 2009; Banwell, 2009; Bartoš, 2009; Swanson and Putman, 2009). Considerable body mass (males: *c.*80 kg; females: *c.*50 kg, Takatsuki, 2009), high fecundity, grouping behaviour, flexible food habits, rapidly increasing population size, as well as distribution range, determine a great potential for sika deer to have a negative impact on ecosystems (Takatsuki, 2009; McCullough *et al.*, 2009b; Pérez-Espona *et al.*, 2009; CABI, 2014). There is some indication that fertility can be greater in female sika deer than in the ecologically similar native red deer, for all age classes (Chadwick *et al.*, 1996; Mayle, 1996), which suggests that the former has the potential to reach greater densities than the latter. In Germany,

Gebhart (1996) observed that 'sika deer is more resilient towards changes in its environment' than red deer. A large variability of timing and duration of rut season has been reported across Europe, indicating a great adaptability of sika deer, which 'may be able to overcome severe environmental conditions sometimes better than the local species' (Bartoš, 2009).

Sika deer may cause considerable economically significant damage to commercial forestry through browsing and bark-stripping in hard winters (Ratcliffe, 1989; Lowe, 1994; Chadwick *et al.*, 1996; Abernethy, 1998; Forestry Commission for Scotland, unpublished data; Swanson and Putman, 2009). Mature trees may also suffer additional damage in some areas through 'bole-scoring' when sika stags gouge deep vertical grooves into the bole of particular trees when marking and advertising mating activities (Larner, 1977; Carter, 1984; Putman and Moore, 1998; Takatsuki, 2009); such damage tends to be more severe on coniferous rather than deciduous trees (Carter, 1984). Grazing by sika deer may also affect plant species composition, favouring the spread of alien plants (Cross, 1981; cf. Parker, 2006a). In open heathland and wetland areas (e.g. reedbeds, saltmarshes), grazing and trampling can induce changes in vegetation structure and species composition, a reduction of vegetation cover, height and volume, the decrease of root biomass, the exposure of soil to erosion, as well as other ecological effects (Diaz *et al.*, 2005; House *et al.*, 2005; Hannaford *et al.*, 2006).

Although these impacts seem to be localised to areas with particularly high densities of sika deer, their fast increase in both numbers and distribution range (McCullough *et al.*, 2009; CABI, 2014) suggests that the problems associated to their presence are going to increase, not only in UK but also in other European countries where they have been introduced.

Reeves' muntjac is native to southeast China and Taiwan (Chapman, 2008) and it has been introduced to the UK, France and the Netherlands (Linnell and Zachos, 2011). In the UK, it is now well established and its populations show rapid expansion of distribution (1972–2002: 8.2%/year; Ward, 2005), which is helped by continuing secondary translocations (Chapman *et al.*, 1994a; CABI, 2014). Muntjac can use small home ranges (*c.*10–15 ha), which overlap among individuals (Chapman *et al.*, 1994a) and they may live throughout the year in relatively small areas. In turn, there is a potential for populations to reach extremely high densities (Cooke, 2004, 2005, 2006), which may be greater than those of the ecologically similar roe deer, when in sympatry (e.g. Chapman *et al.*, 1994a; Hemami *et al.*, 2005).

At present, local negative impacts of muntjac have been reported largely from England (CABI, 2014; horticulture: Putman and Moore, 1998; gardens: Chapman *et al.*, 1994b; damages to forestry, through browsing and stem breakage: Tabor, 1993; Cooke, 1994, 1998, 2006; Cooke and Farrell, 2001). Where this species reaches great densities, it may have significant impacts on woodland shrub and ground layers (e.g. Rackham 1975; Tabor 1993, 1999; Cooke 1994, 2004, 2005, 2006, Cooke and Farrell, 1995, 2001; Diaz and Burton, 1996; Taylor and Woodell 2008; CABI, 2014), with cascade effects on invertebrates and birds (e.g. Pollard and Cooke, 1994; Pollard *et al.*, 1998; Cooke and Farrell, 2001; Cooke, 2004). These changes seem to be only partially reversible, after reduction of muntjac density through culling (Cooke, 2005; cf. Nugent *et al.*, 2001; Coomes *et al.*, 2003; Tanentzap *et al.*, 2009; Forsyth *et al.*, 2010, for case-studies on other deer species, introduced to New Zealand). However, these impacts were recorded at extremely high deer densities, and the extent to which these damages are more widely representative is uncertain (Putman *et al.*, 2011).

Muntjac can also show much lower seed densities per pellet group and lower defecation rates, and they can disperse significantly fewer plant species than sympatric roe deer, suggesting that their action in seed dispersal is not equivalent to those of native, ecologically similar deer (Eycott *et al.*, 2007; Dolman and Wäber, 2008). Thus, these deer have the potential to affect plant dynamics, if they displace native ones (Dolman and Wäber, 2008).

4.2.2 Other species

Other alien species of wild ungulates cause damage to vegetation and habitats in Europe, especially mouflon and aoudad (e.g. Smit *et al.*, 2001; Benes *et al.*, 2006; Gangoso *et al.*, 2006; Nogales *et al.*, 2006; Šídová and Schlaghamerský, 2007; Guidi *et al.*, 2009; Garzón-Machado *et al.*, 2010). Although this review is not concerned with domestic animals, we may parallel the action of feral goats and sheep to that of aoudad and mouflons. Their grazing, browsing and trampling may damage vegetation, in mainland areas, leading to changes of plant species composition and habitat degradation (e.g. Mánzano and Návar, 2000; Zamora *et al.*, 2001), although it is often not possible to disentangle the effects of single species in multi-species assemblies (e.g. Smit *et al.*, 2001; Zamora *et al.*, 2001; Benes *et al.*, 2006; Šídová and Schlaghamerský, 2007, for feral goats or mouflon in Europe). Conversely, worldwide, the introduction of these herbivores on islands has determined severe, negative impacts on native vegetation, with cascade effects on other components of local communities, which are often rich in endemic species (e.g. Scowcroft and Giffin, 1983; Scowcroft and Sakai, 1983; Van Vuren and Coblentz, 1987; Bowden and Van Vuren, 1997; Donlan *et al.*, 2002; Courchamp *et al.*, 2003; Campbell and Donlan, 2005; Gangoso *et al.*, 2006; Garzón-Machado *et al.*, 2010).

Island ecosystems evolved mainly in the absence of large herbivores and, therefore, they may lack defences against browsing or grazing (e.g. Bowden and Van Vuren, 1997; Campbell and Donlan, 2005). In particular, feral goats and sheep can have destructive effects on native vegetation: through intensive browsing, grazing and trampling, they can alter plant communities and forest structure, threatening vulnerable plant species and dependent small animals, increasing soil erosion and, in turn, leading to ecosystem degradation and biodiversity loss (e.g. Coblentz, 1978; Van Vuren and Coblentz, 1987; Campbell and Donlan, 2005). Not surprisingly, both in the Mediterranean and the Canary islands, the negative effects of aoudad, feral goats and mouflon on native vegetation have been documented (e.g. Spagnesi and Toso in Boitani *et al.*, 2003; Gangoso *et al.*, 2006; Nogales *et al.*, 2006; Guidi *et al.*, 2009; Garzón-Machado *et al.*, 2010). Their actions can affect vegetation, including forests, where many endemic species can comprise their diet (e.g. Nogales *et al.*, 2006, for a summary).

On the island of Montecristo (Tuscany, Italy), goats have almost completely prevented regeneration of native holm oak (*Quercus ilex*), opening large areas to the spread of an exotic unpalatable plant, the ailanthus *Ailanthus altissima* (Spagnesi and Toso, in Boitani *et al.*, 2003). Mouflon have been suggested to be a threat for many endemic species of vascular plants, in Teide National Park (Tenerife, Canary, see Nogales *et al.*, 2006, for a summary), whilst damage to natural vegetation has also been reported in Tuscany Archipelago (Guidi *et al.*, 2009). In La Caldera de Taburiente National Park (La Palma Island, Canary) aoudad are considered a threat for many endemic plants (e.g. Rodríguez-Pineiro and Rodríguez-Luengo, 1992; Nogales *et al.*, 2006, for a summary). In pine forests of that area, a negative effect of aoudad and goats (and also European rabbit) has been demonstrated on the abundance of four endemic legumes and the paucity of understorey plant species has been related to the presence of introduced herbivores (Garzón-Machado *et al.*, 2010).

It is hard to resolve whether introduced ungulates affect vegetation and habitats because they are exotic or just because they reach great densities. In fact, they may reach greater densities and/or productivity and, as to muntjac, lower rates of seed dispersal than native, sympatric ungulates, which suggests that their impact is not equivalent to, often greater than, that of native species (cf. Milchunas *et al.*, 1988; Hobbs and Huenneke, 1992; Parker *et al.*, 2006a). Further studies are needed to clarify this issue. Alien ungulates can particularly affect vegetation and habitats of islands, which have evolved mainly in the absence of large herbivores. As these ecosystems often have a comparatively great number of endemic species, the presence of alien herbivores may have serious consequences for the conservation of biological diversity.

4.3 Competition between exotic and native ungulates

Interspecific competition may arise between invasive, introduced ungulates and native species at the same trophic level. Two species compete when they share a scarce, natural resource which is actually limited in supply, and where this ecological overlap results in a decrease of growth, fecundity and/or survivorship in one or both species (e.g. de Boer and Prins, 1990; Putman, 1994, 1996a; Begon *et al.*, 2006). Competition can occur through direct interference, when a resource is actively disputed, or through exploitation, when a competitor uses a resource, reducing its availability for the other (e.g. Putman, 1994; see also Chapter 5). Usually, there is an 'inferior' competitor, which decreases in numbers because of the presence of the other species, and a 'superior' competitor, which may be unaffected by the interaction. From an evolutionary point of view, most species which have the potential to compete have developed adaptations to reduce overlap in the use of resources and thus avoid the effects of expressed competition (Pianka, 1973; Schoener, 1974; Connell, 1980, *inter alia*). Thus, one should expect that competition between species is rare in natural communities. Conversely, when alien species are introduced to an area, (whether wild ungulates or domestic animals; see Chapter 5), they have not had a previous opportunity for coevolution with indigenous taxa and thus may show greater overlap in terms of diet, habitat use and/or activity rhythms to those of native species, resulting in a greater potential for competitive interactions (e.g. Putman and Sharma, 1987; Chapman *et al.*, 1994a; Feldhamer and Anderson, 1993; Latham, 1999; Hemami *et al.*, 2005; Focardi *et al.*, 2006; Dolman and Wäber, 2008; Ferretti *et al.*, 2011b). In the longer term, displacement of native species by invasive exotics may occur.

Actual experimental evidence of competition between wild ungulates is restricted to a set of exclosure experiments conducted in Texas with native white-tailed deer and, respectively, introduced sika, chital and fallow deer (Harmel, 1980, 1992). In particular, these studies showed a decrease of white-tailed deer numbers when in sympatry with sika or chital, in *c.*40-ha enclosures with native vegetation, and a decrease of fallow deer when confined in sympatry with white-tailed deer (Harmel, 1980, 1992). In all these cases, the extinction of one species of each pair occurred within 6–8 years from the beginning of the experiment. Under field conditions, it is difficult to use such an experimental approach, e.g. exclosures or manipulation of population densities, because of logistical, technical and/or ethical problems. However, even if experimental evidence of competition is lacking, information on overlap in the use of resources in relation to availability, obvious shifts in patterns of habitat selection or diet, clear evidence of opposing trends in population size or evidence of behavioural interference interactions support the hypothesis of competition between alien ungulates and native ones (e.g. Putman,1996a; Latham, 1999; Dolman and Wäber, 2008, for reviews). In the next paragraphs, we have summarised the main cases where interspecific competition between exotic and native ungulates has been suggested in Europe (see Table 4.1).

4.3.1 Fallow deer–roe deer

The competitive advantage of fallow deer over native ungulate species has been supported by field data (Putman and Sharma, 1987; Putman, 1996a; Focardi *et al.*, 2006; Ferretti *et al.*, 2011b; Imperio *et al.*, 2012) and by predictive modelling (Kramer *et al.*, 2006; Acevedo *et al.*, 2010). Fallow deer are intermediate feeders (*sensu* Hofmann, 1989), thus their diet could theoretically include food items of roe deer (a 'concentrate selector', *sensu* Hofmann, 1989), determining a potential for exploitative competition.

The idea that fallow deer might outcompete roe deer was first proposed in the last century, mainly on the basis of anecdotal information (e.g. Batcheler, 1960, and references therein). However, data coming from field studies gave somewhat contradictory results (Batcheler, 1960; Putman and Sharma, 1987; Putman, 1996a; Bartoš *et al.*, 2002; Focardi *et al.*, 2006; Ferretti *et al.*, 2011b;

Table 4.1 Reported interactions between introduced and native species of ungulates, in Europe

Introduced species	Native species	Area	Observed relationship	Source
Fallow deer	Roe deer	Drummond Hill Forest, UK	Differences in habitat selection; increase of fallow deer density and decrease of roe deer density.	Batcheler, 1960
Fallow deer	Roe deer	New Forest, UK	High diet/space overlap; increase of fallow deer density and decrease of roe deer density.	Putman, 1986b; 1996a and references therein.
Fallow deer	Roe deer	Roydon Wood, UK	Great spatial overlap.	Putman, 1996a and references therein.
Fallow deer	Roe deer	Dobriš Forest, Czech Republic	No behavioural interactions.	Bartoš *et al.*, 2002
Fallow deer	Roe deer	Castelporziano, Italy	Negative relationship between fallow deer density and (a) spatial overlap; (b) roe deer home range size; (c) roe deer body weight. Increase of fallow deer density and decrease of roe deer density.	Focardi *et al.*, 2006; Imperio *et al.*, 2012
Fallow deer	Roe deer	Maremma Regional Park, Italy	Behavioural interference of fallow deer to roe deer. Great diet and space overlap. Negative effects of fallow deer density on small-scale distribution of roe deer. Increase of fallow deer density and decrease of roe deer density.	Ferretti *et al.*, 2008; 2011a, b; 2012; Ferretti, 2011; Manganelli, 2012
Fallow deer	Red deer	Drummond Hill Forest, UK	Differences in habitat use.	Batcheler, 1960
Fallow deer	Red deer	New Forest, UK	Low habitat overlap but great diet overlap.	Putman, 1986b; 1996a and references therein.

Fallow deer	Red deer	Roydon Wood, UK	Great spatial overlap.	Putman, 1996a and references therein.
Fallow deer	Red deer	Dobriš Forest, Czech Republic	No interference; trend to an increase of time grazing with increasing number of individual deer, irrespectively from species.	Bartoš *et al.*, 2002
Fallow deer	Red deer	Žehušice Deer Park, Czech Republic	At artificial feeding stations, interference by fallow deer to red deer was more frequent than *vice versa*.	Bartoš *et al.*, 1996
Fallow deer	Red deer	Cazorla Sierra, Spain	Moderate diet overlap.	Garcia-Gonzalez and Cuartas, 1992
Fallow deer	Red deer	Doñana National Park, Spain	Low habitat overlap. Avoidance, by groups of female red deer, of meadows with great densities of fallow deer.	Braza and Alvarez, 1987; Carranza and Valencia, 2002
Fallow deer	Red deer	Raciborskie Forest, Poland	Great diet overlap.	Obidziński *et al.*, 2013
Fallow deer	Red deer	Mesola Wood, Italy	Impact of fallow deer on natural food resources.	Mattioli *et al.*, 2003
Muntjac	Roe deer	King's Forest, UK	Low temporal overlap; great spatial overlap; increase of muntjac density and decrease of roe deer density.	Champan *et al.*, 1989; Forde, 1989; Wray, 1994.
Muntjac	Roe deer	Thetford Forest	Great spatial overlap.	Hemami *et al.*, 2004; 2005.
Sika deer	Red deer	New Forest, UK	Low space overlap; great diet overlap.	Putman, 1996a and references therein.

(Continued)

Table 4.1 (*Continued*)

Introduced species	Native species	Area	Observed relationship	Source
Sika deer	Red deer	Roydon Wood, UK	Great spatial overlap.	Putman,1996a and references therein..
Sika deer	Red deer	Killarney, Ireland	Inverse relationships between densities.	Burkitt, 2009.
Sika deer	Roe deer	New Forest, UK	Low space overlap; great diet overlap; increase of sika deer numbers and decrease of roe deer numbers.	Putman, 1996a and references therein.
Sika deer	Roe deer	Roydon Wood, UK	Great spatial overlap.	Putman, 1996a and references therein.
White-tailed deer	Roe deer	Dobriš Forest, Czech Republic	Great diet overlap.	Homolka *et al.*, 2008.
White-tailed deer	Red deer	Dobriš Forest, Czech Republic	Great diet overlap.	Homolka *et al.*, 2008.
Mediterranean mouflon	Alpine chamois	French Alps, France	Great diet and space overlap.	Pfeffer and Settimo, 1973
Mediterranean mouflon	Alpine chamois	Orsiera-Rocciavré Park, Italy	Great diet overlap.	Bertolino *et al.*, 2009
Mediterranean mouflon	Alpine chamois	Bauges, France	Great spatial overlap but fine-scale habitat partitioning.	Darmon *et al.*, 2012
Mediterranean mouflon	Alpine chamois	Central-Eastern Alps, Italy	The selection of feeding sites of chamois was affected by the proximity of mouflon groups.	Chirichella *et al.*, 2013

Mediterranean mouflon	Red deer	Orsiera-Rocciavrè Park, Italy	Great diet overlap.	Bertolino *et al.*, 2009
Mediterranean mouflon	Red deer	Drahanská vrchovina Hills, Czech Republic	Great diet overlap.	Homolka, 1996
Mediterranean mouflon	Red deer	Ciudad Real in Castile-La Mancha, Spain	Diet overlap in summer.	Miranda *et al.*, 2012
Mediterranean mouflon	Roe deer	Orsiera-Rocciavrè Park, Italy	Low diet overlap.	Bertolino *et al.*, 2009
Mediterranean mouflon	Roe deer	Drahanská vrchovina Hills, Czech Republic	Substantial diet overlap.	Homolka, 1996
Aoudad	Iberian ibex	Spain	Moderate habitat overlap.	Acevedo *et al.*, 2007
Alpine chamois	Red deer	Jeseníky Mountains, Czech Republic	Great diet overlap.	Homolka, 1996; Homolka and Heroldová, 2001

Imperio *et al.*, 2012). Batcheler (1960) suggested a low potential for competition because the two species 'occupy different kinds of habitats'. Conversely, in the New Forest (UK) and in Maremma Regional Park (central Italy), interspecific overlap in both habitat use and diet was great between roe and fallow deer (Pianka index = 0.79–0.87, habitat use; 0.55–0.87, diet; Putman, 1986b; 1996a; Ferretti *et al.*, 2011a; Manganelli, 2012), suggesting that exploitation competition could occur.

Inverse population trends between fallow and roe deer have been reported in different study areas, with an increase in numbers of the former and a decrease in numbers of the latter (Drummond Hill Forest, Scotland: Batcheler, 1960; New Forest: Putman and Sharma, 1987; Putman, 1996a; Castelporziano: Focardi *et al.*, 2006; Imperio *et al.*, 2012; Maremma Regional Park: Ferretti *et al.*, 2011b). However, since roe deer prefer early growth stages while fallow deer prefer woodlands of intermediate-to-mature age structure, opposite population trends of these deer, throughout years, could simply be related to opposite reactions to successional change in forest structure (Batcheler, 1960; Putman and Sharma, 1987; Putman, 1996a). In the New Forest, a further suggestion of the lack of competition is the observation that, while roe deer populations remained largely unmanaged, populations of fallow deer (and other species present in the area) were subjected to significant culling pressure, which may have served to keep populations of deer below levels at which shared resources become limiting (Putman, 1996a).

While results of population analyses are inconclusive, it is nonetheless possible that the effect of grazing by fallow deer (and other ungulates: especially ponies and cattle) in this New Forest example was additive to natural forest succession in affecting indirectly the habitat quality for roe deer (R. Putman, pers. comm.). In Maremma Regional Park, a multi-factorial analysis showed that, irrespectively of habitat features, the probability of roe deer presence in sampling plots decreased significantly with increasing local density of fallow deer, suggesting that the former may affect the small scale distribution of roe deer (Ferretti *et al.*, 2011b). In another Mediterranean area (Castelporziano Estate, central Italy), Focardi *et al.* (2006) reported that (1) habitat separation between fallow deer and roe deer increased with increasing fallow deer density and (2) there was a correlation between increasing fallow deer density and larger home ranges and lower body weights of roe deer (*n* = 14 individuals). They suggested that great densities of fallow deer lead to a reduction of habitat quality for roe deer (cf. Putman and Sharma, 1987), which, in turn, may determine a reduction of body weight and an increase of home range size in the latter. Focardi *et al.* (2006) suggested that fallow deer density may explain variation in home range size of roe deer better than habitat composition. They concluded that competition by fallow deer determined a sharp reduction of roe deer population size (by 80%, over 2 years; Focardi *et al.*, 2006).

In the Maremma Regional Park, behavioural observations revealed strong interactions between the two species, with direct aggression by fallow to roe deer (Ferretti *et al.*, 2008, 2011b, 2012; Ferretti, 2011). The proximity of fallow deer determined a significant alteration of feeding in roe deer, which increased the relative time spent in vigilance and decreased the relative time feeding; no effect was recorded in fallow (Figure 4.2). There were frequent displacements of roe deer by fallow (with roe deer moving away to increase the distance between them and any fallow to >50 m, 79% of 294 encounters); this event not uncommonly resulted in the roe deer leaving the feeding area altogether (*c.* 50% of displacement events). Direct aggression by fallow deer to roe deer was observed in 34 cases out of 294 encounters (e.g. Ferretti *et al.*, 2011b). Female roe were displaced relatively more often than males. The rate of intolerance was the greatest in spring (when births and early maternal care occur in roe deer). Competition between these deer may occur not only through resource exploitation, but also through interference.

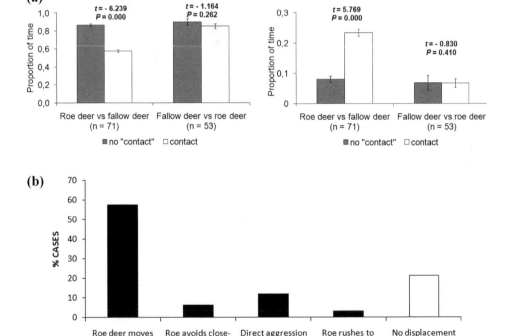

Figure 4.2 Behavioural interference between fallow deer and roe deer in a Mediterranean area (Maremma Regional Park, central Italy: 2006–09). (a) Proportion of time spent feeding and in vigilance (mean ± standard error) by roe deer (left) and by fallow deer (right) without 'contact' compared with those when in 'contact' with the other species. 'Contact': interspecific encounter, when roe and fallow deer were at a mutual distance <50 m. Differences were assessed through *t*-tests. (b) Interspecific behavioural interactions between fallow deer and roe deer, during 'contacts'. Modified from Ferretti *et al.* (2011b).

In conclusion, although competition between fallow and roe deer has not been demonstrated experimentally, there is an increasing amount of evidence that fallow deer may indeed compete with roe deer, on the basis of (i) a demonstration of significant overlap in habitat use and diet (e.g. Putman, 1986b, 1996a; Ferretti *et al.*, 2011a; Manganelli, 2012); (ii) the consistency of inverse population trends across different study areas, over both short (e.g. Focardi *et al.*, 2006; Ferretti *et al.*, 2011b) and long (Putman and Sharma, 1987; Putman, 1996a; Imperio *et al.*, 2012) temporal scales; (iii) the observation of effects of fallow deer density on home range size, body weight and distribution of roe deer (Focardi *et al.*, 2006; Ferretti *et al.*, 2011b); (*iv*) strong evidence of active displacement and aggression between the two species on favoured feeding sites (e.g. Ferretti *et al.*, 2011b). Competition with fallow deer may generate concern for the viability of small populations of Italian roe deer (*C. c. italicus* Festa, 1925), at least at local levels (Focardi *et al.*, 2006; Ferretti *et al.*, 2011b).

4.3.2 Sika deer–other deer species

Sika deer are native to East Asia and have been widely introduced to Europe (see Section 4.2). These highly adaptable, ecologically flexible deer have been reported to outcompete native white-tailed deer in exclosure experiments (Texas, USA: Harmel, 1980) and it has often been claimed that they can compete with native ungulates in the wild (e.g. Feldhamer *et al.*, 1978; Dzieciolowski, 1979; Feldhamer and Armstrong, 1993; Danilkin, 1996; Makovkin, 1999). There are, however, very few definitive studies of the effects of sika on other deer species.

The potential interaction between red and sika deer has been indicated by correlative results in several studies, suggesting suppression of productivity, or geographical displacement of red deer by sika, but competition has not been unequivocally demonstrated (McKelvey, 1959; Dzieciolowski, 1979; Feldhamer and Armstrong, 1993; Abernethy, 1994). Furthermore, some data show lower than expected densities of roe deer in forest inhabited by sika deer (Chadwick *et al.*, 1996).

Only one study has so far been published that explores more objectively patterns of habitat use and diet of sika deer in sympatry with native deer species, in Europe (New Forest, England; Putman, 1986b, 1996a). Overlaps in habitat use were highest with roe deer, most notably in autumn and winter (Pianka index = 0.80–0.90, falling to 0.65–0.75 in spring and summer). Overlaps in diet were highest with red deer (Pianka index = 0.90–0.95), while overlaps with roe deer were lower (between 0.52 and 0.63 in different seasons). Despite this overlap in habitat use and diet, no direct evidence was found to suggest direct competitive effects in terms of population dynamics (Putman and Sharma, 1987), perhaps because population numbers of all species were strongly controlled by (human) management and, thus, kept below levels at which resources become limiting and competition would be apparent. Most recently, Burkitt (2009) has offered suggestive evidence of inverse relationships between densities of sika and red deer in different sympatric populations in Killarney, Ireland.

4.3.3 Reeves' muntjac–roe deer

There is some evidence that where Reeves' muntjac have become established within the UK, they may impose competition upon the native roe deer. Muntjac and roe deer are both 'concentrate selectors' (Hofmann, 1989), which may pre-condition a great potential for interspecific competition. Furthermore, muntjac can reach very high densities, determining heavy negative effects on vegetation, at local scales (see Section 4.2). In turn, availability of food resources for sympatric roe deer could be reduced by heavy browsing.

It has been suggested that increasing muntjac populations may displace and outcompete roe deer: local decreases in roe deer density have been observed, associated with increases in muntjac populations (Forde, 1989; Wray, 1994; Chapman *et al.*, 1994a). Furthermore, a decrease of body weight (17% decrease: fawns; 22% yearlings; 12%: adults) and fertility (66%: yearlings; 23%: adults) of roe deer has been also reported over the period of muntjac increase (Hemami, 2003).

In the King's Forest, a study of ranging behaviour and activity rhythms of sympatric muntjac and roe deer through radio-tracking reported high spatial overlap between the two species, potentially facilitating some level of exploitation competition; by contrast, there appeared to be low temporal overlap between them, suggesting a lower potential for direct interference (Chapman *et al.,* 1994a). In another study area (Thetford Forest), significant habitat overlap was found in the use of forest growth stages (Pianka index = 0.93; Hemami *et al.*, 2004, 2005). Overlap in habitat use increased in winter, when both species concentrated on bramble, once again suggesting a potential for exploitation competition in the limiting season (Hemami *et al.*, 2004).

4.3.4 Other cases

Amongst deer species, fallow deer and red deer may compete: great overlaps have been reported in both habitat use (Putman, 1996a; Braza and Alvarez, 1987) and diet (Putman, 1986b; 1996a; Garcia-Gonzalez and Cuartas, 1992; Obidziński *et al.*, 2013), especially in limiting seasons (e.g. winter, in central-northern Europe). Although aggression by fallow deer to red deer has been documented in captivity (Bartoš *et al.*, 1996) and anecdotally reported in the wild (S. Nicoloso, 2006, video-taped and pers. comm., northern Apennines), observations by Bartoš *et al.* (2002) would suggest some mutualistic relationship between these two species under field conditions, rather than interference, (Bartoš *et al.*, 2002; but see Carranza and Valencia, 2002). Especially in winter, a great diet overlap has been also reported between white-tailed deer and both roe deer and fallow deer (Homolka *et al.*, 2008).

Competitive interactions may also occur between introduced and native ungulates in mountain ecosystems. A potential for competition has been reported between introduced Mediterranean mouflon and both Alpine chamois (Pfeffer and Settimo, 1973; Bertolino *et al.*, 2009; Darmon *et al.*, 2012; Chirichella *et al.*, 2013) and red deer (Homolka, 1996; Bertolino *et al.*, 2009; Miranda *et al.*, 2012), if they share the same feeding grounds. Conversely, the potential for competition seemed to be lower between mouflon and both Alpine ibex (Pfeffer and Settimo, 1973) and roe deer (Homolka, 1996; Bertolino *et al.*, 2009), as these ungulates showed a less substantial diet overlap. In the Jeseníky Mountains (Czech Republic), the diets of introduced Alpine chamois and native red deer showed a great overlap (Homolka 1996; Homolka and Heroldová, 2001). In this case, the alien species (the Alpine chamois) may be at a disadvantage to the native one (the red deer), because of its less generalist food habits with respect to those of the latter (cf. Lovari *et al.*, 2014, for Apennine chamois/red deer interactions).

Native and introduced ungulate species have shown a potential for competition, which has been strongly supported through field data. It would appear that the roe deer is particularly sensitive to the presence of competing species (e.g. fallow deer, muntjac, sika deer, white-tailed deer; cf. Latham, 1999), while of introduced species, the fallow deer seems to have a great potential for competition with other native species, through resource exploitation. As to interference, Ferretti *et al.* (2011b) suggested that fallow deer defend crucial resources as an evolutionary strategy helping this deer to survive in the semi-arid habitats where it evolved. This argument may explain why this deer is relatively intolerant to other ungulates (cf. Bartos *et al.*, 1996, 2002; McGhee and Baccus, 2006). The spread of exotic ungulates in Europe could determine an increase of competitive interactions with native ones: locally, they may pose a threat to small populations of native ungulates, as well as to endemic taxa (e.g. Italian roe deer).

4.4 Hybridisation with native ungulates

Hybridisation with non-native taxa may endanger some European ungulates (see Linnell and Zachos, 2011, for a summary). In particular, hybridisation may occur between different subspecies or even full species within the *Cervus* genus. Hybridisation between red and sika deer has been documented in both the UK and the Republic of Ireland (Harrington, 1973, 1982; Lowe and Gardiner, 1975; Ratcliffe *et al.*, 1992; Goodman *et al.*, 1999; Pemberton *et al.*, 2006; Díaz *et al.*, 2006; McDevitt *et al.*, 2009; Pérez-Espona *et al.*, 2009; Swanson and Putman, 2009) and in the Czech Republic (Bartoš *et al.*, 1981; Bartoš and Žirovnický, 1981; Zima *et al.*, 1990; Bartoš, 2009; Barančeková *et al.*, 2012); there is growing concern about the potential for hybridisation elsewhere (see Linnell and Zachos, 2011). In the UK, genetic analyses suggested that hybridisation occurs through mating between sika stags and red deer hinds (e.g. Goodman *et al.*, 1999; Pérez-Espona *et al.*, 2009; Senn and Pemberton, 2009). The rates of hybridisation and introgression into wild red and sika populations seem to be low (Goodman *et al.*, 1999; Pérez-Espona *et al.*, 2009; Senn and Pemberton, 2009), probably because of the rarity of crossbreeding events and the relatively limited time of contact between the two species (Senn and Pemberton, 2009). Frequency of hybridisation may also be highly variable among sites: in different locations of the Kintyre Peninsula (Scotland), from 0 to 43% of individuals were hybrids (Senn and Pemberton, 2009).

Furthermore, North American wapiti, where introduced, could compromise genetic integrity of native red deer through crossbreeding (Linnell and Zachos, 2011). On the Caucasus mountains, hybridisation between European and American bison among reintroduced populations elicits concern for the conservation of the former (Chapter 3).

The risk of hybridisation could also pose a risk for subspecies of native European ungulates, because of crossbreeding with introduced subspecies (e.g. European roe deer with Italian or south-Iberian subspecies; Alpine chamois, with Tatra chamois, Balkan chamois or Chartreuse chamois; Lovari, 1984; Corlatti *et al.*, 2011; Linnell and Zachos, 2011). As a result, differentiated gene pools may be lost: future translocations should be carefully evaluated in the light of the risk of genetic extinction because of hybridisation.

4.5 Conclusions

Over the last few decades in particular, ungulates, including alien species, have been increasing in numbers and distribution ranges right across Europe, which results in a great potential for impacts on ecosystems. In some cases, clear negative effects of alien species of ungulates have been shown on vegetation, habitats and communities depending on them, especially on island ecosystems, often evolved in absence of large herbivores. However, in many instances it is hard to resolve whether or not the effects of exotic ungulates on vegetation and habitats are specifically due to the fact that they are non-native (and thus not co-evolved) or whether these effects simply result from locally high densities of ungulates of whatever species. Nevertheless, in these cases, the impact of alien ungulates may exacerbate overall effects simply by virtue of adding to the effects imposed by native species.

What may be significant in this regard is that it would appear that non-native species may show a greater productivity or higher effective densities in local areas, as well as different rates of seed dispersal than those typically reached by native ungulate taxa (e.g. Putman and Sharma, 1987; Garcia-Gonzalez and Cuartas, 1992; Chapman *et al.*, 1989; Chadwick *et al.*, 1996; Mayle, 1996; Putman, 1996a; Putman *et al.*, 1996; Mattioli *et al.*, 2003; Hemami *et al.*, 2005; Cooke, 2006; Focardi *et al.*, 2006; Kramer *et al.*, 2006; Eycott *et al.*, 2007; Dolman and Wäber, 2008; Acevedo *et al.*, 2010; Gill and Morgan, 2010; Ferretti *et al.*, 2011a). Mechanisms allowing coexistence between alien ungulates and other natural components of ecosystems may have not developed, thus their impact could be different to that of native ones. If so, one could expect not only that the effects of exotic ungulates be additive to those of native ones, but also that the magnitude of the former be greater than that of the latter (Milchunas *et al.*, 1988; Hobbs and Huenneke, 1992; Parker *et al.*, 2006a). Further studies specifically testing these issues for European ungulates are required.

Alien ungulates could threaten native ones through interspecific competition and hybridisation (see Sections 4.3 and 4.4). Introduced species may show ecological requirements comparable to those of natives, determining a great potential for competition, which may lead up to ecological replacement. Although no study could experimentally demonstrate actual competition between exotic ungulates and European native ones, many studies have highlighted a great potential for it, or strongly supported the competition hypothesis (see Section 4.3). In some cases, there could be reasons of concern for the viability of small populations of native taxa, at least locally (e.g. competition between fallow deer and Italian roe deer; release of hybrids between American bison and European bison, hybridisation between introduced European roe deer and Italian/Spanish roe deer subspecies, as well as between introduced Alpine chamois and native subspecies of northern chamois (Chapter 3).

Thus, there are reasons to consider alien ungulates a threat for native biological diversity, which should require appropriate measures to mitigate – ideally, to prevent/eliminate – their impact. In particular, it would be necessary (1) to prevent the ultimate factors allowing the spread of exotic ungulates, i.e. their release into the wild, and (2) to manage the extant populations of wild exotic ungulates, to reduce their impact. Zoological gardens, safari parks and private enclosures have often been responsible for the establishment of populations of exotic species in the wild, whether by accidental escape or through deliberate release: appropriate legislation, with restrictive measures for importation, detention and maintenance of these species are needed at national and international level (Figure 4.3). In general, a common European policy should promote regulations to prevent introductions, as well as devise effective control measures of alien species when already present.

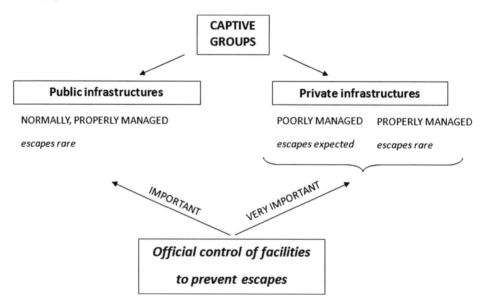

Figure 4.3 Introductions may start with 'escapes' of exotic taxa from captivity. Appropriate official control should be exerted on infrastructures to prevent it, rather than try eradication campaigns later on. In this flow chart, we have attempted to summarise the main factors related to this issue. As to intentional releases, see text.

Thus, an appropriate legislation should be enacted (cf. IUCN, 2000) (1) to avoid the introduction of alien taxa, with the exception of conservation introductions, *sensu* IUCN (2012); (2) to eradicate extant alien species from Europe where this is practicable, or at least to undertake risk assessments and develop feasibility studies to eradicate or limit their further spread (cf. UK policy on Invasive Non-Native Species, DEFRA, 2004); (3) to monitor distribution and population trends of alien taxa, both to evaluate the success of ongoing eradication/control operations, and to detect possible overlaps with ranges of native species.

But not all exotics are by definition damaging and we should beware of stereotyping, or even a potential for some form of 'ethnic cleansing'. In rare cases, introductions ('conservation introductions', *sensu* IUCN, 2012) and importation of exotic animals can have important implications for the conservation of rare taxa, if their native populations decrease or become extinct (for example, the case of Père David's Deer, see Harris and Duckworth, 2008, for a summary, and references therein). Among the alien ungulates presently living in Europe, Chinese water deer are now widely established in the east of England (largely East Anglia) where they were introduced in the latter part of the 19th century. Numbers have increased to an estimated population size of *c*.11,000 individuals. Conversely, numbers in their native China have suffered significant decline in recent decades (Hu *et al.*, 2006) and UK populations represent something of the order of 40% of global populations. A recent risk assessment considered impacts of these deer to be negligible whether to agriculture, forestry or native biodiversity in the UK (Cooke, 2011). In addition it appears that populations in the UK are significantly differentiated from those remaining in China (Fautley *et al.*, in press). Thus from a global perspective, populations of this introduced exotic deer within the UK may represent a significant resource towards conservation of this species worldwide, both in terms of population number and genetics (Fautley *et al.*, in press). Nevertheless, cautionary measures should be implemented to prevent their spread outside their current range. This case highlights the importance of thorough assessments of risks related to the presence of alien ungulates. Yet, exceptions ought not become the rule, as introduced ungulates represent an alteration of the zoogeographical pattern of an area and they may threaten native species, in the short or in the long term. Thus, their management should normally be oriented to population control and mitigation/eradication of their impact.

Not all aliens are equal, but some are worse than others...

Acknowledgements

M. Corazza provided helpful comments on the section of impact on vegetation.

References

Abernethy, K. (1994) *The introduction of sika deer, Cervus nippon nippon, to Scotland.* PhD thesis, University of Edinburgh.

Abernethy, K. (1998) *Sika Deer in Scotland.* Edinburgh: Deer Commission for Scotland, The Stationery Office.

Acevedo, P., Cassinello, J., Hortal, J. and Gortazar, C. (2007) Invasive exotic aoudad (*Ammotragus lervia*) as a major threat to native Iberian ibex (*Capra pyrenaica*): a habitat suitability model approach. *Diversity and Distributions* **13**, 587–597.

Acevedo, P., Ward, A., Real, R. and Smith, G.C. (2010) Assessing biogeographical relationships of ecologically related species using favourability functions: a case study on British deer. *Diversity and Distributions* **16**, 515–528.

Apollonio, M., Andersen, R., and Putman, R.J. (eds) (2010) *European Ungulates and Their Management in the 21st Century.* Cambridge, UK: Cambridge University Press.

Banwell, D.B. (2009) The sika in New Zealand. In D.R. McCullough, S. Takatsuki and K. Kaji (eds), *Sika Deer: Biology, Conservation and Management of Native and Introduced Populations.* Tokyo: Springer, pp. 643–656.

Barančeková, M., Krojerová-Prokešová, J., Voloshina, I.V., *et al.* (2012) The origin and genetic variability of the Czech sika deer population. *Ecological Research*, **27**, 991–1003.

Bartoš, L, 2009. Sika deer in Continental Europe. In D.R. McCullough, S. Takatsuki and K. Kaji (eds), *Sika Deer: Biology, Conservation and Management of Native and Introduced Populations.* Tokyo: Springer, pp. 573–594.

Bartoš, L. and Žirovnický, J. (1981) Hybridisation between red and sika deer. II. Phenotype analysis. *Zoologischer Anzeiger* **207**, 271–287.

Bartoš, L., Žirovnický, J. and Hyanek, J. (1981) Hybridisation between red and sika deer. I. Craniological analysis. *Zoologischer Anzeiger* **207**, 260–270.

Bartoš, L., Vaňková, D., Šiler, J. and Losos, S. (1996) Fallow deer tactic to compete over food with red deer. *Aggressive Behaviour* **22**, 375–385.

Bartoš, L., Vaňková, D. and Miller, K.V. (2002) Interspecific competition between white-tailed, fallow, red, and roe deer. *Journal of Wildlife Management* **66**, 522–527.

Batcheler, C.L. (1960) A study of the relations between roe, red and fallow deer, with special reference to Drummond Hill Forest, Scotland. *Journal of Animal Ecology* **29**, 375–384.

Begon, M., Townsend, C.R. and Harper, J.L. (2006) *Ecology. From Individuals to Ecosystems.* Oxford: Blackwell Publishing.

Benes, J., Cizek, O., Dovala, J. and Konvicka, M. (2006) Intensive game keeping, coppicing and butterflies: the story of Milovicky Wood, Czech Republic. *Forest Ecology and Management* **237**, 353–365.

Bertolino, S., di Montezemolo, N.C. and Bassano, B. (2009) Food-niche relationship within a guild of alpine ungulates including an introduced species. *Journal of Zoology* **277**, 63–69.

Boitani, L., Lovari, S. and Vigna Taglianti, A. (2003) *Mammalia III: Carnivora–Artiodactyla. Fauna d'Italia – Vol. XXXVIII.* Bologna, Italy: Calderini, pp. 403–412.

Bowden, L. and Van Vuren, D. (1997). Insular endemic plants lack defenses against herbivores. *Conservation Biology* **11**, 1249–1254.

Boyle, A.E. (1996) The Rio Convention on biological diversity. In M. Bowman and C. Redgwell (eds), *International Law and the Conservation of Biological Diversity.* London: Kluwer Law International, pp. 33–49.

Braza, F. and Álvarez, F. (1987) Habitat use by red deer and fallow deer in Donana National Park. *Miscellanea Zoologica* **11**, 363–367.

Burkitt, T.D. (2009) *A comparison of ecology between sympatric native red deer (*Cervus elaphus Linnaeus 1758*) and introduced Japanese sika deer (*Cervus nippon nippon Temminck 1836*) populations in southwest Ireland.* PhD thesis, Manchester Metropolitan University, UK.

CABI (2014) *Invasive Species Compendium.* Wallingford, UK: CABI.

Campbell, K.J. and Donlan, C.J. (2005) A review of feral goat eradication on islands. *Conservation Biology* **19**, 1362–1374.

Carnevali, L., Pedrotti, L., Riga, F. and Toso, S. (2009) Banca Dati Ungulati: Status, distribuzione, consistenza, gestione e prelievo venatorio delle popolazioni di Ungulati in Italia. Rapporto 2001–2005. *Biologia e Conservazione della Fauna* **117**, 1–168.

Carranza, J. and Valencia, J. (1999) Red deer females collect on male clumps at mating areas. *Behavioral Ecology* **10**, 525–532.

Carter, N.A. (1984) Bole scoring by sika deer (*Cervus nippon*) in England. *Deer* **6**, 77–78.

CBD Secretariat (2002) *Decision VI/23: Alien species that threaten ecosystems, habitats and species.* Document UNEP/CBD/COP/6/23. Convention on Biological Diversity Secretariat, Montreal.

Chadwick, A.H., Ratcliffe, P.R. and Abernethy, K. (1996) Sika deer in Scotland: density, population size, habitat use and fertility–some comparisons with red deer. *Scottish Forestry* **50**, 8–16.

Challies, C.N. (1985) Establishment, control and commercial exploitation of wild deer in New Zealand. In K. R. Drew and P. F. Fennessy (eds), *Biology of Deer Production. Royal Society of New Zealand Bulletin* **22**, 23–36.

Chapman, N.G. (2008) Reeves muntjac. In S. Harris and D.W. Yalden (eds), *Mammals of the British Isles: Handbook*, 4th edition. London: The Mammal Society, pp. 564–571.

Chapman, N.G. and Chapman, D.I. (1980) The distribution of fallow deer: a world-wide review. *Mammal Review* **10**, 61–138.

Chapman, N.G. and Harris, S. (1996) *Muntjac.* The Mammal Society and British Deer Society, UK.

Chapman, N.G., Harris, A. and Harris, S. (1994b) What gardeners say about muntjac. *Deer* **9**, 302–306.

Chapman, N.G., Harris, S. and Stanford, A. (1994a) Reeves' muntjac *Muntiacus reevesi* in Britain their history, spread habitat selection and the role of human intervention in accelerating their dispersal. *Mammal Review* **24**, 113–160.

Chirichella, R., Ciuti, S. and Apollonio, M. (2013) Effects of livestock and non-native mouflon on use of high-elevation pastures by Alpine chamois. *Mammalian Biology* **78**, 344–350.

Clout, M.N. and Russell, J.C. (2008) The invasion ecology of mammals: a global perspective. *Wildlife Research* **35**, 180–184.

Coblentz, B.E. (1978) The effects of feral goats (*Capra hircus*) on island ecosystems. *Biological Conservation* **13**, 279–286.

Connell, J.H. (1980) Diversity and the coevolution of competitors, or the ghost of competition past. *Oikos* **35**, 131–138.

Cooke, A.S. (1994) Colonisation by muntjac deer *Muntiacus reevesi* and their impact on vegetation. In M.E. Massey and R.C. Welch (eds), *Monks Wood National Nature Reserve. The experience of 40 years 1953–93.* Peterborough, UK: English Nature, pp. 45–61.

Cooke, A.S. (1997) Effects of grazing by muntjac (*Muntiacus reevesi*) on bluebells (*Hyacinthoides non-scripta*) and a field technique for assessing feeding activity. *Journal of Zoology* **242**, 365–369.

Cooke, A.S. (1998) Survival and regrowth performance of coppiced ash (*Fraxinus excelsior*) in relation to browsing damage by muntjac deer (*Muntiacus reevesi*). *Quarterly Journal of Forestry* **92**, 286–290.

Cooke, A.S. (2004) Muntjac and conservation woodland. In C.P. Quine, R.F. Shore and R.C. Trout (eds), *Managing Woodlands and their Mammals: Proceedings of a Joint Mammal Society/ Forestry Commission Symposium.* Edinburgh, UK: Forestry Commission, pp. 65–69.

Cooke, A.S. (2005) Muntjac deer *Muntiacus reevesi* in Monks Wood NNR: their management and changing impact. In C. Gardiner and T. Sparks (eds), *Ten years of change: woodland research at Monks Wood NNR, 1993–2003.* English Nature Research Report **613**, 65–74.

Cooke, A.S. (2006) Monitoring muntjac deer *Muntiacus reevesi* and their impacts in Monks Wood National Nature Reserve. English Nature Research Reports **681**, 1–174.

Cooke, A.S. and Farrell, L. (1995) Establishment and impact of muntjac (*Muntiacus reevesi*) on two national nature reserves. In B.A. Mayle (ed.), *Muntjac Deer. Their Biology, Impact and Management in Britain.* Edinburgh, UK: Forestry Commission, pp. 48–62.

Cooke, A.S. and Farrell, L. (2001) Impact of muntjac deer (*Muntiacus reevesi*) at Monks Wood National Nature Reserve, Cambridgeshire, eastern England. *Forestry* **74**, 241–250.

Coomes, D.A., Allen, R.B., Forsyth, D.M. and Lee, W.E. (2003) Factors preventing the recovery of New Zealand forests following control of invasive deer. *Conservation Biology* **17**, 450–459.

Corlatti, L., Lorenzini, R. and Lovari, S. (2011) The conservation of the chamois *Rupicapra* spp. *Mammal Review* **41**, 163–174.

Côté, S.D., Rooney, T.P., Tremblay, J.P., Dussault, C. and Waller, D.M. (2004) Ecological impacts of deer overabundance. *Annual Reviews of Ecology Evolution and Systematics* **35**, 113–147.

Courchamp, F., Chapuis, J.L. and Pascal, M. (2003) Mammal invaders on islands: impact, control and control impact. *Biological Reviews* **78**, 347–383.

Cross, J.R. (1981) The establishment of *Rhododendron ponticum* in the Killarney oakwoods, S.W. Ireland. *Journal of Ecology* **69**, 807–824.

Danilkin, A.A. (1996) *Behavioural Ecology of Siberian and European Roe Deer.* London: Chapman and Hall.

Darmon, G., Calenge, C., Loison, A., *et al.* (2012) Spatial distribution and habitat selection in coexisting species of mountain ungulates. *Ecography* **35**, 45–53.

de Boer, W.F. and Prins, H.H.T. (1990) Large herbivores that strive mightily but eat and drink as friends. *Oecologia* **82**, 264–274.

DEFRA (2004) *The sustainable management of wild deer populations in England: an action plan.* Department of Environment, Food and Rural Affairs, London.

Diaz, A. and Burton, R.J. (1996) The impact of predation by Muntjac deer *Muntiacus reevesi* on sexual reproduction of the woodland herb, Lords & Ladies *Arum maculatum*. *Deer* **10**, 14–19.

Díaz, A., Hughes, S., Putman, R., Mogg, R. and Bond, J.M. (2006) A genetic study of sika (*Cervus nippon*) in the New Forest and in the Purbeck region, southern England: is there evidence of recent or past hybridization with red deer (*Cervus elaphus*)? *Journal of Zoology* **207**, 227–235.

Diaz, A., Pinn, E.H. and Hannaford, J. (2005) Ecological impacts of sika deer on Poole Harbour saltmarshes. *Proceedings in Marine Science* **7**, 175–188.

Dolman, P.M. and Wäber, K. (2008) Ecosystem and competition impacts of introduced deer. *Wildlife Research* **35**, 202–214.

Donlan, C.J., Tershy, B.R. and Croll, D.A. (2002) Islands and introduced herbivores: conservation action as ecosystem experimentation. *Journal of Applied Ecology* **39**, 235–246.

Dzieciolowski, R. (1979) Structure and spatial organisation of deer populations. *Acta Theriologica* **24**, 3–21.

Eycott, A.E., Watkinson, A.R., Hemami, M.R. and Dolman, P.M. (2007) The dispersal of vascular plants in a forest mosaic by a guild of mammalian herbivores. *Oecologia* **154**, 107–118.

Fautley, R., Coulson, T. and Savoleinen, V. (2012) A comparative analysis of the factors promoting deer invasion, *Biological Invasions* **14**, 2271–2281.

Fautley R., Chen M., Putman R., Coulson T., Zhang E. and Savolainen V. (in press) Genetic structure and diversity of native and introduced Chinese water deer: implications for conservation.

Feldhamer, G.A. and Armstrong, W.E. (1993) Interspecific competition between four exotic species and native artiodactyls in the United States. *Transaction of the North American Wildlife and Natural Resources Conference* **58**, 468–478.

Feldhamer, G.A., Chapman, J.A. and Miller, R.L. (1978) Sika deer and white-tailed deer on Maryland's eastern shore. *Wildlife Society Bulletin* **6**, 155–157.

Feldhamer, G.A. and Demerais, S. (2009) Free-ranging and confined sika deer in North America: current status, biology and management. In D.R. McCullough, S. Takatsuki and K. Kaji (eds), *Sika Deer: Biology, Conservation and Management of Native and Introduced Populations.* Tokyo: Springer, pp. 615–642.

Ferretti, F. (2011) Interspecific aggression between fallow and roe deer. *Ethology, Ecology and Evolution* **23**. 179–186.

Ferretti, F., Sforzi, A. and Lovari, S. (2008) Intolerance amongst deer species at feeding: roe deer are uneasy banqueters. *Behavioural Processes* **78**, 478–491.

Ferretti, F., Bertoldi, G., Sforzi, A. and Fattorini, L. (2011a) Roe and fallow deer: are they compatible neighbours? *European Journal of Wildlife Research* **57**, 775–783.

Ferretti, F., Sforzi, A. and Lovari, S. (2011b) Behavioural interference between ungulate species: roe are not on velvet with fallow deer. *Behavioral Ecology and Sociobiology* **65**, 875–887.

Ferretti, F., Sforzi, A. and Lovari, S. (2012) Avoidance of fallow deer by roe deer may not be habitat-dependent. *Hystrix* **23**, 28–34.

Ferroglio, E., Gortázar, C. and Vicente, J. (2011) Wild ungulate diseases and the risk for livestock and public health. In Putman R., Apollonio, M. and Andersen, R. (eds), *Ungulate Management in Europe: Problems and Practices.* Cambridge, UK: Cambridge University Press, pp. 192–214.

Focardi, S., Aragno, P., Montanaro, P. and Riga, F. (2006) Inter-specific competition from fallow deer *Dama dama* reduces habitat quality for the Italian roe deer *Capreolus capreolus italicus. Ecography* **29**, 407–417.

Forde, P. (1989) *Comparative ecology of muntjac Muntiacus reevesi and roe deer Capreolus capreolus in a commercial coniferous forest.* PhD thesis, University of Bristol.

Forsyth, D.M., Wilmshurst, J.M., Allen, R.B. and Coomes, D.A. (2010) Impacts of introduced deer and extinct moa on New Zealand ecosystems. *New Zealand Journal of Ecology* **34**, 48–65.

Fraser, K.W., Cone, J.M. and Whitford, E.J. (2000) A revision of the established ranges and new populations of 11 introduced ungulate species in New Zealand. *Journal of the Royal Society of New Zealand* **30**, 419–437.

Fuller, R.J. and Gill, R.M.A. (2001) Ecological impacts of increasing numbers of deer in British woodland. *Forestry* **74**, 193–199.

Gangoso, L., Donázar, J.A., Scholz, S., Palacios, C.J. and Hiraldo, F. (2006) Contradiction in conservation of island ecosystems: plants, introduced herbivores and avian scavengers in the Canary Islands. *Biodiversity and Conservation* **15**, 2231–2248.

Garcia-Gonzalez, R. and Cuartas, P. (1992) Food habits of *Capra pyrenaica, Cervus elaphus* and *Dama dama* in the Cazorla Sierra (Spain). *Mammalia* **56**, 195–202.

Garzón-Machado, V., González-Manchebo, J.M., Palomares-Martínez, A., Acevedo-Rodríguez, A., Fernández-Palacios, J.M., Del-Arco-Aguilar, M. and Pérez-de-Paz, P.L. (2010) Strong negative effect of alien herbivores on endemic legumes of the Canary pine forest. *Biological Conservation* **143**, 1685–1694.

Gebhart, H. (1996) Ecological and economic consequences of introductions of exotic wildlife (birds and mammals) in Germany. *Wildlife Biology* **2**, 205–211.

Gill, R.M.A. (1992) A review of damage by mammals in north temperate forests: 1. Deer. *Forestry*, **65** 145–169.

Gill, R.M.A. and Beardall, V. (2001) The impact of deer on woodlands: the effects of browsing and seed dispersal on vegetation structure and composition. *Forestry* **74**, 209–218.

Gill, R.M.A. and Morgan, G. (2010) The effects of varying deer density on natural regeneration in woodlands in lowland Britain. *Forestry* **83**, 53–63.

Goodman, S., Barton, N., Swanson, G., Abernethy, K. and Pemberton, J. (1999) Introgression through rare hybridization: a genetic study of a hybrid zone between red and sika deer (Genus *Cervus*) in Argyll, Scotland. *Genetics* **152**, 355–371.

Gordon, I.J., Hester, A.J. and Festa-Bianchet, M. (2004) The management of wild large herbivores to meet economic, conservation and environmental objectives. *Journal of Applied Ecology* **41**, 1021–1031.

Guidi, T., Foggi, B., Arru, S., Lazzaro, L. and Giannini, F. (2009) Effects of grazer populations on the woody vegetation of the islands of Capraia and Elba (Tuscan Archipelago - Livorno). *Georgofili* **6**, 97–119.

Hannaford, J., Pinn, E.H. and Diaz, A. (2006) The impact of sika deer on the vegetation and infauna of Arne saltmarsh. *Marine Pollution Bullettin* **53**, 56–62.

Harmel, D.E. (1980) *The influence of exotic artiodactyls on white-tailed deer performance and survival.* Performance Report: Job No. 20, Federal Aid Project W-109-R-3, Texas Parks and Wildlife Department, United States.

Harmel, D.E. (1992) *The influence of fallow deer and aoudad sheep on white-tailed deer production and performance and survival.* Performance Report: Job No. 20, Federal Aid Project W-127-R-1, Texas Parks and Wildlife Department, United States.

Harrington, R. (1973) Hybridisation among deer and its implications for conservation. *Irish Forestry Journal* **30**, 64–78.

Harrington, R. (1982) The hybridisation of red deer (*Cervus elaphus* L. 1758) and Japanese sika deer (*C. nippon* Temminck, 1838). *International Congress of Game Biologists* **14**, 559–571.

Harris, R.A. (1981) *Survey of the Fallow Deer Population and its Impact on the Wildlife Conservation Objectives of Castor Hanglands NNR.* Peterborough, UK: Nature Conservancy Counci,.

Harris, R.B. and Duckworth, J.W. (2008) *Hydropotes inermis.* In *IUCN 2013. IUCN Red List of Threatened Species. Version 2013.2.* <www.iucnredlist.org>. Downloaded on 28 January 2014.

Hemami, M.R. (2003) *The ecology of roe deer (*Capreolus capreolus*) and muntjac (*Muntiacus reevesi*) in a forested landscape in eastern England.* PhD thesis, University of East Anglia, Norwich, UK.

Hemami, M.R., Watkinson, A.R. and Dolman, P.M. (2004) Habitat selection by sympatric muntjac (*Muntiacus reevesi*) and roe deer (*Capreolus capreolus*) in a lowland commercial pine forest. *Forest Ecology and Management* **194**, 49–60.

Hemami, M.R., Watkinson, A.R. and Dolman, P.M. (2005) Population densities and habitat associations of introduced muntjac *Muntiacus reevesi* and native roe deer *Capreolus capreolus* in a lowland pine forest. *Forest Ecology and Management* **215**, 224–238.

Heroldová, M. (1996) Dietary overlap of three ungulate species in the Palava Biosphere Reserve. *Forest Ecology and Management* **88**, 139–142.

Hobbs, N.T. (1996) Modification of ecosystems by ungulates. *Journal of Wildlife Management* **60**, 695–713.

Hobbs, R.J. and Huenneke, L.F. (1992) Disturbance, diversity, and invasion: implications for conservation. *Conservation Biology* **6**, 324–337.

Hofmann, R.R. (1989) Evolutionary steps of ecophysiological adaptation and diversification of ruminants: a comparative view of their digestive system. *Oecologia* **78**, 443–457.

Homolka, M. (1996) Foraging strategy of large herbivores in forest habitats. *Folia Zoologica* **45**, 127–136.

Homolka, M. and Heroldová, M. (2001) Native red deer and introduced chamois: foraging habits and competition in a subalpine meadow-spruce forest area. *Folia Zoologica* **50**, 89–98.

Homolka, M., Heroldovà, M. and Bartoš, L. (2008) White-tailed deer winter feeding strategy in area shared with other deer species. *Folia Zoologica* **57**, 283–293.

House, C., May, V. and Diaz, A. (2005). Sika deer trampling and saltmarch creek erosion: preliminary investigation. *Proceedings in Marine Science* **7**, 189–193.

Hu, J., Fang, S.G. and Wan, Q.H. (2006) Genetic diversity of Chinese water deer (*Hydropotes inermis inermis*): Implications for conservation. *Biochemical Genetics* **44**(3–4), 161–172.

Imperio, S., Focardi, S., Santini, G. and Provenzale, A. (2012) Population dynamics in a guild of four Mediterranean ungulates: density-dependence, environmental effects and interspecific interactions. *Oikos* **121**, 1613–1626.

IUCN (1987) *The IUCN Position Statement on Translocation of Living Organisms: introductions, re-introductions and re-stocking.* SSC/Commission on Ecology/Commission on Environmental Policy, Law and Administration, Gland, Switzerland.

IUCN (2000) *Guidelines for the Prevention of Biodiversity Loss Caused by Alien Invasive Species.* Gland, Switzerland: SSC Invasive Species Specialist Group.

IUCN (2012) *IUCN Guidelines for Reintroductions and Other Conservation Translocations.* Gland, Switzerland: SSC Invasive Species Specialist Group.

Jędrzejewski, W., Apollonio, M., Jędrzejewska, B. and Kojola, I. (2011) Ungulate–large carnivore relationships in Europe. In R.J. Putman, M. Apollonio and R. Andersen (eds), *Ungulate Management in Europe: Problems and Practices.* Cambridge, UK: Cambridge University Press, pp. 284–318.

Kirby, K.J. (2001) The impact of deer on the ground flora of British broadleaved woodland. *Forestry* **74**, 219–229.

Kramer, K., Groot Bruinderink, G.W.T.A. and Prins, H.H.T. (2006) Spatial interactions between ungulate herbivory and forest management. *Forest Ecology and Management* **226**, 238–247.

Langbein, J., Putman, R. and Pokorny, B. (2011) Traffic collisions involving deer and other ungulates in Europe and available measures for mitigation. In R.J. Putman, M. Apollonio and R. Andersen (eds), *Ungulate Management in Europe: Problems and Practices.* Cambridge: Cambridge University Press, pp. 215–259.

Larner, J.B. (1977) Sika deer damage to mature woodlands of southwestern Ireland. *Proceedings of the XI11th Congress of Game Biology*, 192–202.

Latham, J. (1999) Interspecific interactions of ungulates in European forest: an overview. *Forest Ecology and Management* **120**, 13–21.

Liberg, O., Bergström, R., Kindberg, J. and von Essen, H. (2010) Ungulates and their management in Sweden. In M. Apollonio, R. Andersen and R.J. Putman (eds), *European Ungulates and their Management in the 21st Century.* Cambridge, UK: Cambridge University Press, pp. 37–70.

Linnell, J.D.C. and Zachos, F.E. (2011) Status and distribution patterns of European Ungulates. In R.J. Putman, M. Apollonio and R. Andersen (eds), *Ungulate Management in Europe: Problems and Practices.* Cambridge, UK: Cambridge University Press, pp. 12–53.

Lovari, S. (1984) *Il popolo delle rocce.* Torino, Italy: Rizzoli.

Lovari, S., Ferretti, F., Corazza, M., *et al.* (2014) Unexpected consequences of reintroductions: competition between increasing red deer and threatened Apennine chamois. *Animal Conservation.* doi: 10.1111/acv.12103

Lowe, R. (1994) *Deer Management: developing the requirements for the establishment of diverse coniferous and broadleaf forests.* Unpublished report, Coilte, Bray, Co. Wicklow.

Lowe, V.P.W. and Gardiner, A.S. (1975) Hybridisation between red deer and sika deer, with reference to stocks in north-west England. *Journal of Zoology* **177**, 553–566.

Makovkin, L.I. (1999) *The Sika Deer of Lazovsky Reserve and Surrounding Areas of the Russian Far East.* Vladivostok, Russia:Almanac Russki Ostrov.

Manganelli, E. (2012) *Daino e capriolo: competizione o neutralismo?* MSc thesis, University of Siena.

Mánzano, M.G. and Návar, J. (2000) Processes of desertification by goats overgrazing in the Tamaulipan thornscrub (matorral) in north-eastern Mexico. *Journal of Arid Environments* **44**, 1–17.

Masseti, M. (1996) The postglacial diffusion of the genus *Dama* Frisch, 1775, in the Mediterranean region. *Supplementi alle Ricerche di Biologia della Selvaggina* **25**, 7–29.

Mattioli, S., Fico, R., Lorenzini, R. and Nobili, G. (2003) Mesola red deer: physical characteristics, population dynamics and conservation perspectives. *Hystrix* **14**, 87–94.

Mayle, B.A. (1996) Progress in predictive management of deer populations in British woodlands. *Forest Ecology and Management* **88**, 187–198.

McCullough, D.R., Jiang, Z.G. and Chun-Wang, L. (2009a) Sika deer in Mainland China. In D.R. McCullough, S. Takatsuki and K. Kaji (eds), *Sika Deer: Biology, Conservation and Management of Native and Introduced Populations.* Tokyo: Springer, pp. 521–540.

McCullough, D.R., Takatsuki, S. and Kaji, K. (2009b) *Sika Deer: Biology, Conservation and Management of Native and Introduced Populations.*Tokyo: Springer.

McDevitt, A.D., Edwards, C.J., O'Toole, P., O'Sullivan, P., O'Reilly, C. and Carden, R.F. (2009) Genetic structure of, and hybridisation between, red (*Cervus elaphus*) and sika (*Cervus nippon*) deer in Ireland. *Mammalian Biology* **74**, 263–273.

McGhee, J.D. and Baccus, J.T. (2006) Behavioural interactions between axis and fallow deer at high-value food patches. *Southwestern Naturalist* **51**, 358–367.

McKelvey, P.J. (1959) Animal damage in North Island protection forests. *New Zealand Science Review* **17**, 28–34.

McLaren, B.E. and Peterson, R.O. (1994). Wolves, moose and tree rings on Isle Royale. *Science* **266**, 1555–1558.

McLaughlan, C., Gallardo, B. and Aldridge, D.C. (in press) How complete is our knowledge of the ecosystem services impacts of Europe's top 10 invasive species? *Acta Oecologica.*

Milchunas, D.G. and Lauenroth, W.K. (1993) Quantitative effects of grazing on vegetation and soils over a global range of environments. *Ecological Monographs* **63**, 327–366.

Milchunas, D.G., Sala, O.E. and Lauenroth, W.K. (1988) A generalized model of the effects of grazing by large herbivores on grassland community structure. *American Naturalist* **132**, 87–106.

Miranda, M., Sicilia, M., Bartolomé, J., Molina-Alcaide, E., Gálvez-Bravo, L. and Cassinello, J. (2012) Contrasting feeding patterns of native red deer and two exotic ungulates in a Mediterranean ecosystem. *Wildlife Research* **39**, 171–182.

Morecroft, M.D., Taylor, M.E., Ellwood, S.A. and Quinn, S.A. (2001) Impacts of deer herbivory on ground vegetation at Wytham Woods, central England. *Forestry* **74**, 251–257.

Nogales, M., Rodríguez-Luengo and Marrero, P. (2006) Ecological effects and distribution of invasive non-native mammals on the Canary Islands. *Mammal Review* **36**, 49–65.

Nugent, G., Fraser, W. and Sweetapple, P. (2001) Top down or bottom up? Comparing the impacts of introduced arboreal possums and 'terrestrial' ruminants on native forests in New Zealand. *Biological Conservation* **99**, 65–79.

Nunez, M.A., Bailey, J.K. and Schweitzer, J.A. (2010) Population, community and ecosystem effects of exotic herbivores: a growing global concern. *Biological Invasions* **12**, 297–301.

Obidziński, A., Kieityk, P., Borkowski, J., Bolibok, L. and Remuszko, K. (2013) Autumn–winter diet overlap of fallow, red, and roe deer in forest ecosystems, Southern Poland. *Central European Journal of Biology* **8**, 8–17.

Parker, J.D., Burkepile, D.E. and Hay, M.E. (2006a) Opposing effects of native and exotic herbivores on plant invasions. *Science* **311**, 1459–1461.

Parker, J.D., Burkepile, D.E. and Hay, M.E. (2006b) Response to Comment on 'Opposing effects of native and exotic herbivores on plant invasions. *Science* **313**, 298.

Pemberton, J., Swanson, G., Barton, N., Livingstone, S. and Senn, H. (2006) Hybridisation between red and sika deer in Scotland. *Deer*, **13**, 22–26.

Pérez-Espona, S., Pemberton, J.M. and Putman, R. (2009) Red and sika deer in the British Isles, current management issues and management policy. *Mammalian Biology* **74**, 247–262.

Pfeffer, P. and Settimo, R. (1973) Deplacements saisonniers et competition vitale entre mouflons, chamois and bouquetins dans la Reserve du Mercantour (Alpes Maritimes). *Mammalia* **37**, 203–219.

Pianka, E.R. (1973) The structure of lizard communities. *Annual Review of Ecology and Systematics* **4**, 53–74.

Pollard, E. and Cooke, A.S. (1994) Impact of muntjac deer *Muntiacus reevesi* on egg-laying sites of the white admiral butterfly *Ladoga camilla* in a Cambridgeshire wood. *Biological Conservation* **70**, 189–191.

Pollard, E., Woiwod, R.C., Greatorex-Davies, J.N., Yates, T.J. and Welch, R.C. (1998) The spread of coarse grasses and changes in numbers of Lepidoptera in a woodland nature reserve. *Biological Conservation* **84**, 17–24.

Post, E., Peterson, R.O., Stenseth, N.C. and McLaren, B.E. (1999) Ecosystem consequences of wolf behavioural response to climate. *Nature* **401**, 905–907.

Putman R.J. (1986a) *Grazing in Temperate Ecosystems; Large Herbivores and their effects on the Ecology of the New Forest.* Beckenham UK: Croom Helm.

Putman, R.J. (1986b) Competition and coexistence in a multi-species grazing system. *Acta Theriologica* **31**, 271–291.

Putman R.J. (1994) *Community Ecology.* London: Chapman and Hall.

Putman, R.J. (1996a) *Competition and Resource Partitioning in Temperate Ungulate Assemblies.* London: Chapman and Hall.

Putman R.J. (1996b) *Deer on National Nature Reserves: problems and practices. English Nature Research Report* **173**, English Nature, Peterborough.

Putman, R.J. and Moore, N.P. (1998) Impact of deer in lowland Britain on agriculture, forestry and conservation habitats. *Mammal Review* **28**, 141–164.

Putman, R.J. and Sharma, S.K. (1987) Long term changes in New Forest deer populations and correlated environmental change. *Symposia of the Zoological Society of London* **58**, 167–179.

Putman, R.J., Edwards, P.J., Mann, J.C.E., Howe, R.C., and Hill, S.D. (1989) Vegetational and faunal changes in an area of heavily grazed woodland following relief from grazing. *Biological Conservation* **47**, 13–32.

Putman, R.J., Langbein, A.J., Hewison, A.J.M. and Sharma, S.K. (1996) Relative roles of density-dependent and density independent factors in population dynamics of British deer. *Mammal Review* **26**, 81–101.

Putman, R.J., Langbein, J. Green, P. and Watson, P. (2011) Identifying threshold densities for wild deer in the UK above which negative impacts may occur. *Mammal Review* **41**, 175–196.

Rackham, O. (1975) *Hayley Wood: Its History and Ecology. Cambridgeshire and Isle of Ely.* Cambridge, UK: Cambridge Naturalists' Trust.

Rackham, O. (2003) *Ancient Woodland its History, Vegetation and Uses in England* (new edition). Colvend, Kirkudbright, UK: Castlepoint Press.

Ratcliffe, P.R. (1989) The control of red and sika deer populations in commercial forests. In R.J. Putman (ed.) *Mammals as Pests.* London: Chapman and Hall, pp. 98–115.

Ratcliffe, P.R., Peace, A.J., Hewison, A.J.M., Hunt, E.J. and Chadwick, A.H. (1992) The origins and characterization of Japanese sika deer populations on Great Britain. In N. Maruyama, B. Bobek, Y. Ono, W. Regelin, L. Bartoš and P.R. Ratcliffe (eds), *International Symposium on Wildlife Conservation: Present Trends and Perspectives for the 21st Century.* Tokyo: Japan Wildlife Research Center, pp. 185–190.

Reimoser, F. and Putman, R. (2011) Impacts of wild ungulates on vegetation: costs and benefits. In R.J. Putman, M. Apollonio and R. Andersen (eds), *Ungulate Management in Europe: Problems and Practices.* Cambridge, UK: Cambridge University Press, pp. 144–191.

Ricciardi, A. and Ward, J.M. (2006) Comment on 'Opposing effects of native and exotic herbivores on plant invasions. *Science* **313**, 298.

Rodríguez-Pineiro, J.C. and Rodríguez-Luengo, J.L. (1992) Autumn food-habits of the Barbary sheep (*Ammotragus lervia* Pallas 1777) on La Palma Island (Canary Islands). *Mammalia* **56**, 385–392.

Schoener, T.W. (1974) Resource partitioning in ecological communities. *Science* **4145**, 27–39.

Scowcroft, P.G. and Giffin, J.G. (1983) Feral Herbivores Suppress Mamane and Other Browse Species on Mauna Kea, Hawaii. *Journal of Range Management* **36**, 638–645.

Scowcroft, P.G and Sakai, H.F. (1983) Impact of Feral Herbivores on Mamane Forests of Mauna Kea, Hawaii: Bark Stripping and Diameter Class Structure. *Journal of Range Management* **36**, 495–498.

Senn, H.V. and Pemberton, J.M. (2008) Variable extent of hybridization between invasive sika (*Cervus nippon*) and native red deer (*Cervus elaphus*) in a small geographical area. *Molecular Ecology* **18**, 862–876.

Šídová, A. and Schlaghamerský, J. (2007) The impact of high game density on enchytraeids in a mixed forest. In K. Tajovský, J. Schlaghamerský and V. Piž (eds), *Contribution to soil zoology in central Europe II.* ISB BC AS CR, v.v.i., České Budějovice.

Sinclair, A.R.E., Mduma, S. and Brashares, J.S. (2003) Patterns of predation in a diverse predator–prey system. *Nature* **425**, 288–290.

Smale, M.C., Hall, G.M.J. and Gardner, R.O. (1995) Dynamics of Kanuka (*Kunzea ericoides*) forest on Wouth Kaipara Spit, New Zealand, and the impact of fallow deer (*Dama dama*). *New Zealand Journal of Ecology* **19**, 131–141.

Smit, C. and Putman, R.J. (2011) Large herbivores as 'environmental engineers'; the manipulation of large herbivore populations as agents of deliberate habitat change. In R.J. Putman, M. Apollonio and R. Andersen (eds), *Ungulate Management in Europe: Problems and Practices.* Cambridge, UK: Cambridge University Press, pp. 260–283.

Smit, R., Bokdam, J., den Ouden, J., Schot-Opschoor, H. and Schrijvers, M. (2001) Effects of introduction and exclusion of large herbivores on small rodent communities. *Plant Ecology* **155**, 119–127.

Spear, D. and Chown, S.L. (2009) Non-indigenous ungulates as a threat to biodiversity. *Journal of Zoology* **279**, 1–17.

Swanson, G.M. and Putman R.J (2009) Sika deer in the British Isles. In D.R. McCullough, S. Takatsuki and K. Kaji (eds), *Sika Deer: Biology and Management of Native and Introduced Populations.* Tokyo: Springer, pp. 595–614.

Tabor, R.C. (1993) Control of deer in a managed coppice. *Quarterly Journal of Forestry* **87**, 308–13.

Tabor, R.C. (1999) The effects of Muntjac deer, *Muntiacus reevesi*, and Fallow deer *Dama dama*, on the oxslip *Primula elatior. Deer* **11**, 14–19.

Takatsuki, S. (2009). Effects of sika deer on vegetation in Japan: a review. *Biological Conservation* **142**, 1922–1929.

Tanentzap, A.J., Burrows, L.E., Lee, W.G., *et al.* (2009) Landscape-level vegetation recovery from herbivory: Progress after four decades of invasive red deer control. *Journal of Applied Ecology* **46**, 1064–1072.

Taylor, K. and Woodell, S.R. (2008) Biological flora of the British Isles: *Primula elatior* (L.) Hill. *Journal of Ecology* **96**, 1098–1116.

Van Vuren, D. and B.E. Coblentz (1987) Some ecological effects of feral sheep on Santa Cruz Island, California USA. *Biological Conservation* **41**, 253–268.

Vazquez, D.P. (2002) Multiple effects of introduced mammalian herbivores in a temperate forest. *Biological Invasions* **4**, 175–191.

Vilà, M., Basnou, C., Pysek, P., Josefsson, M., Genovesi, P., *et al.* (2010) How well do we understand the impacts of alien species on ecosystem services? A Pan-European, cross-taxa assessment. *Frontiers in Ecology and the Environment* **8**, 135–144.

Ward, A.I. (2005) Expanding ranges of wild and feral deer in Great Britain. *Mammal Review*, **35**, 165–173.

Ward, A.I., Etherington, T. and Ewald, J. (2008) Five years of change. *Deer* **14**, 17–20.

Ward, L.K., Clarke, R.T. and Cooke, A.S. (1994) Long term scrub succession deflected by fallow deer at Castor Hanglands National Nature Reserve. *Annual Report of the Institute of Terrestrial Ecology* (1993–4), 78–81.

White, P.J. and Garrott, R.A. (2005) Yellowstone's ungulates after wolves: expectations, realizations, and predictions. *Biological Conservation* **125**, 141–152.

Wray, S. (1994) Competition between muntjac and other herbivores in a commercial coniferous forest. *Deer,* **9**, 237–242.

Zamora, R., Gómez, J.M., Hódar, J.A., Castro, J. and García, D. (2001) Effect of browsing by ungulates on sapling growth of Scots pine in a Mediterranean environment: consequences for forest regeneration. *Forest Ecology and Management* **144**, 33–42.

Zima, J., Kožená, I. and Hubálek, Z. (1990) Non-metrical cranial divergence between *Cervus elaphus, C. nippon nippon* and *C. nippon hortulorum. Acta Scientarum Naturalium, Brno* **24**, 1–41.

Chapter 5

Competition Between Domestic and Wild Ungulates

Roberta Chirichella, Marco Apollonio and Rory Putman

The development of large, relatively permanent, agriculture-based societies was the primary event initiating livestock domestication about 10,000 years ago (Price, 2002). With a few exceptions, domestication of ungulates (e.g. cattle, sheep and goats) mainly began in the Near East (Troy *et al.*, 2001). The presence of livestock in Europe goes back to Neolithic times, with domestic sheep and goats showing up at that time particularly in Mediterranean countries (see, for example, Martín Bellido *et al.*, 2001). Domestic livestock graze more than one-third of the world's land area, often sharing space and resources with native wildlife (de Haan *et al.*, 1997). Although many authors have voiced concern regarding the impact of livestock grazing on native wildlife (Fleischner, 1994; Edwards *et al.*, 1996; Aagesen 2000; Prins, 2000), the issue has remained a matter of considerable debate. While some have argued that extant levels of livestock grazing may not adversely affect wildlife (Smith, 1992 cited in Saberwal, 1996; Homewood *et al.*, 2001), others strongly contest this view (e.g. Mishra and Rawat, 1998; Young *et al.*, 2005).

This is especially true in countries with a large free ranging livestock population. At about 460 million animals, India's livestock population is the largest in the world and 8.8% of the country's geographical area is used for grazing (FAO, http://data.fao.org/). India's wild herbivores, on the other hand, have become increasingly confined to wildlife reserves covering less than 5% of the country's land area (Madhusudan and Karanth, 2002). In a detailed review, Madhusudan (2004) reported three specific findings that have a significant bearing on wild herbivore conservation: (1) grazing by livestock appears to limit resources for wild herbivores, and is responsible for lowered wild herbivore densities and even exclusion in livestock-grazed areas; (2) livestock-mediated resource limitation, in particular, appears to affect those wild herbivore species that are most similar to livestock in terms of body size and feeding ecology; (3) wild herbivores like chital and gaur may benefit considerably from management interventions designed to reduce livestock presence in areas from which wild herbivores may have been previously excluded by livestock.

Even in western North America, livestock grazing is the most widespread land management practice. Seventy per cent of the western United States is grazed, including wilderness areas, wildlife refuges, national forests, and even some national parks. The general ecological cost of this widespread form of land use can be dramatic and may include loss of biodiversity; lowering of population densities for a wide variety of taxa; disruption of ecosystem functions, including nutrient cycling and succession; change in community organisation; and change in the physical characteristics of both terrestrial and aquatic habitats (Fleischner, 1994). Specific potential for direct competitive effects on native ungulates have been suggested through demonstration of substantial overlap in diet of introduced domesticates with native species (e.g. Hansen and Reid, 1975; Hansen *et al.*, 1977; Schwartz and Ellis, 1981; Hanley and Hanley, 1982; Ghosh *et al.*, 1987) and where specific evidence has been sought of possible niche shifts or competitive suppression of population growth, results are generally indicative of some level of competition.

The problem becomes particularly important in semi-arid and arid biomes, and there has, in a similar way, been a considerable discussion about the compatibility of livestock production in the context of the conservation and restoration of large mammalian biodiversity in Africa (e.g. MacMillan, 1986; Prins *et al.*, 2000). In East Africa, the majority of the populations of most large mammal species occur outside protected areas, although this proportion is declining (e.g. Western, 1989). These populations mostly occur on land that is also being used for the production of livestock, either by traditional pastoralists or by large-scale ranching enterprises. There is a widespread belief that grazing wildlife, particularly zebra (*Equus* spp.) and wildebeest (*Connochaetes taurinus*) compete with cattle for grass (MacMillan, 1986; Young *et al.*, 2005).

Given this background elsewhere, livestock husbandry must be regarded as capable of having a strong direct and/or indirect impact on wild ungulate populations, even in European countries. There are examples of strong conflict (e.g. Gordon and Illius, 1989; Acevedo *et al.*, 2007): some extreme cases are to be found in the areas where free range livestock management is still a dominant economic activity, as for instance in Sardinia, Italy. On this island extensive livestock husbandry is a traditional and still widespread form of land use; management of livestock in this more extensive manner is facilitated by both the presence of huge portions of public land devoted to this activity and large pasture areas cleared by deforestation and fires. Over 4 million sheep and some 100,000 free ranging cattle, goats, horses, donkeys and domestic pigs contend pastures with wild ungulates. In particular, mouflon suffer a strong impact from the grazing activity of livestock since the latter are brought onto hilly and mountain pastures, the environment used by mouflon, exactly when pasture productivity is at its highest level and when lambing season starts, displacing the mouflon to suboptimal areas like Mediterranean shrubs and forests. One of the outcomes is that this mouflon population shows a very limited productivity (Ciuti *et al.*, 2009).

5.1 Potential interactions within a multi-species assemblage

Direct impacts of domestic livestock on wild ungulates may result through competition for space and food between wild and domestic species (see, for example, Berdoucou, 1986; Mysterud, 2000; Mussa *et al.*, 2003; La Morgia and Bassano, 2009) and/or by transmission of shared diseases (see, for example, East *et al.*, 2011; Ferroglio *et al.*, 2011). The presence of livestock has been reported to influence the distribution and behaviour of native ungulates (e.g. Rebollo *et al.*, 1993), as the latter can be forced to move from the areas occupied by livestock (e.g. Mattiello *et al.*, 2002).

In general, the herbivore species within any given assembly might have three types of interaction: (i) total ecological independence; (ii) competition; and (iii) facilitation. For ungulates, the most commonly described interaction appears to be competition. Competition can manifest itself in various forms, and for the purposes of this chapter the term refers to all interactions where one species adversely affects the fitness of another. In general, competition can occur through two different mechanisms. Interference competition involves direct social interactions and subtle effects on the environment by one species which reduces its quality for another (e.g. by removing shrub layers and, hence, depriving a second species of the cover; Latham, 1999), while in exploitation competition one species either reduces the availability of a shared resource or uses it more efficiently and therefore depletes the availability of the resource for the other species (species use and compete for a shared resource of food or space, e.g. changing feeding habits and niche breadth after the arrival of the second species; Latham, 1999).

However, overlap in patterns of resource use need not necessary be detrimental. There is potential in such cases for the interaction to be mutualistic or at least of no disadvantage to one partner in the interaction while positively beneficial to the other (Putman, 1996). Gordon (1988) reported a convincing example for such interaction on the Isle of Rum, Scotland, where grazing by cattle was shown to increase quality of grass swards available to red deer.

Even where species show substantial overlap in fundamental niche (i.e. potential for competitive interaction), there are several ways in which stable, or at least persistent, coexistence may be achieved. Thus, coexistence may be accommodated through a change in resource use in interaction, or where resources are in superabundant supply. However, it may be noted that a superabundance of resources would be unlikely to persist for long in the absence of other factors (e.g. a generalist predator acting to reduce populations of all potential competitors below the level at which they would be limited by resource availability).

Because so many of those instances where competition is suggested between domestic and wild ungulates are in effect purely anecdotal, it is important to establish from the outset some formal criteria by which we can asses any given 'claim' of competitive interaction. It is clear for example that competition can, by definition, only exist if resources are actually or potentially limiting (Putman, 1996; Tokeshi, 1999).

In many studies of competitive interactions in large ungulates, authors have considered simply the degree of overlap observed in patterns of resource use. However, the interpretation of measures of niche overlap (i.e. habitat and dietary overlap) in terms of the implications for competitive interaction is extremely problematical and ambiguous. High observed overlap may imply competition, but only if resources are limited; observation of high overlap might equally well be considered indicative of a *lack* of competition – on the basis that if severe competition were being experienced some niche shift would have been expected, resulting in reduction of overlap. By that same token, observation of low levels of overlap in the field may imply lack of competition – but may in fact reflect the end-result of changes in the ecology of some or all of the species as a direct result of competition for shared resources. Only where we may find evidence of a clear shift in resource use of a species in allopatry and sympatry may we suspect a competitive interaction – or better still when clear overlap in resource use is accompanied by an inverse relationship in population sizes of a given species pair (Putman, 1996).

Because of these difficulties, effective analysis of competition must combine some assessment of overlap in resource use (providing the potential for interaction) with analysis of the actual population dynamics of the co-occurring species to give indications of densities and environmental conditions under which interactions occur and to show direct evidence of an impact of one species on another. Such studies fall into two broad types: those of sympatric populations of two species whose densities co-vary with time, and studies of contemporary populations of two species existing under a range of environmental conditions and population densities (Latham, 1999). These studies are by their nature large scale, and hence hard to manipulate experimentally. As a consequence, they may give useful insights into possible effects of interspecific interactions on the populations as a whole, but may give little information on the mechanisms involved. Other considerations, such as information about population trend, behavioural interactions and transmission of diseases may lead evidence in favour or against the occurrence of competition.

Recently, in order to try and overcome the lack of correspondence between niche overlap and competition, Richard *et al.* (2013) proposed an innovative approach for analysing spatial distributions of individuals from two sympatric species. Using the null model approach commonly applied in community ecology, they provided a way of performing a formal test of interspecific competition rooted in explicitly defined hypotheses. However, elaborating null models is far from an easy task and has not been used in population ecology to study interactions between animals in a robust way.

Recognition of the difficulty of establishing competition in the field – amongst any group of organisms - has led various workers to try to develop formal protocols for adducing competitive interaction in natural communities. Wiens (1989) has modified and extended the criteria developed earlier by Reynoldson and Bellamy (1970), suggesting a range of types of evidence which may be sought of different levels of cogency (Table 5.1).

Table 5.1 Increasing strength of evidence for competition as proposed by Wiens (1989)

Weak	1. Observed patterns (of population trend or shifts in resource use between sympatry and allopatry) are consistent with predictions from competition
	2. Species overlap in resource use
	3. Intraspecific competition occurs
Suggestive	4. Resource use by one species reduces availability of resources for another species
	5. One or more species is negatively affected
Convincing	6. Alternative process hypotheses are not consistent with observed patterns

Wiens' criteria are stringent – and satisfaction of all is rarely practicable in the field. An alternative logic is proposed by de Boer and Prins (1990). In a sense, de Boer and Prins argue by converse: seeking not to prove competition, but rather (on the basis that it is often easier to disprove a hypothesis rather than prove it beyond doubt) establishing hurdles of definite disproof. In summary, de Boer and Prins (1990) argue that interspecific competition between any two species is only possible where: (1) there is clear evidence of habitat overlap; (2) there is overlap in forage consumed by the two species within those shared habitats; and (3) the shared dietary resources are limiting. But those conditions are simply minimum conditions to offer even a potential for competition. Wiens (1989) criteria for establishing the actuality of competitive interaction in practice further require that 'one or more species is negatively affected' and that 'observed patterns (of population trend or shifts in resource use between sympatry and allopatry) are consistent with predictions from competition'.

5.2 Potential effects of competition

Any interaction between two species is, by that definition, an interaction, and we might anticipate effects of each species on the other in both directions. The common situation reported in most published studies, however, is that one or several domestic species strongly outnumber the wild ungulates (e.g. Putman, 1986a, b; Martín Bellido, 2001; Fankhauser, 2004; Acevedo *et al.*, 2007; Fankhauser *et al.*, 2008) and thus have a more significant influence. As, in addition, domestic cattle are frequently involved, the domestic species is usually also superior in body size. We may thus anticipate that competitive effects observed may be asymmetrical, with a greater influence of the domestic livestock on the ecological dynamics of the wild population. This is apparent in many examples cited, in that only the wild species adjust habitat use or feeding behaviour, while the domestic species remain indifferent to their wild counterparts (e.g. Acevedo *et al.*, 2007). Such interaction between native species and domestic stock is, further, not a simple, universal phenomenon but is quite clearly structured: cattle and horses, as preferential grazers, tend to show highest overlap, and thus potential for competition, with native species reliant on a bulk-feeding strategy (*sensu* Hofmann 1973, 1985); competition for specialist browsers may be afforded by goats, while sheep are most likely to show high overlap with intermediate feeders.

According to potential effects of competition, the contest for space and food among competitors can have negative consequences on the population dynamics of wild species (Madhusudan, 2004; Mishra *et al.*, 2004), particularly in areas where ungulates may face seasonal nutritional bottlenecks (e.g. winters in European mountain ranges). Sometimes the population trend may suffer drastic and sudden changes. This is especially true when spatial overlap between wild and domestic species may lead to transmission of diseases (Gauthier *et al.*, 1992; Lanfranchi, 1993). Endoparasites, in particular generalist species, can have profound effects on ruminant populations, such as reducing reproductive success, growth rate and survival (Hart, 1990), even at subclinical levels where signs of disease might not be apparent (Zaffaroni *et al.*, 1997; Gunn and Irvine, 2003). Ruminants therefore have evolved behavioural strategies to reduce the impact of parasite load, among them adaptations to minimise the uptake of parasites while feeding (Hart, 1990; Lozano, 1991; Brambilla *et al.*, 2013). One way to achieve this goal is by avoiding the swards around dung heaps because eggs of intestinal parasites are excreted with faeces, and after hatching, larvae spread into the surroundings (Sykes and Coop, 1977; Sykes, 1987; Hart, 1990; Smith *et al.*, 2006). Most studies of parasite effects refer to domestic ruminants, but consequences for wild ungulates are known to be similar (van der Wal *et al.*, 2000; Gunn and Irvine, 2003), as wild and domestic species can suffer from the same viral and bacterial infectious diseases (Nicolet and Freundt, 1975; Mayer *et al.*, 1996) and can host the same parasites (Kutzer, 1988; Roberts *et al.*, 2002). In addition, since livestock may spread disease to wild ungulates, it is therefore not surprising that a considerable number of the bacterial, viral and parasitic diseases of domestic ungulates can be carried by wild ungulates and can cause clinical diseases in domestic animals. The possible role of wild ungulates as a source of diseases for livestock in Europe has been recently reviewed (Froliche *et al.*, 2002; Bohm *et al.*, 2007; Ferroglio *et al.*, 2011). High densities of wild animals as well as range expansion are likely to exacerbate the potential for disease persistence due to the formation of multi-species assemblages, which may act as disease reservoirs (Bohm *et al.*, 2007).

Spatial overlap between wild and domestic species may also result in indirect competition through modification of habitat (both by the livestock themselves and by man in his management efforts to 'improve' habitat quality for those livestock), leading to a reduction of environmental quality for wild species. The adverse effects of the interference caused by man and his stock on the natural environment are well recognised (Wilcove *et al.*, 1998).

5.3 Evidence for competition between domestic and wild ungulates in Europe: Some case studies

5.3.1 Cattle and ponies in the New Forest of southern England

Putman (1986a, b) has demonstrated clear niche overlap in terms of both habitat use and diet, between free-ranging domestic horses and cattle in the New Forest of southern England and red, fallow and sika deer, although a lesser overlap with roe deer.

As we have already noted however, overlap in resource use *per se* is not necessarily indicative of competition. Putman and Sharma (1987) subsequently considered changes in the abundance of each of the herbivore species over a 24-year period in relation to the changing abundance of the various other ungulates, and changes in the Forest's vegetation cover. Correlation analyses considered for each species: the censused population in any year, in separate subareas of the Forest, of roe, fallow, sika and red deer in both the same and the previous year; numbers of cattle and ponies pastured on the Forest in the same year (and a rolling 3-year mean of the number of cattle and ponies pastured on the Forest in the previous 3-year period).

Numbers of roe deer, over that 24-year period, showed significant negative correlation with grazing pressure imposed upon the Forest by domestic stock. Roe deer numbers in any year were found to correlate with the mean numbers of cattle and ponies pastured on the Forest in the preceding 3 years (Putman and Sharma, 1987). Sharma (1994) subsequently updated this analysis of the effects on ungulate population numbers of the abundance of other herbivores within the Forest – and extended the analysis to explore in addition any possible effects of climatic factors on observed population change. Because of concerns about consistency of census methods prior to the 1970s Sharma's re-analysis was based only on data from (and including) 1972, although the data extended until 1988 and thus still span 17 years. In this re-analysis of the period 1972–88, no significant correlations emerged between recorded numbers of roe deer and numbers of cattle or ponies, individually or in combination (Sharma 1994). Fallow deer numbers over the whole Forest from 1972–88, however, showed a significant and persistent negative correlation with numbers of cattle and ponies pastured on the Forest in the previous year (cattle individually: $r = -0.53$; ponies individually: $p = -0.49$, or both combined: $r = -0.62$; $p < 0.05$ in all cases); fallow numbers were also significantly negatively correlated with the accumulated grazing pressure from domestic stock over the previous 3-year period.

In this study, while there may be limited evidence that grazing pressure from cattle and ponies deleteriously affects populations of fallow (Sharma, 1994), and perhaps roe (at least suggested by the analyses of Putman and Sharma over the longer period of 1962–85), there is no consistent evidence for interaction among the populations of the different deer species themselves (although see page 92).

5.3.2 Sheep and chamois

In European mountains, wild chamois or isard coexist over wide areas with domestic sheep kept on alpine pastures during summer. This domestic species usually has some advantages over its wild competitors: their herd densities locally are often far above those of wild species, they are usually released to the best grazing grounds and they may also receive supplementary food from the farmer. For example, there are some 430,000 sheep pastured in the Swiss Alps, and they outnumber chamois by almost five to one on average, while locally the ratio may be considerably higher. La Morgia and Bassano (2009), using faecal microhistological analyses, revealed a high dietary overlap between chamois and domestic sheep during summer (Pianka's index = 0.93–0.99). In particular, a reduction of highly digestible forbs was observed in the chamois diet during August, when both species grazed in the same range. As a consequence of sheep grazing, chamois may have been forced to reduce niche breadth and to change their food habits, increasing percentages of monocotyledons in the diet and feeding mainly on *Cyperaceae*. When chamois fed alone on the same summer range, niche breadth was larger, and more digestible food items, such as forbs, were positively selected.

Spatial segregation between the two species has also been noted by several authors, with displacement of chamois from favoured grazing sites (Fankhauser, 2004) and it is also reported that the presence of sheep also affects the spatial distribution of some populations of Pyrenean chamois (Pépin and N'Da, 1991; Rebollo *et al.*, 1993). Moreover, Chirichella *et al.* (2013) suggested that larger livestock flocks had a stronger impact in forcing chamois groups to move from areas with higher forage abundance. In addition, Fankhauser *et al.* (2008) demonstrated that in Swiss Alps the need to minimise endoparasite uptake from faeces may play a role in driving spatial behaviour of chamois and could result in competitive imbalance between wild and domestic ungulates.

In summary: a negative interaction between chamois and sheep on the basis of observed spatial segregation and feeding habits suggested that competition issues must be taken into account when evaluating compatibility on Alpine meadows. Despite this, the studies of Rüttimann *et al.* (2008) did not reveal a great effect of sheep on chamois in the Swiss Alps, even if competition between the two species could still be occurring over a longer time scale. It is possible that the number of sheep present in the study area (100 sheep in a 18 ha meadow) was not sufficiently high to induce a change in the time spent feeding by chamois (Kie *et al.*, 1991) and that there was also a passive tolerance of sheep by chamois, if sheep did not get too close (Pépin and N'Da, 1991). Chirichella *et al.* (2013), as above, have suggested that larger livestock flocks had a stronger impact in forcing chamois groups to move from areas with higher forage abundance. In this study, the presence of the shepherd's dogs guarding the livestock also seemed to force chamois to stay in areas with lower forage availability.

5.3.3 Goats and Iberian ibex

The status and distribution of the Iberian ibex have been studied by several authors, either in the whole Iberian peninsula (e.g. Granados *et al.*, 2002; Pérez *et al.*, 2002) or in specific areas (e.g. Granados *et al.*, 1998; Palomares and Ruiz-Martínez, 1993; Lasso De La Vega, 1994; Pérez *et al.*, 1994; Gortazar *et al.*, 2000).

Acevedo and Cassinello (2009) report that one of the main threats to conservation of this species is the increasing presence of domestic livestock, which can compete for resources (e.g. Acevedo *et al.*, 2007) and transmit diseases to wild ungulates (see examples in Gortázar *et al.*, 2006). Indeed, much of the range of this species, especially in summer months when exploiting the summer high mountain pastures, is shared with sheep, domestic goats, cattle and horses. This leads to modification of feeding strategy (Fandos, 1991; Martínez, 1992, 2002a, b; Pérez *et al.*, 2002; Moço *et al.*, 2013) and overgrazing, which may become particularly significant in dry years.

It is clear that the presence of livestock has a negative effect on ibex relative abundance, causing the ibex to select areas of poor, sparse vegetation, cultivated lands and forests, whereas in the absence of livestock, the ibex is mainly present in pasture–scrub lands and non-cultivated lands (Acevedo *et al.*, 2007). In addition, Astorga Márquez *et al.* (2014) recently investigated the prevalence of infection with different pathogens in domestic goat and Iberian ibex populations from two neighbouring geographical regions in Spain, where the populations were in sympatry and in allopatry. Their results showed clearly that co-occurrence of the two species was a significant risk factor in relation to incidence of certain infections (e.g. Q fever).

5.4 Final remarks

All the various pieces of information we can piece together suggest that established natural communities are likely to be relatively free of competition. From a purely theoretical standpoint, we may argue that natural selection would be expected to promote clear separation in resource use between regularly interacting sets of species specifically to minimise the loss of fitness incurred through competition. We would expect relatively little evidence of any interaction between species; those recorded are more likely to be facilitative than competitive. Competition should become apparent only when an established system is challenged by some perturbation of species composition or relative density. Indeed those few studies which have demonstrated clear population interactions or have demonstrated niche shift almost all seem to derive from situations where the system is perturbed by introduction of a new 'exotic' (see Chapter 4) or where additional grazing pressure from domestic livestock is imposed upon a natural ecosystem.

Within that latter context, the main goal must be the need to find an acceptable compromise and to promote a balanced land use that may favour both wild ungulates and a moderate traditional livestock husbandry. In particular, research/management efforts should be undertaken to consider different effects of their co-occurrence:

1. *Competition for space and food*
   The success and durability of efforts to reduce wildlife/livestock competition hinge on our ability to design interventions that offer robust reconciliation between the societal benefits of livestock grazing and its ecological impacts.

In this context, increased efforts to analyse and quantify the real occurrence of competition, following a multidisciplinary approach, should be a priority.

2. *Strong potential for the transmission of shared diseases*

   Although many data on infectious diseases are available in various European countries, there is more need for systematic surveillance and coordinated research. Moreover, climatic changes are likely to have a direct impact on the presence and abundance of various pathogens and their vectors, so that with a warming climate, exotic diseases may play a role in future livestock and wildlife disease management. Thus, a monitoring strategy for wildlife diseases should be a priority in European countries where wild and domestic animals coexist, especially at high densities. Some work has already been initiated in this regard (see, for example, East *et al.*, 2011; Ferroglio *et al.*, 2011) but more effort is required.

3. *Environmental modification with a reduction of quality*

   Livestock grazing can have both positive and negative impacts, depending on density. A large proportion of threatened European habitats and their associated species are linked with systems where livestock grazing and mowing are important to maintain an open landscape or to preserve high quality meadows (i.e. to enhance forage quality or to preserve grassland from evolution into shrubland and/or woodland). Thus, for a balanced land use, a moderate traditional livestock husbandry should be considered.

References

Aagesen, D. (2000) Crisis and conservation at the end of the world: sheep ranching in Argentine Patagonia. *Environmental Conservation* **27**, 208–215.

Acevedo, P. and Cassinello, J. (2009) Biology, ecology and status of Iberian ibex *Capra pyrenaica*: a critical review and research prospectus. *Mammal Review* **39**, 17–32.

Acevedo, P., Cassinello, J. and Gortázar, C. (2007) The Iberian ibex is under an expansion trend but displaced to suboptimal habitats by the presence of extensive goat livestock in central Spain. *Biodiversity and Conservation* **16**, 3361–3376.

Astorga Márquez, R.J., Carvajal, A., Maldonado, A., Gordon, S.V., Salas, R., Gómez-Guillamón, F., Sánchez-Baro, A., López-Sebastián, A. and Santiago-Moreno, J. (2014) Influence of cohabitation between domestic goat (*Capra aegagrus hircus*) and Iberian ibex (*Capra pyrenaica hispanica*) on seroprevalence of infectious diseases. *European Journal of Wildlife Research*, **60**, 387–390.

Berdoucou, C. (1986) Spatial and trophic interactions between wild and domestic ungulates, in the French mountain national parks. In: P.J. Joss, P.W. Lynch and O.B. Williams (eds),*Rangelands: A Resource under Siege.* Cambridge: Cambridge University Press, pp. 390–391.

Bohm, M., White, P.C.L., Chambers, J., Smith, L. and Hutchings, M.R. (2007) Wild deer as a source of infection for livestock and humans in the UK. *The Veterinary Journal* **174**, 260–276.

Brambilla, A., von Hardenberg, A., Kristo, O., Bassano, B. and Bogliani, G. (2013) Don't spit in the soup: faecal avoidance in foraging wild Alpine ibex, *Capra ibex. Animal Behaviour* **86**, 153–158.

Chirichella, R., Ciuti, S. and Apollonio, M. (2013) Effects of livestock and non-native mouflon on use of high-elevation pastures by Alpine chamois. *Mammalian Biology* **78**, 344–350.

Ciuti, S., Pipia, A., Grignolio, S., Ghiandai, F. and Apollonio, M., (2009) Space use, habitat selection and activity patterns of female Sardinian mouflon (*Ovis orientalis* musi*mon*) during the lambing season. *European Journal of Wildlife Research* **55**, 589–595.

de Boer, W.E. and Prins, H.H.T. (1990) Large herbivores that strive mightily but eat and drink as friends. *Oecologia* **82**, 264–74.

De Haan, C., Steinfeld, H. and Blackburn, H. (1997) *Livestock and the environment: finding a balance.* Report of study by the Commission of the European Communities, the World Bank and the governments of Denmark, France, Germany, the Netherlands, UK and USA. Wren Media, Eye, Suffolk, UK.

East, M.L., Bassano, B. and Ytreus, B. (2011) The role of pathogens in the population dynamics of European ungulates. In R.J. Putman, M. Apollonio and R. Andersen (eds), *Ungulate Management in Europe: Problems and Practices.* Cambridge, UK: Cambridge University Press, pp. 319–348.

Edwards, G.P., Croft, D.B. and Dawson, T.J. (1996) Competition between red kangaroo (*Macropus rufus*) and sheep (*Ovis aries*) in the arid rangelands of Australia. *Australian Journal of Ecology* **21**, 165–172.

Fandos, P. (1991) *La cabra montés Capra pyrenaica en el Parque Natural de las Sierras de Cazorla, Segura y Las Villas.* Madrid, Spain: ICONA-CSIC.

Fankhauser, R. (2004) *Competition between domestic and wild ungulates: do sheep affect habitat use of chamois?* Degree thesis, ETH Zürich.

Fankhauser, R., Galeffi, C. and Suter, W. (2008) Dung avoidance as a possible mechanism in competition between wild and domestic ungulates: two experiments with chamois *Rupicapra rupicapra. European Journal of Wildlife Research* **54**, 88–94.

Ferroglio, E., Gortazar, C. and Vicente, J. (2011) Wild ungulate diseases and the risk for livestock and public health. In R.J. Putman, M. Apollonio and R. Andersen (eds), *Ungulate Management in Europe: Problems and Practices.* Cambridge, UK: Cambridge University Press, pp. 192–214.

Fleischner, T.L. (1994) Ecological costs of livestock grazing in western North America. *Conservation Biology* **8**, 629–644.

Froliche, K., Thiede, T., Kozikowski, T. and Jakob, W. 2002. A review of mutual transmission of important.infectious diseases between livestock and wildlife in Europe. *Annals of the New York Academy of Sciences* **969**, 4–13.

Gauthier, D., Gibert, P. and Hars, J. (1992) Sanitary consequences of mountain cattle breeding on wild ungulates. In F. Spitz, G. Janeau, G. Gonzalez and S. Aulagnier (eds), *Ongules/ Ungulates 91: Proceedings of an International Symposium.* Paris: SFEPM-IRGM, pp. 621–630.

Ghosh, P.K., Goyal, S.P. and Bohra, H.C. (1987) Competition for resource utilisation between wild and domestic ungulates in the Rajasthan desert. *Tigerpaper* **14**, 2–7.

Gordon, I.J. (1988) Facilitation of red deer grazing by cattle and its impact on red deer performance. *Journal of Applied Ecology* **25**, 1–10.

Gordon, I.J. and Illius, A.W. (1989) Resource partitioning by ungulates on the Isle of Rhum. *Oecologia* **79**, 383–389.

Gortazar, C., Herrero, J., Villafuerte, R. and Marco, J. (2000) Historical examination of the distribution of large mammals in Aragón, Northeastern Spain. *Mammalia* **61**, 411–422.

Gortázar, C., Acevedo, P., Ruiz-Fons, F. and Vicente, J. (2006) Disease risks and overabundance of game species. *European Journal of Wildlife Research* **52**, 81–87.

Granados, J.E., Chirosa, M., Pérez, M.C., Pérez, J.M., Ruiz Martínez, I., Soriguer, R.C. and Fandos, P. (1998) Distribution and status of the Spanish ibex *Capra pyrenaica* in Andalusia, Southern Spain. Proceedings of the 2nd World Conference of Mountain Ungulates, Aosta, pp. 129–133.

Granados, J.E., Soriguer, R.C., Pérez, J.M., Fandos, P. and García-Santiago, J. (2002) *Capra pyrenaica* Schinz, 1838. In: L.J. Palomo and J. Gisbert (eds), *Atlas de los Mamíferos Terrestres de Espana*. Madrid: Dirección General de Conservación de la Naturaleza, SECEM, SECEMU, pp. 326–329.

Gunn, A. and Irvine, R.J. (2003) Subclinical parasitism and ruminant foraging strategies: a review. *Wildlife Society Bulletin* **31**, 117–126.

Hanley, T.A. and Hanley, K.A. (1982) Food resource partitioning by sympatric ungulates on Great Basin rangeland. *Journal of Range Management* **35**, 152–58.

Hansen, R.M. and Reid, L.D. (1975) Diet overlap of deer, elk and cattle in southern Colorado. *Journal of Range Management* **26**, 43–47.

Hansen, R.M., Clark, R.C. and Lawhorn, W. (1977) Foods of wild horses, deer and cattle in the Douglas Mountain area, Colorado. *Journal of Range Management* **30**, 116–18.

Hart, B.L. (1990) Behavioural adaptations to pathogens and parasites: Five strategies. *Neuroscience and Biobehavioral Reviews* **14**, 273–294.

Hofmann, R.R. (1973) *The Ruminant Stomach.* Nairobi: East African Literature Bureau.

Hofmann, R.R. (1985) Digestive physiology of the deer: their morphophysiological specialisation and adaptation. In K.R. Drew and P.F. Fennessy (eds), *Biology of Deer Production. Royal Society of New Zealand, Bulletin* **22**, 393–407.

Homewood, K., Lambin, E.F., Coast, E., Kariuki, A., Kikula, I., Kivelia, J., Said, M., Serneels, S. and Thompson, M. (2001) Long-term changes in Serengeti–Mara wildebeest and land cover: pastoralism, population or policies? *Proceedings of the National Academy of Sciences of the United States of America* **98**, 12544–12549.

Kie, J.G., Evans, C.J., Loft, E.R. and Menke, J.W. (1991) Foraging behaviour by mule deer: the influence of cattle grazing. *Journal of Wildlife Management* **55**, 665–674.

Kutzer, E. (1988) Bedeutung parasitärer Wechselinfektionen bei Hausund Wildwiederkäuern. *Monatshefte fur Veterinarmedizin* **43**, 577–580.

La Morgia, V. and Bassano, B. (2009) Feeding habits, forage selection, and diet overlap in Alpine chamois (*Rupicapra rupicapra* L.) and domestic sheep. *Ecological Research* **24**, 1043–1050.

Lanfranchi, P. (1993) Patrimonio zootecnico e faunistico: interazioni sanitarie e relative implicazioni gestionali. *Atti della Società Italiana di Buiatria* **25**, 147–155.

Lasso De La Vega, B. (1994) Estimación de la población de cabra montés en Sierra Tejeda y Almijara (Málaga). Actas del I Congreso Internacional del Género Capra en Europa, Consejería de Medio Ambiente, Junta de Andalucia, Ronda, pp. 217–218.

Latham, J. (1999) Interspecific interactions of ungulates in European forests: an overview. *Forest Ecology and Management* **120**, 13–21.

Lozano, G.A. (1991) Optimal foraging theory: a possible role for parasites. *Oikos* **60**, 391–395.

MacMillan, S. (1986) *Wildlife/Livestock Interfaces on Rangelands*. Nairobi: Inter-African Bureau for Animal Resources.

Madhusudan, M.D. (2004) Recovery of wild large herbivores following livestock decline in a tropical Indian wildlife reserve. *Journal of Applied Ecology* **41**, 858–869.

Madhusudan, M.D. and Karanth, K.U. (2002) Local hunting and the conservation of large mammals in India. *Ambio* **31**, 49–54.

Martín Bellido, M., Escribano Sánchez, M., Mesías Díaz, F.J., Rodríguez de Ledesma Vega, A. and Pulido García, F. (2001) Sistemas extensivos de producción animal. *Archivos de zootecnia* **50**, 465–489.

Martínez, T.M. (1992) *Estrategia alimentaria de la cabra montés (Capra pyrenaica) y sus relaciones tróficas con los ungulados silvestres y domésticos en S^a Nevada, S^a de Gredos y S^a de Cazorla*. PhD Thesis. Universidad Complutense, Madrid, Spain.

Martínez, T.M. (2002a) Summer feeding strategy of Spanish wild goat *Capra pyrenaica* and domestic sheep *Ovis aries* in south-eastern Spain. *Acta Theriologica* **47**, 479–490.

Martínez, T.M. (2002b) Comparison and overlap of sympatric wild ungulate diet in Cazorla, Segura and Las Villas Natural Park. *Pirineos* **157**, 103–115.

Mattiello, S., Redaelli, W., Carenzi, C. and Crimella C. (2002) Effect of dairy cattle husbandry on behavioural patterns of red deer (*Cervus elaphus*) in the Italian Alps. *Applied Animal Behaviour Science* **79**, 299–310.

Mayer, D., Nicolet, J., Giacometti, M., Schmitt, M., Wahli, T. and Meier, W. (1996) Isolation of Mycoplasma conjunctivae from conjunctival swabs of Alpine ibex (*Capra ibex ibex*) affected with infectious keratoconjunctivitis. *Journal of Veterinary Medicine* **43**, 155–161.

Mishra, C. and Rawat, G.S. (1998) Livestock grazing and biodiversity conservation: comments on Saberwal. *Conservation Biology* **12**, 712–714.

Mishra, C., Van Wieren, S., Ketner, P., Heitkonig, I.M.A. and Prins, H.H.T. (2004) Competition between domestic livestock and wild bharal *Pseudois nayaur* in the Indian trans-Himalaya. *Journal of Applied Ecology* **41**, 344–354.

Moço, G., Serrano, E., Guerreiro, M., Ferreira, A.F., Petrucci-Fonseca, F., Maia, M.J., Soriguer, R.C. and Pérez, J.M. (2013) Seasonal dietary shifts and selection of Iberian wild goat *Capra pyrenaica* Schinz, 1838 in Peneda-Gerês National Park (Portugal). *Galemys* **25**, 13–27.

Mussa, P.P., Aceto, P., Abba, C., Sterpone, L. and Meineri, G. (2003) Preliminary study on the feeding habits of roe deer (*Capreolus capreolus*) in the western Alps. *Journal of Animal Physiology and Animal Nutrition* **87**, 105–108.

Mysterud, A. (2000) Diet overlap among ruminants in Fennoscandia. *Oecologia* **124**, 130–137.

Nicolet, J. and Freundt, E.A. (1975) Isolation of Mycoplasma conjunctivae from chamois and sheep affected with keratoconjunctivitis. *Journal of Veterinary Medicine* **22**, 302–307.

Palomares F. and Ruiz Martínez, I. (1993) Status and conservation perspectives for the Spanish ibex population of Sierra Ma´gina Natural Park, Spain. *Zeitschrift für Jagdwissenschaft* **39**, 87–94.

Pépin, D. and N'Da, L. (1991) Spatial and temporal relationships between sheep and a protected population of Isards (*Rupicapra pyrenaica*) during daytime in summer. In: S. Aulagnier, G. Gonzalez, G. Janeau and F. Spitz (eds), *Ongules/Ungulates 91: Proceedings of an International Symposium*. Paris: SFEPM-IRGM, pp. 331–333.

Pérez, J.M., Granados, J.E. and Soriguer, R.C. (1994) Population dynamics of the Spanish ibex *Capra pyrenaica* in Sierra Nevada Natural Park (southern Spain). *Acta Theriologica* **39**, 289–294.

Pérez, J.M., Granados, J.E., Soriguer, R.C., Fandos, P., Márquez, F.J. and Crampe, J.P. (2002) Distribution, *status* and conservation problems of the Spanish Ibex, *Capra pyrenaica* (Mammalia: Artiodactyla). *Mammal Review* **32**, 26–39.

Price, E.O. (2002) *Animal Domestication and Behaviour*. Wallingford, UK, and New York: CABI Publishing.

Prins, H.H.T. (2000) Competition between wildlife and livestock in Africa. In H.H.T. Prins, J.G. Grootenhuis and T.T. Dolan (eds), *Wildlife Conservation by Sustainable Use*. Boston, MA: Kluwer Academic Publishers, pp. 51–80.

Prins, H.H.T., Grootenhuis, J.G. and Dolan, T.T. (2000) *Wildlife Conservation by Sustainable Use*. Boston, MA: Kluwer Academic Publishers.

Putman R.J. (1986a) Competition and coexistence in a multispecies grazing community: the large herbivores of the New Forest. *Acta Theriologica* **31**, 271–291.

Putman R.J. (1986b) *Grazing in Temperate Ecosystems; Large Herbivores and their Effects on the Ecology of the New Forest*. Beckenham, UK: Croom Helm.

Putman, R.J. (1994) *Community Ecology*. London: Chapman and Hall.

Putman, R.J. (1996) *Competition and Resource Partitioning in Temperate Ungulate Assemblies.* London: Chapman and Hall.

Putman, R.J. and Sharma, S.K. (1987) Long term changes in New Forest deer populations and correlated environmental change. In S. Harris (ed.), *Mammal Population Studies. Symposia of the Zoological Society of London* **58**, 167–79.

Rebollo, S., Robles, L. and Gómez-Sal, A. (1993) The influence of livestock management on land use competition between domestic and wild ungulates: sheep and chamois *Rupicapra pyrenaica parva* Cabrera in the Cantabrian range. *Pirineos* **141–142**, 47–62.

Richard, E., Calenge, C., Saïd, S., Hamann, J.L. and Gaillard, J.M. (2013) Studying spatial interactions between sympatric populations of large herbivores: a null model approach. *Ecography* **36**, 157–165.

Roberts, M.G., Dobson, A.P., Arneberg, P., de Leo, G.A., Krecek, R.C., Manfredi, M.T., Lanfranchi, P. and Zaffaroni, E. (2002) Parasite community ecology and biodiversity. In P.J. Hudson, A. Rizzoli, B.T. Grenfell, H. Heesterbeek and A.P. Dobson (eds), *The Ecology of Wildlife Diseases.* Oxford: Oxford University Press, pp. 63–82.

Rüttimann, S., Giacometti, M. and McElligott, A.G. (2008) Effect of domestic sheep on chamois activity, distribution and aboundance on sub-alpine pastures. *European Journal of Wildlife Reserch* **54**, 110–116.

Saberwal, V.K. (1996) Pastoral politics: gaddi grazing, degradation, and biodiversity conservation in Himachal Pradesh, India. *Conservation Biology* **10**, 741–749.

Schwartz, C.C. and Ellis, J.E. (1981) Feeding ecology and niche separation in some native and domestic ungulates on the shortgrass prairie. *Journal of Applied Ecology* **18**, 343–54.

Sharma, S.K. (1994) *The decline of the roe deer (Capreolus capreolus L.) in the New Forest Hampshire.* PhD thesis, University of Southampton.

Smith, L.A., White, P.C.L. and Hutchings, M.R. (2006) Effect of the nutritional environment and reproductive investment on herbivore–parasite interactions in grazing environments. *Behavioral Ecology* **17**, 591–596.

Sykes, A.R. (1987) Endoparasites and herbivore nutrition. In J.B. Hacker (ed.), *Nutrition of Herbivores.* Marrickville, Australia: Academic, pp. 211–232.

Sykes, A.R. and Coop, R.L. (1977) Chronic parasitism and animal efficiency. *ARC Research Review* **3**, 41–46.

Tokeshi, M. (1999) *Species Coexistence. Ecological and evolutionary perspectives.* Oxford: Blackwell.

Troy, C.S., MacHugh, D.E., Bailey, J.F., Magee, D.A., Loftus, R.T., Cunningham, P., Chamberlain, A.T., Sykes, B.C. and Bradley, D.G. (2001) Genetic evidence for Near-Eastern origins of European cattle. *Nature* **410**, 1088–1091.

Van der Wal, R., Irvine, J., Stien, A., Shepherd, N. and Albon, S.D. (2000) Faecal avoidance and the risk of infection by nematodes in a natural population of reindeer. *Oecologia* **124**, 19–25.

Western, D. (1989) Conservation without parks: wildlife in the rural landscape. In: D. Western and M.C. Pearl (eds), *Conservation for the Twenty-first Century.* Oxford: Oxford University Press, pp. 158–165.

Wiens, J.A. (1989) *The Ecology of Bird Communities. Volume 2: Processes and Variations.* Cambridge, UK: Cambridge University Press.

Wilcove, D.S., Rothstein, D., Dubow, J., Phillips, A. and Losos, E. (1998) Quantifying threats to imperiled species in the United States. *BioScience* **48**, 607–615.

Young, T.P., Palmer, T.M. and Gadd, M.E. (2005) Competition and compensation among cattle, zebras, and elephants in a semi-arid savanna in Laikipia, Kenya. *Biological Conservation* **122**, 351–359

Zaffaroni, E., Citterio, C., Sala, M. and Lauzi, S. (1997) Impact of abomasal nematodes on roe deer and chamois body condition in an alpine environment. *Parasitologia* **39**, 313–317.

Chapter 6

Effects of Selective Harvesting on Ungulate Populations

Atle Mysterud

Selective harvesting of ungulate populations is an old management tradition. Most commonly it is driven by either a desire to shoot males with large 'trophies' such as antlers and horns (Geist, 1986) or avoiding shooting females with offspring to enhance population growth (Milner *et al.*, 2011). Such intentional selective harvesting practices arising from hunter preferences are clearly very important for offtake, and also modified or strengthened by management requirements such as quotas and economic incentives (Mysterud, 2011). Selectivity can also arise due to non-intentional selection due to differences in behaviour among animals making them more or less prone to being harvested. Any level of harvesting of an animal population clearly has an impact on population dynamics by removing individuals and, when selective, it can greatly affect the remaining age and sex structure of the population (Beddington, 1974; Ginsberg and Milner-Gulland, 1994; Langvatn and Loison, 1999). This will obviously again affect population growth, but also a range of other processes.

That some animals are more important than others for population growth has likely been known to humans since prehistoric times. Clearly, only females produce new young and how many depends on their age (Gaillard *et al.*, 2000). This is formalised into the concept of reproductive value describing the extent to which individuals of different age contribute to future population growth (Fisher, 1930; Engen *et al.*, 2007, 2009). There is a huge theoretical literature, using in particular matrix models, to investigate how age-specific harvesting may affect population growth. For a long period of time, studies of the effects of selective harvesting were mainly linked to such direct effects of individual removal on population growth. I will term these the direct demographic effects (Figure 6.1). However, indirect demographic effects may arise due to such age-specific harvesting that also affects dynamics through changes in age structure and sex ratio of the remaining population.

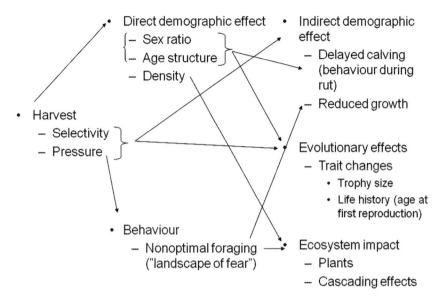

Figure 6.1 A schematic representation of how harvesting might affect ungulate populations both directly and indirectly.

Ungulates are typically polygynous to differing degrees (Loison *et al.*, 1999a). As only female ungulates give birth and alone raise the young, most of these population models, when applied to ungulates, only consider the female population. Males are not regarded as limiting for population growth, and it is presumed that skewing the population sex ratio towards females will largely enhance population growth. However, it is quite clear that at some point sex ratios may become so skewed that males may become limiting, or need longer time periods to inseminate all females, as was first convincingly shown in an influential theoretical paper by Ginsberg and Milner-Gulland (1994). Despite the fact that some form of selective harvesting is nearly universal in the management of ungulate populations, it is only until quite recently that such more indirect effects of selective harvesting became a topical issue (Mysterud *et al.*, 2002).

For a long time, most studies of ungulates and their life history were deliberately focused on populations without harvesting, such as the red deer on Rum (Clutton-Brock *et al.*, 1982) and the Soay sheep of St Kilda (Clutton-Brock and Pemberton, 2003). There has been a renewed interest in studies of the effects of harvesting during the last decade. The patterns of age- and size-specific mortality induced by human harvesting differ markedly from those seen in populations both without harvesting and predation (Gaillard *et al.*, 1998). In theory, such differing mortality pattern will yield markedly different selective pressures from those impinging on populations without harvesting (Proaktor *et al.*, 2007) or with large predators (Bischof *et al.*, 2008). There is indeed huge current interest in how selective harvesting might affect evolutionary responses (Harris *et al.*, 2002; Allendorf *et al.*, 2008; Allendorf and Hard, 2009). For terrestrial systems, this was sparked by an influential study of the genetic consequences of trophy hunting of bighorn (*Ovis canadensis*) rams in Canada (Coltman *et al.*, 2003).

   The aim of this chapter is to review how hunter selectivity might affect ungulate populations beyond direct demographic effects, i.e. indirect demographic effects related to changes in age structure and sex ratio and evolutionary effects. I will consider general mechanisms, but a key issue is to review the empirical evidence for its importance relative to other factors with an emphasis on European ungulate populations.

6.1 Harvesting selectivity

Harvesting or hunter selection can be defined as any non-random offtake of individuals from the population. Hunter selection will thus by definition yield a shift in the structure of the remaining population. The three main categories of harvesting selectivity are those related to selectivity for sex, age and body (or antler/horn) size. For population dynamics, clearly those related to offtake depending on age and sex is the most critical, while body (or antler/horn) size (relative to cohort average) is the more important for directional selection and potential evolutionary effects (more below).

6.1.1 Intentional and non-intentional factors

Some selectivity of harvest may result from both direct and intentional factors such as the influence of hunter preferences (e.g. trophy-hunting), quotas and economic costs, etc., but also as the result of indirect and non-intentional factors through more general impacts on animal behaviour and abundance. Direct effects are likely by far the more important, though the indirect effects are currently little studied. That there is clear selection for age and sex is well established, and most often imposed as the result of quotas set for harvest by regulatory authorities who may determine a required cull of a particular number of males or females and/or may determine a particular age-structure within that cull. The way of giving quotas differs substantially across Europe (Apollonio *et al.*, 2010), and the proportion of different sex and age classes in the harvest differ accordingly. As an example of this, when comparing red deer (*Cervus elaphus*) harvest statistics across 11 European countries (Milner *et al.*, 2006), the proportion of calves in the harvest varied from 10–40%, while males typically accounted for 40–60% of the remainder. Cultural traditions setting management aims are likely a main driver of such variability (Milner *et al.*, 2006).

   After quotas are established, the intentional factors causing harvesting selection are those related to hunter preferences and the opportunities given for being selective. Hunters' preferences differ depending on hunter motivation, whether they are aiming for meat, recreation or trophy; clearly, trophy hunters are the most selective. It has been suggested that several factors causing selection are related to the opportunity to be selective (Mysterud, 2011): animal trait variation, animal behaviour, animal abundance, population structure (sex ratio and age structure) and habitat openness. Hunters have time limitation, and we expect hunters to be less selective if the chance of not getting anything is a worse alternative. Therefore, selectivity will be reduced where there is low population density (Tenhumberg *et al.*, 2004), a skewed sex ratio leading to low density of one sex (Nilsen and Solberg, 2006), a high quota relative to population size (Solberg *et al.*, 2000), a short duration of hunting season or small estate size (Mysterud, 2011). It is less well documented how the level of knowledge and skill (use of guides), cultural background, religion (taboos) and individual ethics affect selectivity, but it is likely to play a role. In Alaska, the largest moose (*Alces alces*) were shot by client hunters where guides were able to find better areas with higher availability of large moose (Schmidt *et al.*, 2007). Competition among hunters can also decrease selectivity if passing a chance increases the risk that the animal may be shot by another hunter. In some cases, quotas are given not to single hunters or teams, but to larger areas. No such case is reported for ungulates, but such a quota system is commonly used for large carnivores (brown bears *Ursus arctos*: Bischof *et al.*, 2009; lynx *Lynx lynx*: Nilsen *et al.*, 2012).

Lastly, it might not be possible to achieve the level of selectivity that would be the optimal choice, due to practical management or implementation constraints (Milner *et al.*, 2011). For example in Scandinavia, it is a management goal for moose populations to harvest a high proportion of yearlings, and the aim is to shoot as many female as male yearlings. It is quite easy to separate a yearling male from other classes of moose, while it is difficult to separate a yearling female from an adult female. Therefore, when given quotas of yearlings, hunters end up shooting 1:4 female:male yearlings and then skewing sex ratio towards females to an undesired degree (Haagenrud and Lørdahl, 1974). Similarly, it might be difficult to separate male from female chamois in some cases. Therefore, empirical analysis of what has been actually harvested is important, but there are currently few such studies.

6.1.2 Empirical evidence for hunter selection

Ideally, evidence of hunter selection should compare the composition of the harvest relative to the structure of the remaining or overall population, or be based on individually marked animals so that selectivity can be determined relative to other marked individuals (Bischof *et al.*, 2009). Unfortunately, most studies of harvest selection are not designed this way due to data limitation. Rather, the majority of published studies available have compared the composition of offtake among hunting methods (drive versus stalking) or different groups of hunters (trophy stalker versus local hunter) to identify differences. Typically, this assumes that since they hunt from the same population, differences will reflect variation in selectivity, which is likely a fair assumption within a given hunting season with low levels of other mortality. Also, several studies compare the harvests taken by human hunt-ers with offtake of natural predators, following the same logic (moose–wolf *Canis lupus*: Sand *et al.*, 2012; roe deer *Capreolus capreolus*–lynx: Andersen *et al.*, 2007; wild boar–wolf: Nores *et al.*, 2008). A recurrent problem with many studies is the lack of separation with selection for age and size (Mysterud, 2011); older males are also larger, but separating these effects has rarely been done.

A very common pattern of selectivity is to avoid harvesting females with off-spring. Female moose 5–10 years old which were without a calf had a 3.2 times larger risk of being harvested (Ericsson, 2001). No selection for size or between single and twin moose calves was found (Moe *et al.*, 2009). For chamois, too, survival was higher for females with offspring due to hunting (Rughetti and Festa-Bianchet, 2011). In some areas, wild boar females with piglets are not allowed to be shot (Keuling *et al.*, 2010). [In such a context, it has been dem-onstrated that orphaned red deer calves do indeed have lower body growth and survival even after weaning (Andres *et al.*, 2013).]

Studies have shown that for red deer in Spain, larger males, overall, were shot during trophy stalking than during management hunting and montería, a particular way of drive hunting in Spain using chasing dogs (Martínez *et al.*, 2005). A special kind of selective montería, with the aim of removing inferior males, showed that smaller males were shot compared to during commercial trophy hunting even after correcting for age (Torres-Porras *et al.*, 2009). For red deer harvesting in Hungary, local hunters shot younger and on average smaller stags than foreign, trophy stalk-ers (Rivrud *et al.*, 2013). Similarly, local roe deer hunters in Poland aimed more for meat and shot smaller males, while foreign hunters aiming for trophies natu-rally shot larger bucks (Mysterud *et al.*, 2006). Part of this was due to the fact that late in the season, fewer large males may be around due to depletion (Mysterud *et al.*, 2006). Selectivity for larger and older males among moose hunters in Nor-way decreased when the opportunity to select was limited due to a female-biased sex ratio and a young male age structure (Nilsen and Solberg, 2006).

Selection may decrease if trait size is similar among groups of animals due to low body size dimorphism or lack of visual secondary sexual characters. If so, we would expect reduced hunter selectivity for monomorphic species like roe deer. However, a study from Italy showed that hunters were surprisingly capable of targeting the larger males even for roe deer with fairly small trophies compared to larger species (Ramanzin and Sturaro, 2013). In this specific case, this may be partly achieved due to the openness of the habitat; habitat openness, which also often promotes gregariousness, is likely a very important factor in increasing the potential for accurate selection, but documentation of this is lacking. Selectivity is also affected by habitat *quality*, as this affects the overall size of antlers (Ramanzin and Sturaro, 2013).

In Portugal, nocturnal hunts by single hunters for wild boar at bait (called 'espera') were highly selective for larger males and avoiding piglets compared to drive ('montaria') hunting (Braga *et al.*, 2010). No evidence of sex-biased hunt-ing of marked piglets was found for wild boar in Germany, where the dominant hunting method was from hides, with some drive hunting (Keuling *et al.*, 2010). For wild boar with drive hunting, the mere size of the animal may be important, due to the difficulty of hitting smaller targets, but data are currently lacking.

Few studies quantify how much animal behaviour affects harvest-related selection. For wapiti in Rocky Mountains, USA, the probability of a female surviving during the hunting season decreased with age, year, extent of space use, cover and human tracks (Webb *et al.*, 2011). Clearly, it is possible that animals with higher energy requirements, for example hinds with male offspring, are more frequently using pastures or ranging over larger areas and hence have an increased likelihood of being shot. Slightly higher selection for male brown bear likely resulted from the fact of their larger home ranges (Bischof *et al.*, 2009). Despite the intuitive logic of some of these relationships, a study of red deer harvesting in Norway found limited support for the notion that behavioural differences were important for the resulting offtake (Rivrud *et al.*, 2014). There is currently no study to document specifically whether the use of dogs might affect selectivity. The use of dogs on a leash or barking in Fennoscandia (Ruusila and Pesonen, 2004) or during drive hunting in continental Europe (Apollonio *et al.*, 2010) is common. Drive hunts with or without dogs are typically less selective, but the role of dogs in affecting this pattern remains unclear. Though some effect of non-intentional selectivity is likely, this is clearly less important than intentional selection.

6.2 Demographic consequences of selective harvesting

Individuals differ greatly in their contribution to population growth rate (Sæther *et al.*, 2007). Hence selective harvesting may have a huge impact on the population growth rate depending on whether actively reproductive individuals are being targeted. Female ungulates raise offspring alone without any help from males. They start reproducing at ages between 2 and 5 years and continue to do so every year or every other year, depending on species, habitat, climate and population density, until they reach a senescent stage (Gaillard *et al.*, 2000). The most obvious link between selective harvesting and population growth rate is thus the number and proportion of adult females in the harvest (e.g. Nilsen *et al.*, 2005); harvesting the adult female population will have the largest impact on future population growth and this is also confirmed empirically for red deer (Milner *et al.*, 2006). It is quite clear that indiscriminate hunting of all age and sex classes is much more detrimental to population growth than selective trophy hunting (Caro *et al.*, 2009). In theory, it is 'easy' to find the optimal way to harvest populations given an explicit and clear aim. In reality, aims are often not so clear; often there are multiple aims and there is a strong cultural component involved in harvest regimes. In addition, selectivity might be difficult to achieve due to practical problems of separating animals of different age and sex, or due to limitations of time as explained above.

Indirect demographic consequences of selective harvesting can be defined as those not linked to the removal of individuals (simply in terms of numbers) or the reproductive value of those remaining in the population. Mainly, indirect effects are linked to how mating is affected by changes in the sex and age structure of the population. Population models only consider the female part of the population, assuming there are always a sufficient number of males. This assumption may not hold for populations with severely skewed sex ratio, and it may depend on the mating system of the species in question (Caro *et al.*, 2009; Schindler *et al.*, 2013). Selective harvesting might skew sex ratios and age structure up to a limit where the role of males might be important (Ginsberg and Milner-Gulland, 1994; review in Mysterud *et al.*, 2002). Such models can be termed behaviourally sensitive models, and there has now long been an aim to link individual behaviour to population dynamics (Sutherland, 1996).

It is clear that without any males, no offspring will be produced. Equally true for polygynous ungulates, a few males can sire many offspring. The interesting question is exactly how skewed sex ratios need to be before male limitation becomes an issue causing reduced productivity, and whether this differs depending on the level of polygyny. Simulations suggest that male capacity to inseminate females can be limiting when the adult sex ratio is severely skewed (Gruver *et al.*, 1984; Ginsberg and Milner-Gulland, 1994), especially for monogamous mating systems (Caro *et al.*, 2009). We can divide effects into several demographic or life history variables, mainly reduced reproductive rate and later, and less synchronous, breeding as a result of changes in sex ratio and age structure of males (review in Mysterud *et al.*, 2002). Note also that hunting during the breeding season in itself might affect populations (Apollonio *et al.*, 2011). Less attention has been paid to the fact that sexual selection processes will change with sex ratio and male age structure. Reduced male competition at skewed sex ratio may affect mass loss in males (Mysterud *et al.*, 2005, 2008) and in consequence reduce mortality. With lower male harassment of females, there might be less mass loss also in females.

What is the evidence there are simply too few males to inseminate all females? Is there evidence that fewer males take longer to inseminate all females? Are females more reluctant to mate younger males? Can the presence of males induce ovulation (the male effect)? Does male harassment affect mass loss or female distribution?

6.3 Potential indirect effects of population age and sex structure on demography

Ungulates in Europe face a strongly seasonal environment. Short growing seasons impose a severe constraint on the life histories of ungulates, as, especially in higher latitudes, there is only a comparatively short time to raise an offspring that will be of a sufficiently large size to survive the winter. Plant quality and quantity often peak early in the season (Bunnell, 1982; Côté and Festa-Bianchet, 2001). Therefore timing of mating is important. If mating is delayed, it will normally affect the birth dates, as duration of gestation is flexible only within relatively narrow limits (Kiltie, 1982, 1988; Clutton-Brock *et al.*, 1982). A late ovulation provides the next year's offspring with a poor start in life, as they will have less time to grow before winter and may also be born after peak protein levels (Hogg *et al.*, 1992).

Late-born offspring are more likely to die (bighorn sheep: Festa-Bianchet, 1988) and have lower social status later in life (American bison *Bison bison*: Green and Rothstein, 1993). Small body mass in autumn is known to cause higher overwinter mortality (red deer: Loison *et al.*, 1999b) and it may take an additional year to reach the size required to mature for females (Langvatn *et al.*, 1996, 2004).

Despite the fact that gestation periods are comparatively inflexible, earlier mated females, being typically older and larger, have somewhat longer gestation period than females mated later in the season (bison: Berger, 1992; semi-domestic reindeer *Rangifer tarandus*: Mysterud *et al.*, 2009a). In a manipulative study using vasectomised males to 'trick' females into reovulation, the late conceiving females had 10 days shorter gestation times (Holand *et al.*, 2006a). Late-breeding females gave birth to offspring of lower mass, and they were substantially smaller in autumn.[1] However, a late calving may also affect the females next reproductive cycle (Hogg *et al.*, 1992; Langvatn *et al.*, 2004), reducing future fertility. In red deer, a 1% reduction in fertility was reported for every day past the date of median conception in previous year (Clutton-Brock *et al.*, 1987; see also for caribou: Cameron *et al.*, 1993).

There is no study reporting how much the reduced synchrony in birth, due to more variable breeding induced by selective harvesting, may affect calf survival. Many calves are taken by predators during the first few week of life (Linnell *et al.*, 1995; Aanes and Andersen, 1996; Jarnemo *et al.*, 2004; Panzacchi *et al.*, 2008) and, in general, synchrony of birth is considered an important strategy to reduce predation rates through a swamping effect (Estes, 1976; Leuthold, 1977). In bighorn sheep, no difference was found between birth dates and synchrony in two founder populations with either sex ratio 1:7.5 adult male (≥2-year olds):female (≥3.5-year olds) and in a group with 0:12, but with 1.5 year old males (Whiting *et al.*, 2008).

Time limitation leading to later mating can potentially be linked to several processes: (1) if prime-age males are few, it may take them longer to inseminate all females, either due to the requirement for courtship or simply because they become exhausted (courtship), (2) females may be reluctant to mate if there are only young males around early in the rut (female choice), (3) presence of (prime-aged) males may itself stimulate oestrus directly (male effect), and (4) female distribution may differ (mate search). There may be *sperm depletion* during late rut, even for feral sheep with comparatively large testicles (Preston *et al.*, 2001), but currently there is no evidence that this is a severe limitation for breeding in selectivity harvested populations.

6.3.1 Female choice, search and competition

In general, female ungulates clearly compete more for resources needed for successful reproduction rather than for mating opportunities (Clutton-Brock, 2009).

---

[1]In contrast, female condition was similar among the groups, providing evidence of the conservative reproductive tactics used by ungulates (Gaillard and Yoccoz, 2003).

However, among topi, females were shown to compete actively for preferred males (Bro-Jørgensen, 2002, 2007). Female red deer prefer the roars of larger males (Charlton *et al.*, 2007) and selected males with a high roaring frequency (McComb, 1991). In bison, too, it was shown that females in groups with large males (≥5-year olds) were more likely to copulate (Bowyer *et al.*, 2007). A reluctance of females to mate with younger males was reported in fallow deer (Komers *et al.*, 1999). A considerable energy cost of mate sampling was also documented in female pronghorn (Byers *et al.*, 2005). That this was a costly behaviour was also suggested since effort for mate search was reduced in a dry summer when there was less forage available (Byers *et al.*, 2006). Such mate search may be to obtain better genes (Byers and Waits, 2006) or act as fertility insurance (fallow deer: Briefer *et al.*, 2013). And in evidence that a reduction in availability of mature males may have a significant effect on reproductive behaviour and dynamics, Milner-Gulland *et al.* (2003) suggested that increased female aggression and competition for the remaining few males was an important factor driving the reproductive collapse observed in saiga antelope, although such assertion is largely based on anecdotal evidence.

6.3.2 Male effect

That the presence of a male may stimulate female reproduction is well documented in animal husbandry and termed the male effect (see more generally in McClintock, 1983). Presence of rams stimulates ewes to oestrus with an elevation in plasma luteal hormone (LH) secretion (Rosa *et al.*, 2006). The pre-ovulatory pulse of LH was 7 hours earlier in ewes housed in the same pen as a ram, and thus able to interact directly, compared to those with a ram housed in a separate pen situated between every two ewes (Abecia *et al.*, 2002). A longer exposure to rams increased the chances of ewe lambs starting breeding (Kenyon *et al.*, 2008). In goats, it was shown that higher ranked females responded first to males (Alvarez *et al.*, 2003, 2009).

There is also some evidence from cervids. In reindeer, when males were introduced to groups of females late, this resulted in a 10-day later onset of ovarian activity (Shipka *et al.*, 2002). Females are stimulated by both sound and smell of males. Recorded roaring by red deer stags induced oestrus in females, but smell and/or vision seemed also to play a role (McComb, 1987). Male urine contains odiferous compounds that may induce oestrus (Whittle *et al.*, 2000). Cow moose aggressively compete for access to bull urine (Miquelle, 1991). The concentration of some of these compounds are related to male age (Miller *et al.*, 1998), and therefore not only male presence, but also to age of available males (fallow deer: Komers *et al.*, 1999).

6.3.3 Male search and female distribution

Distribution patterns of females during the rut may also vary with sex ratio and male age structure and hence affect the likelihood of a synchronous calving season. Recent studies documented variation in female spatial organisation during the peak of the rut depending on the availability of males. In a manipulated herd of semi-domestic reindeer with an even sex ratio and male age structure, the female distribution was best described as one large floating harem (L'Italien *et al.*, 2012). Most of the females were easily accessible for the males and vice versa during the main rut. With a skewed sex ratio, females gathered in several distinct harems throughout the main rut and quite a few females were always in transit between harems although the frequency decreased during peak rut. In fallow deer, a switch to lekking was found with a high density of mature males (Schaal and Bradbury, 1987; Langbein and Thirgood, 1989; Thirgood *et al.*, 1999). Therefore, the presence of high quality, prime-aged males, may lead to females being less dispersed and hence increasing the probability of a more synchronous rut. Reindeer males with higher rank control larger and more stable groups; however, mating group size and stability decreased with more male-based sex ratio (L'Italien *et al.*, 2012). Prime-aged male reindeer devoted more and young males less energy with more females in the mating group (Tennenhouse *et al.*, 2011). Under conditions of high female aggregation in red deer, males were unable to monopolise whole female groups (Pérez-González and Carranza, 2011), but mainly when the population sex ratio and male age structure was even so the proportion of competitive males was high.

6.3.4 Male harassment

Females may have energetic costs related to mate search, but may also incur additional costs by being harassed by males (Clutton-Brock *et al.*, 1982). Therefore, change in rutting activity related to changes in sex ratio and age structure of males imposed by selectivity harvesting may also affect female body condition. An excess of males may lead to more aggression towards adult females (Le Galliard *et al.*, 2005). Male harassment of females may be exacerbated or reduced by selective harvesting; female biased sex ratios might be expected to cause less intense mate competition and harassment. By contrast, in populations where there are few mature males and the population is dominated by an excess of younger males, these may have a less well-developed social structure and heightened levels of harassment of females (Valdez *et al.*, 1991; Komers *et al.*, 1999; but see Shackleton, 1991.

The strongest evidence for a high cost of rutting comes from studies of feral sheep in Kerguelen sub-Antarctic archipelago, where male harassment caused direct mortality of females at a very male-biased sex ratio (Réale *et al.*, 1996). Female fallow deer in captivity confined with young males lost more weight than when confined with older males during rut (Komers *et al.*, 1999). In semi-domestic reindeer, female mass loss was highest where both young and old males were present (Holand *et al.*, 2006b), suggesting again that more female-biased sex ratios may relieve such costs. In all these studies, the loss of body mass of females was linked to the harassment frequency and resulted in reduced time for feeding.

Females in larger herds were less frequently harassed (red deer: Clutton-Brock *et al.*, 1982; McComb and Clutton-Brock, 1994).

6.3.5 Sex ratio adjustments

A controversial theory in life-history evolution is the possibility of adaptive sex ratio adjustment (Hewison and Gaillard, 1999). There are two manipulative studies simulating effects of selective harvesting on offspring sex ratio producing largely the same results. In moose, when skewing population structure towards females and with young males, offspring sex ratio was skewed towards females (Sæther *et al.*, 2004). Similarly in studies of semi-domestic reindeer, offspring sex ratio was markedly skewed towards females when fewer and younger males were present (Røed *et al.*, 2007). For Scandinavian populations of red deer (Mysterud *et al.*, 2000) and moose (Sæther *et al.*, 2004), there are quite marked effects in offspring sex ratios observed in the harvest statistics, and such an increasing production of female offspring will positively affect population productivity.

6.4 Actual effects of selective harvesting on demography

Most studies reporting the number of females a male can inseminate come from animal production, and must be seen as upper thresholds for what is possible. At highly skewed sex ratios, there is often also a younger male age structure (Ginsberg and Milner-Gulland, 1994). Younger males do not have the same capacity to inseminate a large number of females as prime-aged males (Haigh and Hudson, 1993). For Scandinavia, a lower proportion of primiparous females were found to be breeding in moose when sex ratios were skewed and male age structure was young (Solberg *et al.*, 2002). Reduced fertility was registered at approximately 1:3 male:female, and currently management of moose in Norway often aims to keep sex ratios no more extreme than 1:2.5. In the wild, the only reported case of a dramatic decline in reproductive rates of females is for the saiga antelope at 1:50 male:female (Milner-Gulland *et al.*, 2003; above). Though reproductive rates may be compromised in extreme cases, it is rather an exception that females are not mated at all. Much more common is that rutting is delayed (Mysterud *et al.*, 2002).

There are still comparatively few studies providing evidence for how skewed sex ratios need to be before there may be an effect on reproductive success or how important this is for actual population growth. The only quantified study of the effects of unintentional selective harvesting for population dynamics is for red grouse (and thus an example outside of ungulates; Bunnefeld *et al.*, 2011). With integral projection models (Coulson, 2012), it should be possible to model the population effect of reduced body growth and so forth, but before such modelling studies are done, we cannot really determine how important changes in timing of mating on offspring survival and future reproduction really are in terms of their effects on overall population dynamics.

6.5 Evolutionary consequences

Strong directional selection for specific phenotypic traits, such as is typical of trophy hunting, can potentially lead to long-term evolutionary consequences (Harris *et al.*, 2002; Festa-Bianchet, 2003; Allendorf *et al.*, 2008; Coltman, 2008; Allendorf and Hard, 2009; Darimot *et al.*, 2009). There is little doubt trophy hunting is directionally selective. However, in such analysis it is important to remember that many populations are often not only harvested by trophy hunters and that in practice trophy males usually make up a very small proportion of the total harvest from any population. Moreover, management regulations often restrict hunters from following their preferences.

The level of directional selection for hunters targeting meat, subsistence, recreation or population control rather than trophies is not well documented. Reported evidence of directional selection acting on size is weak, as most such analyses only compare age classes differing in size (Mysterud, 2011), while the age-specific size within a population (the unit for evolution) remain uncertain. The degree of size selection may strongly differ between the age classes that are targeted in both males (Mysterud and Bischof, 2010) and females (Proaktor *et al.*, 2007). Directional selection is sometimes reduced by counter selection pressure on small, young males (Mysterud and Bischof, 2010). In addition, trophy males are often shot at the age of trophy culmination, so that they have done most of their breeding before getting shot (Apollonio *et al.*, 2010). It therefore cannot be taken for granted that the mere presence of trophy hunting always induces strong directional selection as a result of hunter preferences for large trophies. A lower level of selection will affect the expected rate of evolutionary response.

Further, we must recognise that any selection in harvest will always operate alongside forces of natural selection working either in the same or opposite direction (Ratner and Lande, 2001). It is also clear that increasing knowledge on how genetic architecture may limit evolutionary responses and of contrasting selection pressures in males and females (Johnston *et al.*, 2013) will move our understanding forward in years to come. A critical point, in particular, is the extent to which antler size is heritable from father to son (Kruuk *et al.*, 2002) and consistent from young to late age in deer during ontogeny, which remains debated (Bartos *et al.*, 2007; Koerth and Kroll, 2008, 2010; Demarais and Strickland, 2010). The latter assumption is more likely to hold for horns that are not shed every year.

For small populations there might be other challenges related to selective harvesting (Hard *et al.*, 2007; Steenkamp *et al.*, 2007). Effective population size may become lower (Sæther *et al.*, 2009). Similarly, under such conditions, it has been shown that it is the combination of mating system and sex ratio that is a driver of the variation in population growth (Lee *et al.*, 2011). In small populations of highly polygynous species, there are only a few males to inseminate a large number of females. Any random event that affects the males ability to do their job may markedly affect population productivity (Lee *et al.*, 2011). This may be the case, for example, in small alpine populations of bovids being target of trophy hunting.

Indeed, harvesting may remove particularly important individuals from the population. For example in elephants, old mothers know migration routes (McComb et al., 2001). There are limited studies of how removal of top ranking males might affect the social structure, but anecdotal evidence from red deer in low density populations suggest removal of the top stag may lead to local redistribution of females. For large populations, regions with limited harvesting might buffer selective effects of harvesting through migration (Tenhumberg et al., 2004).

6.5.1 Empirical evidence of trends in trophy size

For mouflon in Massif Central, France, a decline in overall size and width of horns could be related to trophy hunting (Garel et al., 2007). In alpine chamois, horn length appeared to have a limited role in male reproductive success and hunter selection was regarded as unlikely to yield an evolutionary response in males (Rughetti and Festa-Bianchet, 2010) or females (Rughetti and Festa-Bianchet, 2011). In red deer in Hungary, based on data from trophy exhibitions from 1880–2008, there was no clear overall trend in trophy size (Rivrud et al., 2013; Figure 6.2). Note that trophy sizes from Hungary were not age-corrected and variation in trophy sizes among periods was most likely related to overharvesting connected to the World Wars.

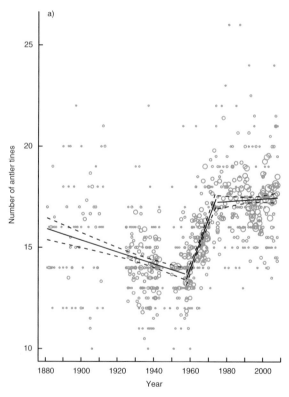

Figure 6.2a Trophy trends of red deer in Hungary (from Rivrud et al., 2013). Figure is reproduced from the *Journal of Applied Ecology* by kind permission of John Wiley and Sons (Wiley-Blackwell Publishers) and shows number of antler tines per trophy through time.

Analysing the trophy scoring of North American ungulates also showed that age structure seems to be the main driver of size (Monteith *et al.*, 2013). For Iberian wild goat (Iberian ibex) and the aoudad or Barbary sheep in south-eastern Spain, an observed decline in horn length over an 18-year period was suggested to be due in part to effects of density, but also to removal of large males by trophy hunters (Pérez *et al.*, 2011). Clearly, temporal variation in phenotypic traits is not very strong evidence of evolutionary change, as it may reflect phenotypic plasticity. In three populations in Vosges in eastern France, however, changes in allele frequency coding for the number of tines on the antler in red deer was linked to the level of selective hunting (Hartl *et al.*, 1991).

The limited empirical evidence should not be taken as argument that trophy hunting cannot generate evolutionary effects. Rather, one might consider taking a precautionary principle and apply either spatial and temporal refuges, compensatory culling, saving stags until prime age culmination and demanding higher prices for larger trophies (Rivrud *et al.*, 2013). Due to the economic incentives and demands for larger trophies, there is a risk that supplementary feeding, selective breeding through introduction of males with specific, desired, characteristics, and other management actions for hunted populations of ungulates become so interventive that they take animals close to a semi-domestic stage (Mysterud, 2010). I will argue this latter development might be important to consider in conservation.

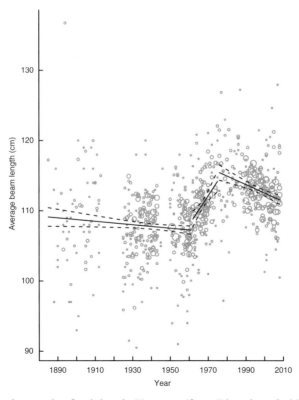

Figure 6.2b Trophy trends of red deer in Hungary (from Rivrud *et al.*, 2013). Figure is reproduced by kind permission of John Wiley and Sons (Wiley-Blackwell Publishers) and shows changes in average beam length (cm) through time.

6.5.2 Reduced sexual selection among males

Growth patterns of male moose suggested that in populations where there is a strong bias towards females, males may show lower body and trophy sizes, presumably as a consequence of lower levels of sexual selection (Mysterud *et al.*, 2005; Tiilikainen *et al.*, 2010). In red deer it is clearly documented that such changes in sex ratio and age structure affect the degree of polygyny (Pérez-González and Carranza, 2011). Also in red deer, male-biased dispersal is the most common pattern of dispersal. However, a transition to female-biased dispersal was found in areas with selective harvesting, causing sex ratios strongly biased towards females and with a high proportion of young males (Pérez-González and Carranza, 2009). Under these conditions, male–male competition for mating opportunities was lower, and this reduction of competition made it possible for young males to remain, which in turn may have meant that females were more likely to disperse as a response to male philopatry. In Spain, selective harvesting of red deer documented lower genetic diversity in hunting estates than within a national park without hunting, arguably due to fewer males and limited competition among males (Martínez *et al.*, 2002). It is clear from such examples that selective hunting might affect social dynamics within a harvested population, but compared to the extent of such hunting, studies are still few.

6.5.3 Female life history and hunting

Theoretical models for red deer females suggest that under heavy harvest pressure, individuals that begin reproduction at a young age and at a light weight have a greater chance of reproducing at least once compared with those that begin reproduction at heavier weights later in life (Proaktor *et al.*, 2007). This will be a trade-off against the cost of early reproduction, which is often substantial. No trend towards earlier maturation was found for red deer in populations where a high proportion of non-breeding juveniles are harvested (Mysterud *et al.*, 2009b). This can be seen as support of the theoretical model for how selective harvesting might affect life history (Proaktor *et al.*, 2007), as hunting is mainly expected to affect life history if hunting pressure is most heavy after the animals reach reproductive maturity. Differences in harvesting pressure among populations of wild boar correlated with the proportion of juveniles reproducing; generation time was 3.5 years in a lightly hunted population and 2.3 years when heavily hunted (Servanty *et al.*, 2009). Wild boar started to reproduce at an earlier age under heavy hunting pressure (Gamelon *et al.*, 2011). Hunting might thus lead to evolution for earlier age of first reproduction in ungulates. The tradition of saving females with offspring might in theory also lead to evolution for more frequent reproduction (Ericsson, 2001).

6.6 Conclusion

Hunting is usually selective for age, sex or size. Selectivity is driven mainly by intentional factors such as harvest quota and hunter preferences, though cases of non-intentional selection occur. Theoretical models give clear motivation and directions on how to harvest populations given clear management aims, but they have often ignored that a certain proportion of males are needed. Effects of selective harvesting may be difficult to predict as there might be multiple goals, such as both aiming for high quality trophy males as well as controlling the population size and using the meat.

From a management perspective, there are many positive effects of being selective. Selective harvesting can increase productivity when saving adult females. However, there is a limit to how low the proportion of males within the populations can be without affecting productivity and, in particular, delaying breeding. Further, skewed sex ratios and age structure in harvested populations may lead to a relaxation of sexual selection processes. These processes are currently understudied. Still, there is limited empirical evidence to demonstrate convincingly how important such evolutionary processes are for trait development among European ungulates.

References

Aanes, R. and Andersen, R. (1996) The effects of sex, time of birth, and habitat on the vulnerability of roe deer fawns to red fox predation. *Canadian Journal of Zoology* **74**, 1857–1865.

Abecia, J.A., Forcada, F. and Zuñiga, O. (2002) A note on the effect of individual housing conditions on LH secretion in ewes after exposure to a ram. *Applied Animal Behaviour Science* **75**, 347–352.

Allendorf, F.W. and Hard, J.J. (2009) Human-induced evolution caused by unnatural selection through harvest of wild animals. *Proceedings of the National Academy of Sciences, USA* **106**, 9987–9994.

Allendorf, F.W., England, P.R., Luikart, G., Ritchie, P.A. and Ryman, N. (2008) Genetic effects of harvest on wild animal populations. *Trends in Ecology and Evolution* **23**, 327–337.

Alvarez, L., Martin, G.B., Galindo, F. and Zarco, L.A. (2003) Social dominance of female goats affects their response to the male effect. *Applied Animal Behaviour Science* **84**, 119–126.

Alvarez, L., Ramos, A.L. and Zarco, L. (2009) The ovulatory and LH responses to the male effect in dominant and subordinate goats. *Small Ruminant Research* **83**, 29–33.

Andersen, R., Karlsen, J., Austmo, L.B., Odden, J., Linnell, J.D.C. and Gaillard, J.-M. (2007) Selectivity of Eurasian lynx *Lynx lynx* and recreational hunters for age, sex and body condition in roe deer *Capreolus capreolus*. *Wildlife Biology* **13**, 467–474.

Andres, D., Clutton-Brock, T.H., Kruuk, L.E.B., Pemberton, J.M., Stopher, K.V. and Ruckstuhl, K.E. (2013) Sex differences in the consequences of maternal loss in a long-lived mammal, the red deer (*Cervus elaphus*). *Behavioral Ecology and Sociobiology* **67**, 1249–1258.

Apollonio, M., Andersen, R. and Putman, R. (2010) *European Ungulates and their Management in the 21st Century*. Cambridge, UK: Cambridge University Press.

Apollonio, M., Putman, R., Grignolio, S. and Bartos, L. (2011) Hunting seasons in relation to biological breeding seasons and the implications for the control or regulation of ungulate populations. In R. Putman, R. Andersen and M. Apollonio (ed.), *Ungulate Management in Europe: Problems and Practices*. Cambridge, UK: Cambridge University Press, pp. 80–105.

Bartos, L., Bahbouh, R. and Vach, M. (2007) Repeatability of size and fluctuating asymmetry of antler characteristics in red deer (*Cervus elaphus*) during ontogeny. *Biological Journal of the Linnean Society* **91**, 215–226.

Beddington, J.R. (1974) Age structure, sex ratio and population density in the harvesting of natural animal populations. *Journal of Applied Ecology* **11**, 915–624.

Berger, J. (1992) Facilitation of reproductive synchrony by gestation adjustment in gregarious mammals: a new hypothesis. *Ecology* **73**, 323–329.

Bischof, R., Mysterud, A. and Swenson, J.E. (2008) Should hunting mortality mimic the patterns of natural mortality? *Biology Letters* **4**, 307–310.

Bischof, R., Swenson, J.E., Yoccoz, N.G., Mysterud, A. and Gimenez, O. (2009) The magnitude and selectivity of natural and multiple anthropogenic mortality causes in hunted brown bears. *Journal of Animal Ecology* **78**, 656–665.

Bowyer, R.T., Bleich, V.C., Manteca, X., Whiting, J.C. and Stewart, K.M. (2007) Sociality, mate choice, and timing of mating in American bison (*Bison bison*): effects of large males. *Ethology* **113**, 1048–1060.

Braga, C., Alexandre, N., Fernández-Llario, P. and Santos, P. (2010) Wild boar (*Sus scrofa*) harvesting using the *espera* hunting method: side effects and management implications. *European Journal of Wildlife Research* **56**, 465–469.

Briefer, E.F., Farrell, M.E., Hayden, T.J. and McElligott, A.G. (2013) Fallow deer polyandry is related to fertilization insurance. *Behavioral Ecology and Sociobiology* **67**, 657–665.

Bro-Jørgensen, J. (2002) Overt female competition and preference for central males in a lekking antelope. *Proceedings of the National Academy of Sciences, USA* **99**, 9290–9293.

Bro-Jørgensen, J. (2007) Reversed sexual conflict in a promiscuous antelope. *Current Biology* **17**, 2157–2161.

Bunnefeld, N., Reuman, D.C., Baines, D. and Milner-Gulland, E.J. (2011) Impact of unintentional selective harvesting on the population dynamics of red grouse. *Journal of Animal Ecology* **80**, 1258–1268.

Bunnell, F.L. (1982) The lambing period of mountain sheep: synthesis, hypothesis, and tests. *Canadian Journal of Zoology* **60**, 1–14.

Byers, J.A. and Waits, L. (2006) Good genes sexual selection in nature. *Proceedings of the National Academy of Sciences, USA* **103**, 16343–16345.

Byers, J.A., Wiseman, P.A., Jones, L. and Roffe, T.J. (2005) A large cost of female mate sampling in pronghorn. *American Naturalist* **166**, 661–668.

Byers, J.A., Byers, A.A. and Dunn, S.J. (2006) A dry summer diminishes mate search effort by pronghorn females: evidence for a significant cost of mate search. *Ethology* **112**, 74–80.

Cameron, R.D., Smith, W.T., Fancy, S.G., Gerhart, K.L. and White, R.G. (1993) Calving success of female caribou in relation to body weight. *Canadian Journal of Zoology* **71**, 480–486.

Caro, T.M., Young, C.R., Cauldwell, A.E. and Brown, D.D.E. (2009) Animal breeding systems and big game hunting: models and application. *Biological Conservation* **142**, 909–929.

Charlton, B.D., Reby, D. and McComb, K. (2007) Female red deer prefer the roars of larger males. *Biology Letters* **3**, 382–385.

Clutton-Brock, T.H. (2009) Sexual selection in females. *Animal Behaviour* **77**, 3–11.

Clutton-Brock, T.H. and Pemberton, J. (2003) *Soay Sheep: Dynamics and Selection in an Island Population.* Cambridge: Cambridge University Press.

Clutton-Brock, T.H., Guinness, F.E. and Albon, S.D. (1982) *Red deer: Behaviour and Ecology of Two Sexes.* Edinburgh, UK: Edinburgh University Press, Edinburgh.

Clutton-Brock, T.H., Iason, G.R. and Guinness, F.E. (1987) Sexual segregation and density-related changes in habitat use in male and female red deer (*Cervus elaphus*). *Journal of Zoology* **211**, 275–289.

Coltman, D.W. (2008) Molecular ecological approaches to studying the evolutionary impact of selective harvesting in wildlife. *Molecular Ecology* **17**, 221–235.

Coltman, D.W., O'Donoghue, P., Jorgenson, J.T., Hogg, J.T., Strobeck, C. and Festa-Bianchet, M. (2003) Undesirable evolutionary consequences of trophy harvesting. *Nature* **426**, 655–658.

Côté, S.D. and Festa-Bianchet, M. (2001) Birthdate, mass and survival in mountain goat kids: effects of maternal characteristics and forage quality. *Oecologia* **127**, 230–238.

Coulson, T. (2012) Integral projections models, their construction and use in posing hypotheses in ecology. *Oikos* **121**, 1337–1350.

Darimot, C.T., Carlson, S.M., Kinnison, M.T., Paquet, P.C., Reimchen, T.E. and Wilmers, C.C. (2009) Human predators outpace other agents of trait changes in the wild. *Proceedings of the National Academy of Sciences, USA* **106**, 952–954.

Demarais, S. and Strickland, B.K. (2010) White-tailed deer antler research: a critique of design and analysis methodology. *Journal of Wildlife Management* **74**, 193–197.

Engen, S., Lande, R., Sæther, B.-E. and Festa-Bianchet, M. (2007) Using reproductive value to estimate key parameters in density-independent age-structured populations. *Journal of Theoretical Biology* **244**, 308–317.

Engen, S., Lande, R., Sæther, B.-E. and Dobson, F.S. (2009) Reproductive value and the stochastic demography of age-structured populations. *American Naturalist* **174**, 795–804.

Ericsson, G. (2001) Reduced cost of reproduction in moose *Alces alces* through human harvest. *Alces* **37**, 61–69.

Estes, R.D. (1976) The significance of breeding synchrony in the wildebeest. *East African Wildlife Journal* **14**, 135–152.

Festa-Bianchet, M. (1988) Birthdate and lamb survival in bighorn lambs (*Ovis canadensis*). *Journal of Zoology* **214**, 653–661.

Festa-Bianchet, M. (2003) Exploitative wildlife management as a selective pressure for life-history evolution of large mammals. In M. Apollonio and M. Festa-Bianchet (eds), *Animal Behavior and Wildlife Conservation*. Washington DC: Island Press, pp. 191–207.

Fisher, R.A. (1930) *The Genetical Theory of Natural Selection*. Oxford: Oxford University Press.

Gaillard, J.-M. and Yoccoz, N.G. (2003) Temporal variation in survival of mammals: a case of environmental canalization? *Ecology* **84**, 3294–3306.

Gaillard, J.-M., Festa-Bianchet, M. and Yoccoz, N.G. (1998) Population dynamics of large herbivores: variable recruitment with constant adult survival. *Trends in Ecology and Evolution* **13**, 58–63.

Gaillard, J.-M., Festa-Bianchet, M., Yoccoz, N.G., Loison, A. and Toïgo, C. (2000) Temporal variation in fitness components and population dynamics of large herbivores. *Annual Review of Ecology and Systematics* **31**, 367–393.

Gamelon, M., Besnard, A., Gaillard, J.-M., Servanty, S., Baubet, E., Brandt, S. and Gimenez, O. (2011) High hunting pressure selects for earlier birth date: wild boar as a case study. *Evolution* **65**, 3100–3112.

Garel, M., Cugnasse, J.-M., Maillard, D., Gaillard, J.-M., Hewison, A.J.M. and Dubray, D. (2007) Selective harvesting and habitat loss produce long–term life history changes in a mouflon population. *Ecological Applications* **17**, 1607–1618.

Geist, V. (1986) Super antlers and pre-world war II European research. *Wildlife Society Bulletin* **14**, 91–94.

Ginsberg, J.R. and Milner-Gulland, E.J. (1994) Sex biased harvesting and population dynamics in ungulates: implications for conservation and sustainable use. *Conservation Biology* **8**, 157–166.

Green, W.C.H. and Rothstein, A. (1993) Persistent influences of birth date on dominance, growth and reproductive success in bison. *Journal of Zoology* **230**, 177–186.

Gruver, B.J., Guynn, D.C. and Jacobsen, H.A. (1984) Simulated effects of harvest strategy on reproduction in white-tailed deer. *Journal of Wildlife Management* **48**, 535–541.

Haagenrud, H. and Lørdahl, L. (1974) *Rettet avskytning i elgbestander* [*Selective harvesting of moose populations*]. Preliminary report No. 3. Agricultural University, Ås.

Haigh, J.R. and Hudson, R.J. (1993) *Farming Wapiti and Red Deer.* St. Louis, MI: Mosby-Year Book Inc.,

Hard, J.J., Mills, L.S. and Peek, J.M. (2007) Genetic implications of reduced survival of male red deer *Cervus elaphus* under harvest. *Wildlife Biology* **12**, 427–439.

Harris, R.B., Wall, W.A. and Allendorf, F.W. (2002) Genetic consequences of hunting: what do we know and what should we do? *Wildlife Society Bulletin* **30**, 634–644.

Hartl, G.B., Lang, G., Klein, F. and Willing, R. (1991) Relationships between allozymes, heterozygosity and morphological characters in red deer (*Cervus elaphus*), and the influence of selective hunting on allele frequency distributions. *Heredity* **66**, 343–350.

Hewison, A.J.M. & Gaillard, J.-M. (1999) Successful sons or advantaged daughters? The Trivers-Willard model and sex-biased maternal investment in ungulates. *Trends in Ecology and Evolution* **14**, 229–234.

Hogg, J.T., Hass, C.C. and Jenni, D.A. (1992) Sex-biased maternal expenditure in Rocky Mountain bighorn sheep. *Behavioral Ecology and Sociobiology* **31**, 243–251.

Holand, Ø., Mysterud, A., Røed, K.H., Coulson, T., Gjøstein, H., Weladji, R.B. and Nieminen, M. (2006a) Adaptive adjustment of offspring sex ratio and maternal reproductive effort in an iteroparous mammal. *Proceedings of the Royal Society of London, Series B* **273**, 293–299.

Holand, Ø., Weladji, R.B., Røed, K.H., Gjøstein, H., Kumpula, J., Gaillard, J.-M., Smith, M.E. and Nieminen, M. (2006b) Male age structure influences females' mass change during rut in a polygynous ungulate: the reindeer (*Rangifer tarandus*). *Behavioral Ecology and Sociobiology* **59**, 694–703.

Jarnemo, A., Liberg, O., Lockowandt, S., Olsson, A. and Wahlström, L.K. (2004) Predation by red fox on European roe deer fawns in relation to age, sex, and birth date. *Canadian Journal of Zoology* **82**, 416–422.

Johnston, S.E., Gratten, J., Berenos, C., Pilkington, J.G., Clutton-Brock, T.H., Pemberton, J.M. and Slate, J. (2013) Life history trade-offs at a single locus maintain sexually selected genetic variation. *Nature* **502**, 93–95.

Kenyon, P.R., Morel, P.C.H., Morris, S.T. and West, D.M. (2008) A note on the effect of vasectomised rams and short-term exposures to entire rams prior to the breeding period on the reproductive performance of ewe lambs. *Applied Animal Behaviour Science* **110**, 397–403.

Keuling, O., Lauterbach, K., Stier, N. and Roth, M. (2010) Hunter feedback of individually marked wild boar *Sus scrofa* L.: dispersal and efficiency of hunting in northeastern Germany. *European Journal of Wildlife Research* **56**, 159–167.

Kiltie, R.A. (1982) Intraspecific variation in the mammalian gestation period. *Journal of Mammalogy* **63**, 646–652.

Kiltie, R.A. (1988) Gestation as a constraint on the evolution of seasonal breeding in mammals. In M.S. Boyce (ed.), *Evolution of Life Histories of Mammals. Theory and Pattern.* New Haven, CT: Yale University Press, pp. 257–289.

Koerth, B.H. and Kroll, J.C. (2008) Juvenile-to-adult antler development in white-tailed deer in South Texas. *Journal of Wildlife Management* **72**, 1109–1113.

Koerth, B.H. and Kroll, J.C. (2010) White-tailed deer antler research: a response to Demarais and Strickland. *Journal of Wildlife Management* **74**, 198–202.

Komers, P.E., Birgersson, B. and Ekvall, K. (1999) Timing of estrus in fallow deer is adjusted to the age of available mates. *American Naturalist*, **153**, 431–436.

Kruuk, L.E.B., Slate, J., Pemberton, J.M., Brotherstone, S., Guinness, F. and Clutton-Brock, T.H. (2002) Antler size in red deer: heritability and selection but no evolution. *Evolution* **56**, 1683–1695.

L'Italien, L., Weladji, R.B., Holand, Ø., Røed, K.H., Nieminen, M. and Côté, S.D. (2012) Mating group size and stability in reindeer *Rangifer tarandus*: the effects of male characteristics, sex ratio and male age structure. *Ethology* **118**, 783–792.

Langbein, J. and Thirgood, S.J. (1989) Variation in mating systems of fallow deer (*Dama dama*) in relation to ecology. *Ethology* **83**, 195–214.

Langvatn, R. and Loison, A. (1999) Consequences of harvesting on age structure, sex ratio and population dynamics of red deer *Cervus elaphus* in central Norway. *Wildlife Biology* **5**, 213–223.

Langvatn, R., Albon, S.D., Burkey, T. and Clutton-Brock, T.H. (1996) Climate, plant phenology and variation in age at first reproduction in a temperate herbivore. *Journal of Animal Ecology* **65**, 653–670.

Langvatn, R., Mysterud, A., Stenseth, N.C. and Yoccoz, N.G. (2004) Timing and synchrony of ovulation in red deer constrained by short northern summers. *American Naturalist* **163**, 763–772.

Le Galliard, J.-F., Fitze, P.S., Ferriére, R. and Clobert, J. (2005) Sex ratio bias, male aggression, and population collapse in lizards. *Proceedings of the National Academy of Sciences, USA* **102**, 18231–18236.

Lee, A.M., Sæther, B.-E. and Engen, S. (2011) Demographic stochasticity, Allee effects, and extinction: the influence of mating system and sex ratio. *American Naturalist* **177**, 301–313.

Leuthold, W. (1977) African ungulates. *Zoophysiology and Ecology* **8**, 1–307.

Linnell, J.D.C., Aanes, R. and Andersen, R. (1995) Who killed Bambi? The role of predation in the neonatal mortality of temperate ungulates. *Wildlife Biology* **1**, 209–223.

Loison, A., Gaillard, J.-M., Pélabon, C. and Yoccoz, N.G. (1999a) What factors shape sexual size dimorphism in ungulates? *Evolutionary Ecology Research* **1**, 611–633.

Loison, A., Langvatn, R. and Solberg, E.J. (1999b) Body mass and winter mortality in red deer calves: Disentangling sex and climate effects. *Ecography* **22**, 20–30.

Martínez, J.G., Carranza, J., Fernández-García, J.L. and Sánchez-Prieto, C.B. (2002) Genetic variation of red deer populations under hunting exploitation in southwestern Spain. *Journal of Wildlife Management* **66**, 1273–1282.

Martínez, M., Rodríquez, V., Jones, O.R., Coulson, T. and San Miguel, A. (2005) Different hunting strategies select for different weights in red deer. *Biology Letters* **1**, 353–356.

McClintock, M.K. (1983) Pheromonal regulation of the ovarian cycle: enhancement, suppression, and synchrony. In J.G. Vandenbergh (ed.), *Pheromones and Reproduction in Mammals*. New York: Academic Press, pp. 113–149.

McComb, K. (1987) Roaring by red deer stags advances the date of oestrus in hinds. *Nature* **330**, 648–649.

McComb, K. and Clutton-Brock, T.H. (1994) Is mate choice copying or aggregation responsible for skewed distributions of females on leks? *Proceedings of the Royal Society of London, Series B* **255**, 13–19.

McComb, K., Moss, C., Durant, S.M., Baker, L. and Sayialel, S. (2001) Matriarchs as repositories of social knowledge in African elephants. *Science* **292**, 491–494.

McComb, K.E. (1991) Female choice for high roaring rates in red deer, *Cervus elaphus*. *Animal Behaviour* **41**, 79–88.

Miller, K.V., Jemiolo, B., Gassett, J.W., Jelinek, I., Wiesler, D. and Novotny, M. (1998) Putative chemical signals from white-tailed deer (*Odocoileus virginianus*): Social and seasonal effects on urinary volatile excretion in males. *Journal of Chemical Ecology* **24**, 673–683.

Milner, J.M., Bonenfant, C., Mysterud, A., Gaillard, J.-M., Csányi, S. and Stenseth, N.C. (2006) Temporal and spatial development of red deer harvesting in Europe: biological and cultural factors. *Journal of Applied Ecology* **43**, 721–734.

Milner, J.M., Bonenfant, C. and Mysterud, A. (2011) Hunting bambi: evaluating the basis for selective harvesting of juveniles. *European Journal of Wildlife Research* **57**, 565–574.

Milner-Gulland, E.J., Bukreeva, O.M., Coulson, T., Lushchekina, A.A., Kholodova, M.V., Bekenov, A.B. and Grachev, I.A. (2003) Reproductive collapse in saiga antelope harems. *Nature* **422**, 135.

Miquelle, D.G. (1991) Are moose mice? The function of scent urination in moose. *American Naturalist* **138**, 460–477.

Moe, T., Solberg, E.J., Herfindal, I., Sæther, B.-E., Bjørneraas, K. and Heim, M. (2009) Sex ratio variation in harvested moose (*Alces alces*) calves: does it reflect population calf sex ratio or selective hunting? *European Journal of Wildlife Research* **55**, 217–226.

Monteith, K.L., Long, R.A., Bleich, V.C., Heffelfinger, J.R., Krausman, P.R. and Bowyer, R.T. (2013) Effects of harvest, culture, and climate on trends in size of horn-like structures in trophy ungulates. *Wildlife Monographs* **183**, 1–26.

Mysterud, A. (2010) Still walking on the wild side? Management actions as steps towards 'semi-domestication' of hunted ungulates. *Journal of Applied Ecology* **47**, 920–925.

Mysterud, A. (2011) Selective harvesting of large mammals: How often does it result in directional selection? *Journal of Applied Ecology* **48**, 827–834.

Mysterud, A. and Bischof, R. (2010) Can compensatory culling offset undesirable evolutionary consequences of trophy hunting? *Journal of Animal Ecology* **79**, 148–160.

Mysterud, A., Yoccoz, N.G., Stenseth, N.C. and Langvatn, R. (2000) Relationships between sex ratio, climate and density in red deer: the importance of spatial scale. *Journal of Animal Ecology* **69**, 959–974.

Mysterud, A., Coulson, T. and Stenseth, N.C. (2002) The role of males in the population dynamics of ungulates. *Journal of Animal Ecology* **71**, 907–915.

Mysterud, A., Solberg, E.J. and Yoccoz, N.G. (2005) Ageing and reproductive effort in male moose under variable levels of intra-sexual competition. *Journal of Animal Ecology* **74**, 742–754.

Mysterud, A., Bonenfant, C., Loe, L.E., Langvatn, R., Yoccoz, N.G. and Stenseth, N.C. (2008) The timing of male reproductive effort relative to female ovulation in a capital breeder. *Journal of Animal Ecology* **77**, 469–477.

Mysterud, A., Tryjanowski, P. and Panek, M. (2006) Selectivity of harvesting differs between local and foreign roe deer hunters – trophy stalkers have the first shot at the right place. *Biology Letters* **2**, 632–635.

Mysterud, A., Røed, K.H., Holand, Ø., Yoccoz, N.G. and Nieminen, M. (2009a) Age-related gestation length adjustment in a large iteroparous mammal at northern latitude. *Journal of Animal Ecology* **78**, 1002–1006.

Mysterud, A., Yoccoz, N.G. and Langvatn, R. (2009b) Maturation trends in red deer females over 39 years in heavily harvested populations. *Journal of Animal Ecology* **78**, 595–599.

Nilsen, E.B. and Solberg, E.J. (2006) Patterns of hunting mortality in Norwegian moose (*Alces alces*) populations. *European Journal of Wildlife Research* **52**, 153–163.

Nilsen, E.B., Pettersen, T., Gundersen, H., Milner, J.M., Mysterud, A., Solberg, E.J., Andreassen, H.P. and Stenseth, N.C. (2005) Moose harvesting strategies in the presence of wolves. *Journal of Applied Ecology* **42**, 389–399.

Nilsen, E.B., Brøseth, H., Odden, J. and Linnell, J.D.C. (2012) Quota hunting of Eurasian lynx in Norway: pattern of hunter selection, hunter efficiency and monitoring accuracy. *European Journal of Wildlife Research* **58**, 325–333.

Nores, C., Llaneza, L. and Alvarez, M.A. (2008) Wild boar *Sus scrofa* mortality by hunting and wolf *Canis lupus* predation: an example in northern Spain. *Wildlife Biology*. **14**, 44–51.

Panzacchi, M., Linnell, J.D.C., Serrao, G., Eie, S., Odden, M., Odden, J. and Andersen, R. (2008) Evaluation of the importance of roe deer fawns in the spring–summer diet of red foxes in southeastern Norway. *Ecological Research* **23**, 889–896.

Pérez, J.M., Serrano, E., González-Candela, M., León-Vizcaino, L., Barberá, G.G., de Simón, M.A., Fandos, P., Granados, J.E., Soriguer, R.C. and Festa-Bianchet, M. (2011) Reduced horn size in two wild trophy-hunted species of Caprinae. *Wildlife Biology* **17**, 102–112.

Pérez-González, J. and Carranza, J. (2009) Female-biased dispersal under conditions of low male mating competition in a polygynous mammal. *Molecular Ecology* **18**, 4617–4630.

Pérez-González, J. and Carranza, J. (2011) Female aggregation interacts with population structure to influence the degree of polygyny in red deer. *Animal Behaviour* **82**, 957–970.

Preston, B.T., Stevenson, I.R., Pemberton, J.M. and Wilson, K. (2001) Dominant rams lose out by sperm depletion. A waning success in siring counters a ram's high score in competition for ewes. *Nature* **409**, 681–682.

Proaktor, G., Coulson, T. and Milner-Gulland, E.J. (2007) Evolutionary responses to harvesting in ungulates. *Journal of Animal Ecology* **76**, 669–678.

Ramanzin, M. and Sturaro, E. (2013) Habitat quality influences relative antler size and hunters selectivity in roe deer. *European Journal of Wildlife Research*, in press.

Ratner, S. and Lande, R. (2001) Demographic and evolutionary responses to selective harvesting in populations with discrete generations. *Ecology* **82**, 3093–3104.

Réale, D., Boussès, P. and Chapuis, J.-L. (1996) Female-biased mortality induced by male sexual harassment in a feral sheep population. *Canadian Journal of Zoology* **74**, 1812–1818.

Rivrud, I.M., Sonkoly, K., Lehoczki, R., Csányi, S., Storvik, G.O. and Mysterud, A. (2013) Hunter selection and long term trend (1881–2008) of red deer trophy sizes in Hungary. *Journal of Applied Ecology* **122**, 137–145.

Rivrud, I.M., Meisingset, E.L., Loe, L.E. and Mysterud, A. (2014) Interaction effects between weather and space use on harvesting effort and -selection in red deer. submitted ms.

Røed, K.H., Holand, Ø., Mysterud, A., Tverdal, A., Kumpula, J. and Nieminen, M. (2007) Male phenotypic quality influences offspring sex ratio in a polygynous ungulate. *Proceedings of the Royal Society of London, Series B* **274**, 727–733.

Rosa, H.J.D., Silva, C.C. and Bryant, M.J. (2006) The effect of ram replacement and sex ratio on the sexual response of anoestrous ewes. *Small Ruminant Research* **65**, 223–229.

Rughetti, M. and Festa-Bianchet, M. (2010) Compensatory growth limits opportunities for artificial selection in alpine chamois. *Journal of Wildlife Management* **74**, 1024–1029.

Rughetti, M. and Festa-Bianchet, M. (2011) Effects of early horn growth on reproduction and hunting mortality in female chamois. *Journal of Animal Ecology* **80**, 438–447.

Ruusila, V. and Pesonen, M. (2004) Interspecific cooperation in human (*Homo sapiens*) hunting: the benefits of a barking dog (*Canis familiaris*). *Annales Zoologica Fennica* **41**, 545–549.

Sæther, B.-E., Solberg, E.J., Heim, M., Stacy, J.E., Jakobsen, K. and Olstad, R. (2004) Offspring sex ratio in moose *Alces alces* in relation to paternal age: an experiment. *Wildlife Biology* **10**, 51–57.

Sæther, B.-E., Engen, S., Solberg, E.J. and Heim, M. (2007) Estimating the growth of a newly established moose population using reproductive value. *Ecography* **30**, 417–421.

Sæther, B.-E., Engen, S. and Solberg, E.J. (2009) Effective size of harvested ungulate populations. *Animal Conservation* **12**, 488–495.

Sand, H., Wikenros, C., Ahlqvist, P., Strømseth, T.H. and Wabakken, P. (2012) Comparing body condition of moose (*Alces alces*) selected by wolves (*Canis lupus*) and human hunters: consequences for the extent of compensatory mortality. *Canadian Journal of Zoology* **90**, 403–412.

Schaal, A. and Bradbury, J.W. (1987) Lek breeding in a deer species. *Biology of Behaviour* **12**, 28–32.

Schindler, S., Neuhaus, P., Gaillard, J.-M. and Coulson, T. (2013) The influence of nonrandom mating on population growth. *American Naturalist* **182**, 28–41.

Schmidt, J.I., Ver Hoef, J.M. and Bowyer, R.T. (2007) Antler size of Alaskan moose *Alces alces gigas*: effects of population density, hunter harvest and use of guides. *Wildlife Biology* **13**, 53–65.

Servanty, S., Gaillard, J.-M., Toïgo, C., Serge, B. and Baubet, E. (2009) Pulsed resources and climate-induced variation in the reproductive traits of wild boar under high hunting pressure. *Journal of Animal Ecology* **78**, 1278–1290.

Shackleton, D.M. (1991) Social maturation and productivity in bighorn sheep: are young males incompetent? *Applied Animal Behaviour Science* **29**, 173–184.

Shipka, M.P., Rowell, J.E. and Ford, S.P. (2002) Reindeer bull introduction affects the onset of the breeding season. *Animal Reproduction Science* **72**, 27–35.

Solberg, E.J., Loison, A., Sæther, B.-E. and Strand, O. (2000) Age-specific harvest mortality in a Norwegian moose *Alces alces* population. *Wildlife Biology* **6**, 41–52.

Solberg, E.J., Loison, A., Ringsby, T.H., Sæther, B.-E. and Heim, M. (2002) Biased adult sex ratio affects fecundity in primiparous moose *Alces alces*. *Wildlife Biology* **8**, 117–128.

Steenkamp, G., Ferriera, S. M. and Bester, M. N. (2007) Tusklessness and tusk fractures in free-ranging African savanna elephants (*Loxodonta africana*). *Journal of the South African Veterinary Association* **78**, 75–80.

Sutherland, W. J. (1996) *From Individual Behaviour to Population Ecology.* Oxford: Oxford University Press.

Tenhumberg, B., Tyre, E.J., Pople, A.R. and Possingham, H.P. (2004) Do harvest refuges buffer kangaroos against evolutionary responses to selective harvesting? *Ecology* **85**, 2003–2017.

Tennenhouse, E.M., Weladji, R.B., Holand, Ø., Røed, K.H. and Nieminen, M. (2011) Mating group composition influences somatic costs and activity in rutting dominant male reindeer (*Rangifer tarandus*). *Behavioral Ecology and Sociobiology* **65**, 287–295.

Thirgood, S., Langbein, J. and Putman, R.J. (1999) Intraspecific variation in ungulate mating strategies: the case of the flexible fallow deer. *Advances in the Study of Behaviour* **28**, 333–361.

Tiilikainen, R., Nygren, T., Pusenius, J. and Ruusila, V. (2010) Variation in growth patterns of male moose *Alces alces* after two contrasted periods of hunting. *Annales Zoologica Fennica* **47**, 159–172.

Torres-Porras, J., Carranza, J. and Pérez-González, J. (2009) Selective culling of Iberian red deer stags (*Cervus elaphus hispanicus*) by selective montería in Spain. *European Journal of Wildlife Research* **55**, 117–123.

Valdez, R., Cardenas, M. and Sanchez, J. (1991) Disruptive mating behavior by subadult Armenian wild sheep in Iran. *Applied Animal Behaviour Science* **29**, 165–171.

Webb, S.L., Dzialak, M.R., Wondzell, J.J., Harju, S.M., Hayden-Wing, L.D. and Winstead, J.B. (2011) Survival and cause-specific mortality of female Rocky Mountain elk exposed to human activity. *Population Ecology* **53**, 483–493.

Whiting, J.C., Bowyer, R.T. and Flinders, J.T. (2008) Young bighorn (*Ovis canadensis*) males: can they successfully woo females? *Ethology* **114**, 32–41.

Whittle, C.L., Bowyer, R.T., Clausen, T.P. and Duffy, L.K. (2000) Putative pheromones in urine of rutting male moose (*Alces alces*): evolution of honest advertisement? *Journal of Chemical Ecology* **26**, 2747–2762.

Chapter 7

The Management of Urban Populations of Ungulates

Rory Putman, Jochen Langbein, Peter Watson,
Peter Green and Seán Cahill

Right across Europe, the majority of species of ungulates are increasing in numbers and expanding their ranges (Apollonio *et al*., 2010). This appears in large part to be a response to change in availability of suitable habitat: with increased planting of both native and exotic woodlands, intensification of agriculture and agricultural abandonment of more marginal areas, as well as (re)introductions to new or previously occupied areas from which they had disappeared. All such changes are accompanied also by an increased urbanisation of human populations (Putman, 2011) and an expansion of the urban landscape itself, with, in consequence, a dramatic increase in the contact zone between urban areas and wildlife habitat, often referred to as the wildland–urban interface (Radeloff *et al*., 2008).

A European Environment Agency analysis of land-cover change across 36 European countries showed a change in land-cover type between 2000 and 2006 for 68,353 km^2 (1.3% of total land of 5.42 million km^2; Meiner *et al*., 2010). This analysis concluded that land-use specialisation, namely urbanisation, agricultural intensification and land abandonment plus natural afforestation, is currently a very strong trend in Europe which is expected to continue in the future. Urban development in particular is the most rapidly expanding land use change in Europe, and it is predicted that urban expansion will continue at a rate of 0.5–0.7% per year, which is over 10 times higher than any other land use change (Piorr *et al*., 2011).

Such expansion of the urban landscape produces conflicts between human activities and wildlife. Whether or not directly managed by humans, wild animals suffer significant impacts from our activities in direct loss of habitat, loss of ecological connectivity through the proliferation of road systems or other transport infrastructure, impacts on food quality and so forth; and increasing numbers of animals are killed and injured on our roads. But at the same time, expansion of towns and cities also means that many, more opportunistic, species are increasingly becoming established actually within urbanised areas. The definition of 'urban' can vary considerably from country to country. For the purposes of this review, we would (after Dandy *et al.*, 2009) make a clear distinction between urban areas (the centre or suburbs of cities, towns and other conurbations) and peri-urban areas, which are predominantly semi-rural, but 'infiltrated', as it were, by suburban extension, 'out of town' developments (commonly industrial or retail) and the associated infrastructure of the urban fringe (roads, etc.).[1]

Within the UK, as also in Germany, the phenomenon of urban foxes has been recognised, publicised and widely studied for many years (Harris and Rayner, 1986; König, 2008). In contrast, ungulates have only more recently been recognised first as occasional visitors to our cities (Chapman, 1991) and then as inhabitants (e.g. Prior, 2000). There has been increasing colonisation of larger towns and cities in the UK over recent decades by, in particular, muntjac (*Muntiacus reevesi*) and roe deer (*Capreolus capreolus*). Deer of one or the other of these species are now established well within the centre of cities such as Bristol, Edinburgh, Glasgow, London, Aberdeen and Southampton, with the most closely-documented history of colonisation reported for Sheffield (McCarthy and Rotherham, 1994; McCarthy *et al.*, 1996, 1999). There are also well-publicised incidences of larger deer species, including fallow (*Dama dama*) and red deer (*Cervus elaphus*), entering urban or suburban areas both within the UK and other European countries, causing disruption and serious deer–vehicle collisions.

Wild boar populations have also increased dramatically in many European countries during recent decades (Sáez-Royuela and Tellería, 1986; ELO and Polytechnic University of Madrid, 2012), to such an extent that a formal question was raised in 2012 in the European Parliament on the 'overpopulation of wild boar in Europe' (Franco and Mathieu, 2012). Among other aspects, this parliamentary question makes specific reference to problems regarding wild boar in European urban areas. Wild boar have become increasingly habituated to human activity and have begun to become established both in peri-urban and truly urban areas within and around many European cities (e.g. Kotulski and König, 2008; Cahill *et al.*, 2012; Podgórski *et al.*, 2013). Cahill *et al.* (2012) indeed report from an internet search that, up until 2010, at least 44 cities or towns in 15 countries (Belgium, China, France, Germany, India, Israel, Italy, Japan, Korea, Poland, Singapore, Spain, UK, United States and Romania) had reports of incidents concerning wild boar/feral pigs. Of these cities, 36.4% had experienced just one or two incidents, while the majority (63.6%) had already reported several or many cases.

Within fully urban areas there is concern in relation to damage by ungulates to gardens or urban parks (see, for example, Chapman *et al.*, 1994; Coles, 1997), golf courses, structural damage to fences, increased risk of road traffic accidents involving ungulates (Langbein *et al.*, 2011), as well as concern about the potential implication of deer and boar in the transmission of disease to both humans,

---

[1]Piorr *et al.* (2011) describe 'peri-urban' as the area between urban settlement areas and their rural hinterland, and larger peri-urban areas can include towns and villages within an urban agglomeration.

livestock and domestic pets (e.g. Simpson, 2002; Jansen *et al.*, 2007; Nidaira *et al.*, 2007; Meng *et al.*, 2009; Schielke *et al.*, 2009). Habituated wild boar can even become a nuisance for people due to their pestering for food (Bobek *et al.*, 2011). Welfare concerns extend to the physical condition of wild ungulates established within urban sites, which in some circumstances may be poorer than that of animals within more natural habitat (Green, 2008). However, (peri-)urban areas can also provide ungulates with significant anthropogenic food sources, thus representing attractive habitat for them (Cahill *et al.*, 2012), as well as other resources such as refuge from hunting or natural predators, and even more benign temperatures during winter (Podgórski *et al.*, 2013).

In this chapter we review available data on the distribution of urban and peri-urban ungulates and the conflicts which may result, as well as considering options available for control or mitigation.

7.1 Urbanised ungulates in Europe

As above, just as in more rural areas, wild ungulates in both urban and peri-urban contexts may be implicated in collisions with vehicles, in damage to agriculture and especially horticulture: market gardens or orchards in the peri-urban area; damage to amenity plantings, farm woodlands or community forests. The presence of established wild ungulate populations may raise issues of their potential as reservoirs or vectors in the transmission of diseases to humans or livestock. On the other hand, some view the establishment of such animals near or within urban areas as being of amenity benefit, and enjoy seeing them as part of their urban landscape.

While deer of a range of different species are now widely reported as occasional transients or actual residents within a number of conurbations within the UK, a thorough literature review (through two academic search engines) failed to discover any published literature on urban deer or their management in any other European countries, although clearly there are regular media reports of deer and other wildlife within urban areas. We therefore undertook our own survey of colleagues (wildlife biologists and management biologists) in a number of different European countries in order to assemble what unpublished information may be available specifically in relation to problems and management of deer within urban environments. Responses from Austria, Germany, the Czech Republic, Hungary, Belgium, the Netherlands, France, Portugal, Spain, Italy, Slovenia, Sweden, Finland and Norway suggest that while deer are frequently present in peri-urban areas, the presence of deer within major towns or cities is not generally considered a widespread problem in the majority of European countries outside the UK and in general no specific management or specialist management approach is developed to deal with the problem (details in Watson *et al.*, 2009).

It should be noted however that the fact that urbanisation of deer is not generally recognised as a problem in these other countries should not be taken to imply that it is not a potential problem for the future, or that the perception of a recently growing problem in the UK is an overreaction or in some other regard factually incorrect.

Many European countries have lived with roe deer within their residential areas for many years and do not therefore perceive them as a 'new problem', having designed and adapted policies to cope with what they consider normal and inevitable deer presence since the middle of the 20th century. Locally, however, conflicts are starting to appear in response to the changing characteristics of certain urban developments: in areas of the Costa del Sol (Malaga, Spain) there are now new extensive low-density residential resorts in close proximity to woodlands, and which also incorporate fragments of these natural areas within the developments in combination with irrigated landscaped areas such as gardens and golf courses (Duarte *et al.*, 2012). These residential areas now experience conflicts due to the presence of ungulates such as red deer and fallow deer (as well as wild boar, below).

It is also clear that wild boar are more regularly being encountered within both peri-urban and urban areas: as examples, wild boar have been present in peri-urban areas of Berlin for the past two decades (Georgii *et al.*, 1991). This city is undoubtedly the one which suffers most problems in Europe as a result of wild boar presence, considering the length of time that problems have been occurring, the frequency of incidents, and the number of wild boar that have to be removed from its urban areas. An estimated 5,000 wild boar live within Berlin's urban and suburban areas (Jansen *et al.*, 2007), although this may be a considerable underestimate, and they are now considered as a serious nuisance in this city (Kotulski and König, 2008). More than 2,300 were shot between 2008 and 2009 (Marco Apollonio, pers. comm.). In Spain, Barcelona and certain dormitory towns and neighbourhoods on the outskirts of Madrid also experience frequent incidents as a result of wild boar presence, and in Seraing, Belgium, 141 wild boar were removed by hunting from the periphery of its urban area in little over a year (Morel, 2011).

Updated results obtained from the internet search of Cahill *et al.* (2012), mentioned in our Introduction, have more recently been combined with those from an international consultation using a questionnaire survey on wild boar and feral pigs in (peri-)urban areas (Licoppe *et al.*, 2013) This survey indicates that up to 2012, at least 104 cities and towns in 18 European countries have registered incidents concerning the presence of wild boar within their (peri-)urban area (Figure 7.1). The majority of these locations experience problems on a regular basis (63% with 10 or more incidents per year), with five countries – Spain, France, Germany, the Netherlands and Poland – making up over three quarters (79%) of the 66 areas reporting incidents more than 10 times per year.

Wild boar are opportunistic omnivores with a very broad dietary spectrum in Europe (Schley and Roper, 2003), as well as having the highest reproductive rates among ungulates, with the capability of more than doubling local populations (Massei and Genov, 2004). Moreover, pigs are noted for their high cognitive capacity (Broom *et al.*, 2009), a trait which is also likely to aid exploitation by wild boar of novel artificial environments such as those found in urban settings. Such characteristics may make this particular synanthrope an even more complicated challenge for wildlife managers in urban areas than might be the case with other ungulates such as deer.

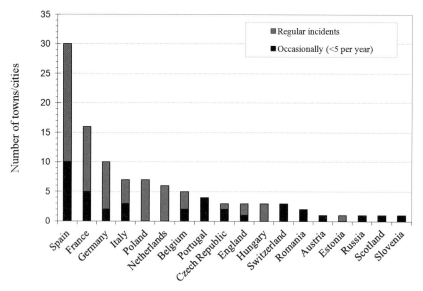

Figure 7.1 Number of towns and cities reporting incidents with wild boar within their (peri-) urban area by 2012. *Source:* From Licoppe *et al.* (2013), with permission.

7.2 Potential conflicts

In their global survey on wild boar and feral pigs, Licoppe *et al.* (2013) report a qualitative evaluation of the importance of a dozen potential impacts and conflicts relating to the presence of this species in (peri-)urban areas. Results obtained from this survey for European countries rank collisions with vehicles as the most important impact of wild boar presence, followed by damage to agriculture/horticulture within the (peri-)urban area, as well as damage to private and public gardens and parks (unpublished data from Licoppe *et al.*, 2013). Damage to biodiversity by wild boar was considered to be the impact of least concern by respondents. This might be due to either low impacts, or unknown impacts, of wild boar on biodiversity in these areas, as boar can indeed have a significant impact on their environment, both negative and positive (see review by Massei and Genov, 2004; Sims, 2005).

7.2.1 Collisions with vehicles

Although much information has been reported on ungulate–vehicle collisions in Europe (see, for example, Groot-Bruinderink and Hazebroek, 1996; Langbein *et al.*, 2011; Lagos *et al.*, 2012), data are lacking on their specific extent and corresponding trends in (peri-)urban areas. Nevertheless, they are known to be an issue of growing concern and, for example, were identified as the most important impact relating to the presence of wild boar in urban areas in the survey undertaken by Licoppe *et al.*, (2013). Although human casualties are relatively rare, occurring usually in only from 1 to 3% of incidents, most collisions involving ungulates in urban areas do cause material damage, whilst their mere presence can disrupt traffic on roads and railways.

A study of the patterns of distribution of deer–vehicle collisions (DVCs) in relation to urban areas in the UK shows that among a sample of over 30,700 DVCs
logged in a nationwide DVC project carried out in England, 21% occurred within
that land area classed as 'urban' by the Department for the Environment, Food and
Rural Affairs, with thus 21% of all incidents occurring within 8% of the total land
area (Langbein, 2008). If we define a peri-urban zone as a ring of 1.6 km around
that urban land classification, then the published figures show that a further 23%
of all recorded DVCs fall inside this extremely conservative definition of the peri-
urban zone and in total 44% of all recorded accidents fall within 1 mile of built up
areas (Langbein, 2008).

While inadequate past data are available to assess changes in the number of deer
established within urban areas over the last 50 years or so, there is some evidence
from DVCs to suggest, at least for the UK, that the overall split between numbers
in rural and peri-urban or urban areas may not have changed very significantly, at
least over the last decade. Records of call-outs to injured live deer reported to the
two largest animal welfare organisations called out to live deer injured in traffic
collisions in the UK, (the Royal Society for the Prevention of Cruelty to Animals
in England and the Scottish Society for the Prevention of Cruelty to Animals in
Scotland), show a significant increasing trend in total numbers of incidents dealt
with between 2001–2007, with any recent levelling off likely related to a slow
in traffic growth as a consequence of the economic recession (Langbein, 2011).
However, an analysis of changes in the annual proportion of DVCs dealt with by
the above animal welfare organisations in urban rather than rural areas between
2001–2010 showed no significant shift in favour of urban areas over that period.
(Langbein, unpublished).

The large concentration of DVCs in the peri-urban and urban sector is the
combined result of the much greater levels of road traffic, human habitation
and disturbance, compared with truly rural areas, and the significant extent to
which deer have colonised the sub-urban fringes over the past century. Deer
in these areas are essentially exposed to a much greater risk of collisions than
a deer population at similar density residing in more rural or remote regions.
In the case of deer in North America, DeNicola and Williams (2008) noted
that excessive numbers of DVCs were one of the main reasons why local governments had implemented lethal deer management programs in the United
States.

7.2.2 Damage to horticulture/market gardens

Within the UK, as elsewhere, there have also been growing numbers of reports
of damage to horticulture/market gardens. Between 1987 and 1989, 10.5%
of complaints relating to deer received by the COSTER database overall
(Computerised Summary of Technical Reports) maintained by the former Ministry of Agriculture, Fisheries and Foods' Wildlife Services Branch (WSB)
related to damage to horticulture or nursery crops (Putman and Moore, 1998;
Putman, 2004).

Impacts from ungulates tend to be highly localised rather than widespread (Putman and Kjellander, 2003; Reimoser and Putman, 2011); however, it is notable that in the urban or peri-urban area there is a greater concentration of more intensively managed horticultural enterprises, or orchards and soft fruit, as well as nurseries of trees or garden plants. In these high-value cash crops, damage from ungulates may therefore be of greater significance than impacts on field crops or pastures in more rural areas (for wider consideration of the impacts of ungulates on rural agriculture, see Reimoser and Putman, 2011). We have been unable to find any published material from the UK and little from Europe on the extent of damage caused by deer to nurseries, market gardens, top fruit crops or other horticultural enterprises. There have however been a number of studies in the United States on deer damage to horticulture – especially damage to orchard crops. These mainly involve white-tailed deer (*Odocoileus virginianus*) a species ecologically very similar to the European roe. In forested counties with high deer densities, deer damage may even exceed that caused by birds (Matschke *et al.*, 1984). The majority of damage caused was due to browsing of vegetative growth rather than damage to tree bark/stems or fruits. Young trees appear to be more at risk than older specimens and certain fruit tree species appear more susceptible than others (Matschke *et al.*, 1984). Apple and peach trees were the most severely affected with damage less severe for pears and cherries (Scott and Townsend, 1985). In Ohio, in a survey involving 2236 fruit growers and nurseries, 41% of fruit orchards reported deer damage, of whom 40% reported that damage to be moderate to severe (Scott and Townsend, 1985). However, as with other forms of damage from ungulates, orchard damage tends to be quite localised (Beckworth and Stith, 1968; Harder, 1970) with the extent of damage related to proximity and amount of woodland cover (Conover, 1989).

Schley and Roper (2003) found that crops represent an important component of the diet of wild boar throughout its Western European range, and that they cause significant agricultural damage. In this sense, crops in (peri-)urban areas are no exception, and may even be more vulnerable to damage due to either legal restrictions or practical difficulties regarding the hunting of wild boar in nearby natural habitats. For example, non-hunted woodlands in proximity to agricultural crops have been shown to create a refuge effect for this species in Italy, thus leading to increased damage in intensively cultivated areas (Amici *et al.*, 2012). Vegetable gardens, fruit trees and vineyards were the most affected cultivation types cited by Arqués *et al.* (2009) for the Marina Baja area of Alicante (Spain), where wild boar are present in high densities (15–20 per km^2), even in highly human-modified agricultural areas with very little natural vegetation cover on the periphery of urban settlements. In this study, 65% of hunting grounds in the area cited damage to market gardens and horticultural areas.

Although few formal studies have been found in the scientific literature regarding damage by wild boar to agriculture in peri-urban areas, this conflict is indeed reported in local press. For example, extensive damage has been reported (Manrique, 2012) to numerous market gardens in several municipalities within the important horticultural area surrounding the city of Tudela (Navarra, Spain). This author reports payments by local hunting associations of €10,000–15,000 per year in compensation for damage to horticulture in areas close to Tudela. Wild boar are thought to reach these horticultural areas via the riparian woodland corridor along the Ebro river (op. cit.), and in relation to this, Amici (2012) identified significantly higher damage to intensively cultivated areas within a 1 km buffer of riparian corridors in Italy. As mentioned above, this author also identified the refuge effect created in adjacent non-hunting areas as a cause of increased agricultural damage, and local media in Spain report urban areas as suffering regular damage by wild boar to vegetable gardens, allotments and horticulture in areas where hunting is not usually permitted (for examples from Catalonia, see El 9 Nou, 2012). In response, municipal authorities sometimes grant special temporary hunting permits in order to try and reduce wild boar presence and damage to crops.

7.2.3 Urban ungulates as agents for transmission of disease

This is perhaps the area where public concern is greatest. Ungulates may be implicated in the transmission of disease to humans, their livestock or their pets. Such a potential epidemiological role is economically more important in the rural environment, where the transmission of tuberculosis, bluetongue disease, internal parasites and other pathogens may have significant impact (Froliche *et al.*, 2002; Delahay *et al.*, 2007; Ward *et al.*, 2008; Ward and Smith, 2012; Ferroglio *et al.*, 2011). It is important to note that deer may themselves be unaffected by harbouring organisms that cause disease in other animals or people. For example, recent work has confirmed that deer will carry and multiply bluetongue virus types 1 and 8, but remain apparently disease free (Gortazar *et al.*, 2009; East *et al.*, 2011; Ferroglio *et al.*, 2011).

In the urban and peri-urban environments in Europe, there appears to be little direct or confirmed evidence of the presence of deer or other ungulates increasing disease prevalence among humans or domestic animals. Nevertheless, several studies do highlight the potential for conflicts in these areas and there is also evidence that there is a public perception that increasing numbers of deer may be associated with an increased risk of diseases such as Lyme borreliosis (Barbour and Fish, 1993). In a similar way, the survey by Licoppe *et al.* (2013) suggested that the conflicts ranked by urban residents as being the most feared and the second most feared possible future impacts associated with wild boar in urban areas were attacks on people, followed by sanitary or public health problems, although neither impacts were currently considered to be significant (Licoppe *et al.*, 2013; see also Goulding and Roper, 2002; Wilson, 2005).

Lyme disease (*Borellia burgdorferi*) is certainly on the increase in many countries in Europe: Health Protection Scotland indicate that reported cases of Lyme disease in Scotland rose from 37 to 605 over the period from 2000 to 2010. In a study in the United States (where mean incidence of Lyme disease is 4.73 cases per 100,000 population; Maes *et al.*, 1998) the cost of illness associated with the cases of Lyme disease in the United States was estimated at $2.5 billion over 5 years, equivalent to $500 million per year, including both direct medical and indirect costs. Given an estimated 13,750 cases per annum, the cost of illness associated with each case was $36,372, or £24,248 (£1.00 = approximately $1.50). However, doubts are increasingly being expressed regarding the true significance of deer abundance in the ecology of Lyme disease (Ostfeld, 2011; Levi *et al.*, 2012). While deer may be loosely implicated in the maintenance of populations of both disease and vector in the wider environment, both tick and disease organism can persist in the absence of deer (or any other large mammal host), which are in fact a non-competent host for the Lyme spirochaete (Telford *et al.*, 1988; Jaenson and Tällerklint, 1992; Kurtenbach *et al.*, 1998, 2006).

There is also growing concern regarding emerging tick-borne diseases (Raoult *et al.*, 2002), in which wild boar could potentially play an important epidemiological role (Ortuño *et al.*, 2007). For example, *Rickettsia slovaca* is considered to be the etiological agent of tick-borne lymphadenopathy (TIBOLA), an emerging disease transmitted by tick bites. In Catalonia, *Dermacentor marginatus* is the main tick vector and wild boar are the main wild host (Ortuño *et al.*, 2007). Wild boar there are exposed to *R. slovaca* infection and this pathogen is well established in *D. marginatus* (op. cit.). However, Antón *et al.* (2008) found no difference in seroprevalence of *R. slovaca* infection among humans from rural, suburban and urban dwellings. In relation to the risk of tick-borne diseases within more urbanised areas, Dautel and Kahl (1999) found that the most limiting biotic factor for *Ixodes ricinus* in urban environments was the availability of medium-sized and large mammals, as hosts of the adult stage. They conclude that, although the natural risk in urban tick locations is often lower than in non-urban forest, clearly the level of human exposure can be high.

Navarro-Gonzalez *et al.* (2013a) studied the presence of food-borne zoonotic pathogens and antimicrobial resistance in urban wild boar from Collserola Park, Barcelona, Spain, and report the presence of *Salmonella enterica* in 5.0% of boar and *Campylobacter coli* in 4.9%, although neither *E. coli* O157:H7 nor *C. jejuni* were detected in sampled animals. They also found antimicrobial resistance in several indicator bacteria and report the first case of resistance to linezolid[2] in bacteria carried by wildlife. Notably, antimicrobial resistance was considerably more frequent in urban boars than in boar occupying natural habitat (also in north-eastern Spain) where other ungulates such as the Iberian ibex (*Capra pyrenaica*) and free-ranging livestock were present (Navarro-Gonzalez *et al.*, 2013b). Such findings might raise concern for possible public health implications, considering the close contact between humans and boar in some urban areas.

Leptospirosis (*Leptospira* spp.) is a serious infection which is potentially fatal in humans and which registers about 50 cases per year in Germany, mostly related to outdoor activities such as recreation, gardening and so forth (Jansen *et al.*, 2005). It is reported to be increasing in urban areas, and whilst infections are often

---

[2]Linezolid is a synthetic antibiotic used in treating certain serious bacterial infections resistant to other antibiotics.

associated with dogs and rodents (op. cit.), Jansen *et al.* (2007) report the presence of antibodies to leptospires in 18% of 141 wild boar from urban Berlin and conclude that this species represents a potential source of human leptospirosis in urban environments.

In urban areas, wild boar often scavenge food destined for domestic pets, as well as from refuse containers. Indeed this was reported in some areas as the major impact of concern. The proximity of wild boar to cats and dogs may pose a risk of increased prevalence of certain parasitic zoonoses, such as toxoplasmosis or echinococcosis, given that European wild boar are known to be infected by *Toxoplasmosis gondii* (Antolová *et al.*, 2007; Richomme *et al.*, 2010; Opsteegh *et al.*, 2011; Beral *et al.*, 2012) and *Echinococcus granulosus/multilocularis* (Boucher *et al.*, 2005; Martín-Hernando *et al.*, 2008; Onac *et al.*, 2013). Although urban wild boar might also be of concern from an epidemiological point of view in relation to certain viral infections, studies to date have in fact shown lower prevalence rates among urban boar of both hepatitis E (Schielke *et al.*, 2009) and Aujeszky's disease (Pannwitz *et al.*, 2012) in comparison with wild boar from rural populations.

7.2.4 Direct welfare issues

Welfare concerns are significant–and not simply in relation to constraints on management methods which may be considered (for example, see issues raised in relation to urban deer management at sites such as www.deerfriendly.com). The physical condition of deer established within urban sites is often poor by comparison to deer within more natural habitat, and thus urban populations may be of concern purely from the point of view of welfare of the deer themselves (Green, 2008). By contrast, (peri)urban areas can sometimes offer more resources to wild boar than are available in the outlying natural–rural areas, particularly during specific periods or situations of scarcity (e.g. irrigated areas when natural habitat may be drought stressed). Cahill *et al.* (2012) found that habituated female wild boar from urban areas of Collserola Park, Barcelona, were significantly heavier than non-habituated individuals from the surrounding non-urban woodlands, and Podgórski *et al.* (2013) suggest that the urban environment of Cracow, Poland, provides wild boar with more stable, season-independent, food resources, and also a more benign climate during winter (Podgórski *et al.*, 2013). Nevertheless, despite possible benefits of urban areas for some ungulates, it is clear that welfare issues are indeed present: there is an increase in the number of reports received of attacks by domestic dogs on deer, deer trapped in fences, canals, industrial premises etc. Capture and other methods used to control urban populations can themselves generate welfare concerns (see Section 7.3). Poaching activities sometimes leave deer and boar badly wounded, and there is concern regarding the use of bow-hunting on account of incidences of individuals roaming loose impaled with arrows. Even direct cruelty can become an issue. These incidents may also expose any human rescuers to risk of injury. Likewise, people and pets can occasionally be injured by habituated ungulates in urban areas.

7.3 Management approaches

The broad diversity of conflicts encountered with ungulates in urban areas, of public opinions regarding them, and overlapping of, or absence of, administrative responsibilities, can all complicate and even dictate management options for ungulates in urban areas. In many European situations, conflicts with ungulates have arisen quite recently, and neither expertise nor legal frameworks have had time to develop and be applied within effective management strategies. Nonetheless, it is important that effective management strategies be applied as early as possible, especially in the case of wild boar due to their high reproductive rates and their matriarchal group structure, which facilitates the rapid transmission of habituated behaviour to large numbers of offspring. When problems arise with wild boar in urban and peri-urban areas, their incidence can grow quickly: for example in Collserola Park, Barcelona, captures of habituated boars grew from fewer than five cases a year to almost 50 per year over a period of just 5 years (Llimona *et al.*, 2005).

In rural areas, culling with a high-powered rifle is the generally accepted method of controlling numbers of wild ungulates in the context of wider management. Indeed in our own survey of problems associated with deer in urban or peri-urban areas in Europe the majority of respondents suggested that traditional methods of lethal control were adequate to contain numbers of wild ungulates, at least in peri-urban areas. However, discharge of high-velocity weapons poses significant problems of safety in built-up, truly urban environments and is in consequence illegal in many countries. In addition, lethal control of ungulate populations in towns and cities often encounters considerable opposition from the wider public who can be fond of their presence (Section 7.4). Duarte *et al.* (2012) describe these kinds of difficulties for example in relation to the management of ungulates on the Costa del Sol, Spain.

At present, shooting using rifles by conventional hunters is currently the most widely employed method of deliberate control in the peri-urban environment and in peri-urban areas in general. We believe traditional methods of control can continue to be effective – whether through control of impacts *per se* (by fencing, or for example deployment of individual tree guards in amenity plantings or farm woodlands; Putman, 1996, 2004) or by control of populations themselves. It appears to us that the main problem of management in peri-urban contexts is that it commonly tends to be reactive rather than pro-active. Conflicts are therefore often more difficult to resolve without significant expense and multi-agency involvement.

In the fully urban and occasionally in the peri-urban environment methods other than traditional hunting are used at times by a variety of agencies and others such as local authorities, animal welfare organizations and the emergency services, but in a reactive ad-hoc manner. These include:

- trapping for translocation and release elsewhere
- trapping for humane dispatch
- darting for translocation and release
- darting for humane dispatch
- 'sharpshooting' by police and other marksmen.

But it is our opinion that methods currently deployed are insufficient to address the increasing number of incidents involving ungulates within the urban environment. In peri-urban areas in general, while we believe traditional methods of control may be employed, it appears to us that a major problem of such management is that it tends to be reactive rather than pro-active.

Management in truly urban areas is necessarily reactive to local problems but there are significant constraints on accepted forms of control in areas of high human concentration, where there may be significant problems associated with use of, for example, high-powered rifles, and where effectiveness of capture and translocation is unproven. In addition, there is often strong pressure from the general public against any form of control and especially lethal methods of control.

7.3.1 Non-lethal methods

7.3.1.1 Capture and translocation

Relatively little has been published on the effectiveness of live-capture and translocation of deer in Europe. In Collserola Park, Barcelona, the radiotracking of translocated (habituated) wild boar showed this to be an ineffective solution, given that the translocated animals quickly returned to the same sites where they had been causing problems, despite having been released several kilometres away (Cahill, 2004). In Las Rozas, Madrid, trapping with cages has been used to remove wild boar from residential areas: during 2009, a total of 112 wild boar were captured using 18 traps deployed over 15 days (López *et al.*, 2010). Captured animals were submitted to veterinary control and healthy individuals were subsequently translocated to hunting estates, whilst unfit animals were euthanased.

Research on translocation of feral swine in the United States indicated that animals had to be moved to suitable habitat at least 15–20 km away to avoid them returning (Lewis, 1966; Barrett, 1978). It is clearly an appropriate method only where relatively small numbers of non-habituated animals are causing a specific problem in an identifiable local area, and it is less likely to be widely applicable to more generalised problems or ongoing population control. DeNicola and Williams (2008) note that translocation of deer is no longer an option in most States because of high costs, unsuitability of release sites, risks of transmission of disease within deer populations (e.g. Chronic Wasting Disease) and concerns over stress caused to captured deer (on capture and relocation: Ishmael and Rongstad, 1984; O'Bryan and McCullough, 1985; Witham and Jones, 1990; Conover, 2002). Conover, for example, notes that very few relocated deer survive even as long as a year after relocation.

7.3.1.2 Immunocontraception

Immunocontraception as a technique is widely advocated and perhaps favoured as a first recourse by the general public. Two immunocontraceptive vaccines have been developed and tested in the United States to the point of applying for product registration. These are Spay-Vac™, an adjuvanted vaccine based upon porcine zona pellucida glycoprotein (PZP), and GonaCon™, an immunocontraceptive

GnRH vaccine. Both have been proven to be effective in single-dose delivery to deer and wild boar.

The majority of research and field trials have taken place in North America, although some of the species on which trials have been undertaken are species which occur within Europe (red, sika, fallow and muntjac: e.g. Kirkpatrick *et al.*, 1996; Fraker *et al.*, 2002). Limited investigation indicates however that muntjac, one of the species most likely to present problems in urban UK situations, are much less responsive to manipulation of endocrine function and require frequent booster doses of GnRH agonists (Asa and Boutelle, 2007; Penfold *et al.*, 2007).

Fertility control clearly can be effective in some circumstances (e.g. *inter alia:* Kirkpatrick *et al.*, 1997; Rudolph *et al.*, 2000; Curtis *et al.*, 2002; Rutberg *et al.*, 2004; Rutberg and Naugle, 2008a). However, actual population reduction is comparatively slow and dependent on natural mortality, or some additional agent of imposed mortality. Further its effectiveness is likely to be limited to small closed populations where it is practically feasible to catch and treat a high proportion of the female population and where there is limited, or no, immigration (e.g. Seagle and Close, 1996; Rudolph *et al.*, 2000; Merrill *et al.*, 2007). In addition, there can be significant welfare issues associated with the use of these techniques (see Putman, 1997, 2004; Green, 2007). Injection both of PZP and GonaCon™ may be associated with significant lesions and abscessing at the point of injection. Further, female deer treated with PZP immunocontraceptive vaccines continue to cycle and show normal reproductive behaviour thus extending the rut (Fraker *et al.*, 2002). It has been widely speculated that this extension could lead to exhaustion of males and increased DVCs, although there are no empirical data to support such contentions (Rutberg and Naugle, 2008b). For a more detailed exploration of the potential of immunocontraceptive approaches, however, see Chapter 10.

Surgical sterilisation is also sometimes advocated by the general public as a population control technique for ungulates in urban areas. Although it has been used with some success for controlling white-tailed deer in a few cities of North American (see, for example, Matthews *et al.*, 2005), its very high economic costs usually make it unviable for generalised usage (Boulanger *et al.*, 2012).

7.3.2 Lethal methods

7.3.2.1 Capture for culling

In some instances, control of numbers has been attempted through trapping, or darting of animals prior to euthanasia. This approach can in principle be used in specific areas where there is some concern or legal prohibition about discharge of firearms in close proximity to human habitation, and also where animals are habituated to people and thus easy to dart; however, such an approach has been assessed or considered only in a few locations (Jordan *et al.*, 1995; DeNicola *et al.*, 2000). Capture using tranquiliser darts is the main method for removal by wardens of habituated wild boar from urban areas of Collserola Park, Barcelona, Spain, where it has been used on a systematic basis since 2004 (for details see Cahill *et al.*, 2009, 2012).

Other capture methods have included trapping with box traps, drop nets or rocket nets, but as with capture for relocation, costs of currently available methods for these approaches tend to be prohibitively high (DeNicola *et al.*, 2000; Doerr *et al.*, 2001).

7.3.2.2 Conventional hunting with rifles

In the United States, state wildlife agencies have the legal responsibility to manage wildlife so that their abundance and occurrence are compatible with habitat and consistent with public interest (Messmer *et al.*, 1997b). The state determines policy (and population limits) and culling is controlled by issue to hunters of individual licences. Under such a system, licences may be issued up to the quota desired by the appropriate wildlife agency. Thus some control over population levels or overabundance, at least in peri-urban areas, can be achieved simply through the device of increasing the number of licences (or licensees) authorised under conventional hunting systems (see, for example, Decker *et al.*, 2002, 2004; Kilpatrick *et al.*, 2007).

While similar state-licensing systems are operative in *some* parts of Europe, such as France and parts of Switzerland (Putman, 2011) within other countries the legal right to take deer is vested in the landowner or leaseholder of a hunting ground or *revier* and the state has comparatively limited powers of intervention. In effect, therefore, landowners or leaseholders (often Hunters' Associations; Putman 2011) determine policy and practice. In consequence, it is much more difficult to effect any increase in cull levels whether at a landscape or more local level, since this involves voluntary persuasion of individual landowners as well as coordination of the efforts of many different stakeholders over a large area.

Further, as noted, use of high velocity rifles within truly urban areas is unlikely to prove acceptable from the viewpoint of public safety. In the majority of cases, it would appear, control of urban and peri-urban deer populations in the United States is carried out by experienced, highly trained and often professional, riflemen operating in safe zones within truly urban areas or, commonly, within urban or country parks (e.g. Deblinger and Rimmer, 1995; Jones and Witham, 1995; Stradtmann *et al.*, 1995; Doerr *et al.*, 2001; DeNicola and Williams, 2008). In Berlin, control of boar is undertaken by specially trained volunteer hunters (37 in 2012), specifically trained to shoot within a city area. In the UK, (amateur) deer stalkers have also recently published recommendations on best practice measures aimed at the safe shooting of urban deer, and thus try to avoid possible conflicts with local residents (Quarrell, 2012).

7.3.2.3 Bow-hunting

In some US states, although not the majority, it is legal to hunt deer with a bow. In Pennsylvania and Maryland, use of compound bows was legalised specifically to overcome problems associated with the discharge of high velocity rifles in urban areas in the management of urban deer populations (Kilpatrick *et al.*, 2007). Use of bows may be accepted as part of a strategy of general increased hunter licensing (Kilpatrick *et al.*, 2007) or as part of an approach based on controlled hunts within parks and other open areas (e.g. ver Steeg *et al.*, 1995; Kilpatrick and Walter, 1999). The use of bows to kill ungulates is not however currently permitted in the majority of European countries (Putman, 2011).[3] Bow-hunting was authorised during a brief 'trial' period in order to control wild boar in peri-urban areas of Barcelona, Spain, but the measure was subsequently withdrawn within just a few days, mainly as a result of public and media controversy about its use there.

7.3.3 Costs of control

Doerr *et al.* (2001) consider the relative cost and effectiveness of four different population management methods used in Bloomington, Minnesota, to reduce white-tailed deer populations citywide. Deer removal methods evaluated were:

- controlled hunts within large parks and refuges (as above)
- opportunistic sharpshooting by conservation officers on patrol
- sharpshooting over bait by park rangers in a county park
- sharpshooting over bait by police officers on small public land tracts.

The controlled hunt was the only method that generated revenue (fees charged to hunters) but nonetheless cost $117/deer killed to operate (1993 prices). Costs per deer killed using conservation officers and park rangers as sharpshooters were similar: $108 and $121, respectively. The highest cost ($194/deer killed) occurred when police officers were used as sharpshooters but the highest kill rate (0.55 deer/hour) was achieved when 'sharpshooters' shot deer over bait.

For wild boar in Europe, almost three-quarters of sites surveyed by Licoppe *et al.* (2013) report hunting in some form or other as a control method in their peri-urban area. The methods reported, from most to least applied, are: shooting from high seats and/or at baiting sites, either drive hunts or stalking with dogs, and bow-hunting. Designated sharpshooters have also been applied to control wild boar in conflict situations in urban Seraing, Liége, Belgium (Bovy *et al.*, 2013), and 'municipal hunters' are used to remove wild boar from suburban areas of Berlin (Jansen *et al.*, 2007; Arms, 2011).

7.4 Public acceptability of different forms of control

Hesse (2010) provides a comprehensive overview of urban ungulate conflict management for British Colombia and Canada, and notably highlights the importance of human dimensions in this issue and the need to engage communities and the general public at an early stage and in an honest fashion in order to achieve successful resolution of conflicts. Walter *et al.* (2010) provide a recent review on the

---

[3]Bow-hunting is permitted for roe deer in Finland, Denmark and for other species in France and Italy (bow), Portugal (bow, cross-bow, spear). In some cases there may be a restriction on bow power (as, for example, in Finland where the bow must have a minimum power of 180 N).

management of damage by wapiti (*Cervus canadensis*) in North America, and they also stress the need for consideration of public perception by agencies that seek feasible methods of population control. At present in Europe, wild boar probably generate more conflicts in urban areas than cervids do, and studies on public attitudes regarding their management reveal a high degree of public ambiguity regarding possible solutions (Kotulski and König, 2008).

Lethal control of ungulate populations in towns and cities often encounters considerable opposition from the wider public who often actively enjoy seeing these animals in their urban landscape. In their survey on attitudes of Berliners towards wild boar, Kotulski and König (2008) found that 23% of residents objected to their presence, 37% were in favour and 36% had ambivalent opinions. Despite the fact that 44% of people believed the numbers of wild boar should be reduced, 67% of these were against lethal methods of control. Similar results are reported in UK by Chapman *et al.* (1994), Philip and Macmillan (2003), Wilson (2003), Dandy *et al.* (2009) as in the US (e.g. Kellert, 1988; Decker and Richmond, 1995; McAninch, 1995; Messmer *et al.*, 1997a) in relation to management of urban deer populations.

The press and media can also have an impact on management decisions, and in relation to wild boar in England, Goulding and Roper (2002) conclude that management and conservation programmes are likely to encounter adverse reactions from the press, and they advise taking this into account in implementing such programmes. In a survey of public preferences Dandy *et al.* (2011) found that lethal culling had almost no support as a first management option for conflict resolution in relation to the presence of deer in peri-urban areas of Scotland. However, many respondents to their survey supported it as a third preference management measure, and they suggest that culling would be acceptable as an additional management alternative in peri-urban areas when other preferred management methods had been tried and failed. Public perceptions regarding the management of ungulates in (peri-)urban areas will also clearly depend on the level of conflict that actually exists. For example, under situations with low densities of deer populations in peri-urban areas of central Scotland with few or no associated conflicts, local residents considered that no management measures were necessary or justified and, for the moment, deer in these peri-urban environments were considered to be a positive aspect of the local area (Ballantyne, 2012). Many urban residents enjoy seeing wildlife in residential areas (Connelly *et al.*, 1987; Decker and Gavin, 1987; Conover *et al.*, 1995) and thus experience some conflict in understanding the need for control. Managing wildlife populations can thus be especially problematic in urban situations, both because hunting-based management strategies may not be feasible in many urban–suburban areas, but also because of active public opposition to lethal methods of control (Kellert, 1988; Decker and Richmond, 1995; McAninch, 1995; Messmer *et al.*, 1997a).

An apparent increase in public opposition to lethal control may lead managers to perceive that stakeholders are more likely to accept non-lethal than lethal techniques to reduce the damage associated with overabundant deer or other ungulates in urban environments (Curtis *et al.*, 1993, Wright, 1993; Messmer *et al.*, 1997a).

However, as noted, few non-lethal methods are especially effective, as well as being costly, and general consensus among wildlife managers is that lethal methods of control are required to control ungulate populations in urban situations. In practice, recent evidence suggests that as damage levels in urban environments increase, tolerance towards the ungulates responsible declines (Decker and Purdy, 1988; Kilpatrick *et al.*, 1996; Loker *et al.*, 1999; Dandy *et al.*, 2009) and residents are more likely to accept lethal population control techniques (McAninch, 1995; Loker *et al.*, 1999; Kilpatrick and LaBonte, 2003).

Likewise, Siemer *et al.* (2004) found that homeowner acceptance of lethal management strategies for urban deer populations increased if residents were themselves directly affected by damage. In a similar way, Dandy *et al.* (2009) have investigated deer–human interactions in a peri-urban area in Scotland. They established that for members of the local community culling is initially the least favoured management method; however, after discussion it has widespread support at a general level, although this is strongly contingent upon key additional criteria being fulfilled. These were that:

• all other practical management options have been attempted (i.e. culling is a 'last resort')
• there is an existing and problematic overabundance of animals in area in which the cull is to take place,
• and any culling activities are selective, humane and legal.

Further to this, other participants in discussion groups expressed the opinion that the 'natural' alternatives, such as predation, could actually be a worse option than management by humans, particularly in terms of welfare. Indeed welfare arguments are, in fact, commonly deployed by community members in support of selective, 'professional' and humane culling.

Involvement of community groups and local stakeholders in discussions about control of urban ungulates is also essential in order to raise awareness regarding both private and public responsibilities and participation in mitigating the presence of ungulates in urban areas and reducing associated conflicts: for example, adequate fencing of properties, importance of not feeding deer and boar, careful disposal of domestic refuse, modified landscaping techniques and so forth.

7.5 Conclusions

While wild ungulates in urban areas of Europe have received only quite limited scientific study to date, it is clear from our review of both published and unpublished reports, as well as direct consultations with ungulate researchers across a wide range of countries, that issues relating in particular to deer and wild boar populations encroaching onto cities and suburban areas are becoming increasingly common throughout much of central, southern and western Europe. In the United States, the phenomenon of rising peri-urban deer numbers, and the conflicts and differing challenges these entail in comparison with management of more rural ungulate populations, have been recognised and studied since the mid-1980s (e.g. Ishmael and Rongstad, 1984; Connelly *et al.*, 1987; Decker and Gavin, 1987; Kellert, 1988).

Since then, it would appear that in the United States most problems (or conflicts) with 'urban' deer have remained mainly associated with suburban, or peri-urban areas, although as in parts of Europe, deer have also become established within a number of truly urban parks (see for example papers in McAninch, 1995; Baker and Fritsch, 1997; Decker *et al.*, 2004; Kilpatrick and LaBonte, 2007 for overview). Primary concerns relate to increased risk of road-traffic accidents (collisions with vehicles) and risk of transmission of disease – especially concerns that large peri-urban deer populations may maintain an increased population of ticks and therefore be implicated in the epidemiology of tick-borne diseases such as Lyme disease, babesiosis and ehrlichiosis (Decker and Gavin, 1987; Stout *et al.*, 1993; Conover, 1995; Kilpatrick *et al.*, 1996, Kilpatrick and LaBonte, 2007). There are also concerns about transmission of TSEs (Transmissible Spongiform Encephalopathies) to pets or domestic livestock, since deer in a number of areas in the US are now infected with Chronic Wasting Disease (CWD) (Williams *et al.*, 2001; Doherr, 2007; Sigurdson and Aguzzi, 2007).

In most cases, management effort involves control or reduction of local deer population abundance. There is a growing literature on the relative efficacies of different lethal and non-lethal approaches to population control (e.g. Jordan *et al.*, 1995; Stradtmann *et al.*, 1995; ver Steeg *et al.*, 1995; DeNicola *et al.*, 1997, 2000; Doerr *et al.*, 2001) and the public reaction to these different styles of management (e.g. Stout *et al.*, 1997; Messmer *et al.*, 1997a; West and Parkhurst, 2002; Kilpatrick *et al.*, 2007). Whilst there is continuing debate about the potential of non-lethal techniques, such as capture and translocation, or immunocontraception, in the vast majority of instances, only lethal methods (increase in conventional hunting pressure, sharp-shooting by rifle or bow, or capture for euthanasia) would appear to be generally effective (e.g. DeNicola and Williams, 2008), although local successes with immunocontraception have been reported (Rutberg and Naugle, 2008b).

Because of greater public 'involvement' with wildlife in the urban area, and because so many urban residents identify with deer or other wild animals which colonise urban and peri-urban areas, some measure of community involvement is strongly advocated by many authors in order to confer some measure of involvement and ownership of the decision-making process (Messmer *et al.*, 1997a; Siemer *et al.*, 2000, 2004; Decker *et al.*, 2002, 2004; Raik *et al.*, 2006; Kilpatrick and LaBonte, 2007; Decker *et al.* 2002, 2004; Hesse, 2010).

7.5.1 Options for control of ungulate populations in urban and peri-urban areas in Europe

This review concentrates upon actual and potential methods of dealing with wild ungulates when they are causing problems within residential and commercial areas. There may be ethical considerations in terms of required management since it was human activity that reduced the area of habitat formerly available to them and/or created the new urban landscapes that ungulates have now colonised. Although there may therefore be some need for proportional reduction in numbers to compensate for loss of habitat, some people may believe that any lethal control is unacceptable (Chapman *et al.*, 1994; Philip and Macmillan, 2003; Wilson, 2003).

In both urban and peri-urban contexts we are aware that there are a number of potential methods which might be acceptable elsewhere in the world that would not be acceptable within a European context–in particular, use of toxins and poisons. Therefore, in this review we have deliberately restricted consideration to methods we believe are, or could be, generally acceptable. Management techniques often proposed such as diversionary feeding, fencing, scarers, road design, landscaping and so forth are not addressed, though we acknowledge that some of these may, in particular circumstances, have some role to play in manipulating behaviour and potentially reducing the conflict with human activity; such methods however are generally most useful where employed in combination with methods aimed at controlling ungulate numbers within agreed limits and in mitigation of any impacts which then remain.

Our review leads us to conclude that in any approach to the management of ungulates within the urban and peri-urban environment it would not be feasible to seek eradication of ungulate populations or totally prevent colonisation. Options available to address potential problems in urban areas should start with precautionary measures in the surrounding peri-urban area. If this is not done, there is a real risk that if management is only reactive (i.e. when negative impacts are high), then it is already too late and much more stringent reactive management will be required. Such prophylactic action using conventional hunting techniques would seek to take action to minimise colonisation of urban areas by deer or other ungulates in the first place (or reduce additional recruitment to existing populations). This could be done by control operations in the urban fringe or peri-urban areas to create a zone of reduced ungulate density which would in turn reduce the number of available colonists. The prophylactic element of this approach requires ongoing management, at an appropriate scale. This element of any management policy should be continuous and integrated into routine management of ungulate populations in the wider landscape. We believe that this can only be effectively achieved as part of a formal, planned, management strategy agreed in advance by all the relevant stakeholders.

Even with prophylactic action, it is likely that additional management effort would be needed to respond to individual local issues as and when these occur. The reactive element of any strategy will, by necessity, be responsive to particular individual problems. However, approaches and methods to be used should have been previously debated and appropriate methods agreed in advance, to facilitate a rapid response in case of need. Evidence from the United States (Messmer *et al.*, 1997a; Siemer *et al.*, 2000, 2004; Decker *et al.*, 2002, 2004; Raik *et al.*, 2006; Kilpatrick and LaBonte, 2007) and Scotland (Dandy *et al.*, 2009) highlights the importance of community involvement in the decision-making process, even if not in the actual control to be carried out.

No other methods appear to offer significant advantages over the current suite of options in use in Europe or the United States. The only adaptation is that of 'sharpshooting' which modifies conventional hunting techniques. These adaptations should include use of modern night vision equipment and incorporate advances in ammunition technology to address additional public safety issues pertaining in peri-urban and urban areas. Immunocontraception may still offer possible options for the future on a local scale, but it is unlikely to be available in the short term for control of ungulates in most European countries, and cost may be prohibitive for wide-scale controls. Further it is clear that 'contraception is not a substitute for hunting' as a means of reducing populations (Fagerstone *et al.*, 2008). Trapping, although well developed in many countries for scientific purposes, has become largely ignored in the UK, but is used elsewhere to good effect. Trapping advice and training need to be updated and could then be used for a wider range of applications subject to legislative changes.

In the United States and Europe the most effective control method appears to be the use of accurate 'sharpshooters', with cull rates improved further by using baits to attract the target animals (see, for example, Deblinger and Rimmer, 1995; Jones and Witham, 1995; Stradtmann *et al.*, 1995; Doerr *et al.*, 2001; DeNicola and Williams, 2008). There is no doubt that the use of firearms and ammunition necessary to kill large ungulates humanely is problematic in built-up areas, but recent developments may alter the margin of safety through the use of sound moderators and frangible ammunition.[4] In relation to this, guidelines have recently been proposed regarding the safe shooting of deer in urban areas of the UK (Quarrell, 2012).

We believe there are a number of areas that require further development if the methods above are to be used in a coherent manner to ensure that the highest animal welfare and human safety requirements are achieved and any activity receives the necessary public acceptance.

We would suggest further work is required in the following areas:

- Coordination and standardisation of approaches, including identification of relevant skills in organisations and individuals to increase the capacity to address the growing issues posed by habituation of ungulates to the urban environment
- Developing humane live-capture techniques for individuals and groups of animals and adapting current legislation and Best Practice to enable the use of these techniques
- Examining developments in firearms and ammunition to establish suitable criteria to improve 'sharpshooting' capability, and adapting current legislation and Best Practice to enable the use of these techniques
- Ensuring local community involvement in the decision-making process, even if not in the actual control to be carried out, is likely to be required to reduce public anxiety if trapping or culling is to be undertaken in an urban environment.

---

[4]Frangible, or 'soft,' rounds are designed to break apart when they hit walls or other hard surfaces to prevent ricochets. They were originally designed for close-quarter combat or law-enforcement applications.

Acknowledgements

In presenting this chapter we would like to thank the many wildlife biologists and researchers throughout Europe who responded to our requests for information especially Professor Marco Apollonio (Italy), Drs Miroslava Barančeková and Jarmila Prokešová (Czech Republic), Dr Jörg Beckmann (Germany), Professor Juan Carranza (Spain), Dr Jim Casaer (Belgium), Dr Stefano Focardi (Italy), Dr Carlos Fonseca (Portugal), Dr Mark Hewison (France), Dr Petter Kjellander (Sweden), Drs Ilpo Kojola and Vesa Ruusila (Finland), Dr Erik Lund (Norway), Professor Andras Nahlik (Hungary), Dr Bostjan Pokorny (Slovenia), Professor Fritz Reimoser (Austria), Drs Sip van Wieren and Geert Groot Bruinderink (Netherlands).

We are also extremely appreciative of the help we received from colleagues and contacts in the United States in explaining the situation there to us. In many cases, even though we were not previously acquainted, people were impressively generous with the time and effort they offered to assist us. We would particularly like to acknowledge the tremendous help received from Professor Dan Decker, Dr Tony DeNicola, Professor Howard Kilpatrick, Professor Terry Messmer, Dr Mark Fraker, Dr Lowell Miller, Dr Marty Vavra and Dr Bruce Johnson. Of colleagues in the UK, we would particularly like to thank Dr David Cowan, Dr Giovanna Massei and Dr Piran White.

We also wish to sincerely thank Dr Alain Licoppe, Dr Céline Prévot, Marie Heymens, Céline Bovy and Dr Jim Casaer from Belgium for facilitating access to unpublished aspects from their international survey on wild boar/feral swine in (peri-)urban areas (Licoppe *et al.*, 2013). Our thanks also to colleagues Francesc Llimona, Lluís Cabañeros, Francisco Calomardo, Francisco Javier García, Àngel Mateo, Jordi Piera and Carles Sobrino from Collserola Park (Barcelona).

This chapter is based on a review initially commissioned by the Department of Food, Environment and Rural Affairs (UK). We thank them for permission to re-draft it as a chapter here, but should make it clear that opinions expressed or conclusions drawn in this paper are those of the authors and do not represent any Government view or position.

References

Amici, A., Serrani, F., Rossi, C.M. and Primi, R. (2012) Increase in crop damage caused by wild boar (*Sus scrofa* L.): the 'refuge effect'. *Agronomy for Sustainable Development* **32**, 683–692.

Antolová, D., Miterpáková, M., Reiterová, K. and Dubinský, P. (2006) Influence of anthelmintic baits on the occurrence of causative agents of helminthozoonoses in red foxes (*Vulpes vulpes*). *Helminthologia* **43**, 226–231.

Antón, E., Nogueras, M.M., Pons, I., Font, B., Muñoz, T., Sanfeliu, I. and Segura, F. (2008) *Rickettsia slovaca* infection in humans in the Northeast of Spain: Seroprevalence study. *Vector-Borne and Zoonotic Diseases* **8**, 689–694.

Apollonio, M., Andersen, R. and Putman, R.J. (eds; 2010) *European Ungulates and their Management in the 21st Century.* Cambridge: Cambridge University Press.

Arms, S. A. (2011) Berlin plans hunt of wild boar invaders. *The Guardian,* November 30. Available at: http://www.theguardian.com/environment/2011/nov/30/berlin-hunt-wild-boar.

Arqués, J., Belda, A., Martínez, J.E., Peiró, V., Jiménez, D. and Seva, E. (2009) Análisis de encuestas como herramienta de gestión sostenible de especies cinegéticas en agrosistemas del este de la provincia de Alicante (Marina Baja): Estudio del caso del jabalí *Sus scrofa* Linnaeus, 1758. *Galemys* **21**, 51–62.

Asa, C.S. and Boutelle, S. (2007) The AZA Wildlife Contraception Centre collaborative trial of the GnRH agonist Suprelorin® in North America. *Proceedings of the 6th International Symposium on Fertility Control in Wildlife,* York, UK.

Baker, S.V. and Fritsch, J.A. (1997). New territory for deer management: human conflicts on the suburban frontier. *Wildlife Society Bulletin* **25**, 404–407.

Ballantyne, S. (2012) Urban biodiversity: successes and challenges: human perceptions towards peri-urban deer in Central Scotland. *The Glasgow Naturalist* **25**, 1–3.

Barbour, A.G. and Fish, D. (1993) The biological and social phenomenon of Lyme disease. *Science* **260**, 1610–1616.

Barrett, R.H. (1978) The feral hog on the Dye Creek Ranch, California. *Hilgardia* **46** (9), 283–355.

Beckworth, S.L. and Stith, L.G. (1968) Deer damage to citrus groves in South Florida. *Proceedings of the Annual Congress of the Southeastern Association of Game and Fish Commissioners* **21**, 32–38.

Beral, M., Rossi, S., Aubert, D., Gasqui, P., Terrier, M.-E., Klein, F., Villena, I., Abrial, D., Gilot-Fromont, E., Richomme, C., Hars, J. and Jourdain, E. (2012) Environmental factors associated with the seroprevalence of *Toxoplasma gondii* in wild boars (*Sus scrofa*), France. *EcoHealth* **9**, 303–309.

Bobek, B., Frąckowiak, W., Furtek, J., Merta, D. and Orłowska, L. (2011) Wild boar population at the Vistula Spit: management of the species in forested and urban areas. *Julius-Kühn-Archiv* **432**, 226–227.

Boucher, J.M., Hanosset, R., Augot, D., Bart, J.M., Morand, M., Piarroux, R., Pozet–Bouhier, F., Losson, B. and Cliquet, F. (2005) Detection of *Echinococcus multilocularis* in wild boars in France using PCR techniques against larval form. *Veterinary Parasitology* **129**, 259–266.

Boulanger, J.R., Curtis, P.D., Cooch, E.G. and DeNicola, A.J. (2012) Sterilization as an alternative deer control technique: a review. *Human–Wildlife Interactions* **6**, 273–282.

Bovy, C., Libois, R. and Licoppe, A. (2013) Urban wild boar management: a resource selection analysis based on eradication data. *31st Congress of the International Union of Game Biologists,* Brussels, Belgium. Abstracts, p. 248.

Broom, D.M., Sena, H. and Moynihan, K.L. (2009). Pigs learn what a mirror image represents and use it to obtain information. *Animal Behaviour* **78**, 1037–1041.

Cahill, S. (2004) *Estudi de la població de senglar* (Sus scrofa) *al Parc de Collserola. Memòria de treballs de 2002–2003.* Unpublished report, Estació Biològica de Can Balasc, Consorci del Parc de Collserola, Barcelona.

Cahill, S., Llimona, F., Cabañeros, L. and Calomardo, F. (2009) Habituation of wild boar (*Sus scrofa*) in a metropolitan area. Characterisation, conflicts and solutions in Collserola Park, Barcelona. In A. Náhlik and T. Tari (eds), *Proceedings of the 7th International Symposium on Wild Boar (Sus Scrofa) and on Sub-order Suiformes.* Sopron, Hungary: Institute of Wildlife Management and Vertebrate Zoology, University of West Hungary, pp. 25–27.

Cahill, S., Llimona, F., Cabañeros, L. and Calomardo, F. (2012) Characteristics of wild boar (*Sus scrofa*) habituation to urban areas in the Collserola natural park (Barcelona) and comparison with other locations. *Animal Biodiversity and Conservation* **35**(2), 221–233.

Chapman, N.G. (1991) Chinese muntjac (*Muntiacus reevesi*). In G.B. Corbet and S. Harris (eds), *The Handbook of British Mammals.* Oxford: Blackwell, Oxford, pp. 526–532.

Chapman, N.G., Harris, A. and Harris, S. (1994) What gardeners say about muntjac. *Deer* **9**, 302–306.

Coles, C.L. (1997) *Gardens and Deer: A Guide to Damage Limitation.* Shrewsbury, UK: Swan Hill Press.

Connelly, N.A., Decker, D.J. and Wear, S. (1987) Public tolerance of deer in a suburban environment. *Proceedings of the Eastern Wildlife Damage Control Conference* **3**, 207–218.

Conover, M.R. (1989) Relationships between characteristics of nurseries and deer browsing. *Wildlife Society Bulletin* **17**, 414–18.

Conover, M.R. (1995) What is the urban deer problem and where did it come from? In J.B. McAninch (ed.), *Urban Deer: A Manageable Resource?* Proceedings of a Symposium of the 55th Midwest Fish and Wildlife Conference, The Wildlife Society, Bethesda, USA, pp. 11–18.

Conover, M.R., Pitt, W.C., Kessler, K.K., Dubow, T.J. and Sanborn, W.A. (1995) Review of human injuries, illnesses, and economic losses caused by wildlife in the United States. *Wildlife Society Bulletin* **23**, 407–414.

Conover, M.R. (2002) *Resolving Human–Wildlife Conflicts: The science of wildlife damage management.* Boca Raton, FL: Lewis.

Curtis, P.D., Knuth, B.A., Myers, L.A. and Rockwell, T.M. (1993) Selecting deer management options in a suburban environment: a case study from Rochester, New York. *Transactions North American Wildlife and Natural Resources Conference* **58**, 102–116.

Curtis, P.D., Pooler, R.L., Richmond, M.E., Miller, L.A., Mattfield, G.F., Quimby, F.W. (2002) Comparative effects of GnRH and porcine zona pellucida (PZP) immunocontraceptive vaccines for controlling reproduction in white-tailed deer (*Odocoileus virginianus*). *Journal of Reproduction and Fertility, Supplement* **60**, 131–141.

Dandy, N., Ballantyne, S., Moseley, D., Gill, R. and Quine, C. (2009) *The management of roe deer in peri-urban Scotland.* Forest Research Report, Forestry Commission, UK.

Dandy, N., Ballantyne, S., Moseley, D., Gill, R., Peace A. and Quine, C. (2011) Preferences for wildlife management methods among the peri-urban public in Scotland. *European Journal of Wildlife Research* **57**, 1213–1221.

Dautel, H. and Kahl, O. (1999) Ticks (Acari: Ixodoidea) and their medical importance in the urban environment. In W.H. Robinson, F. Rettich and G.W. Rambo (eds), *Proceedings of the 3rd International Conference on Urban Pests.* Prague, Czech Republic, pp. 73–82. Available at: http://www.icup.org.uk/reports/ICUP481.pdf.

Deblinger, R.D. and Rimmer, D.W. (1995) Development and implementation of a deer management program in Ipswich, Massachusetts. In J.B. McAninch (ed.), *Urban Deer: A Manageable Resource?* Proceedings of a Symposium of the 55th Midwest Fish and Wildlife Conference, The Wildlife Society, Bethesda, USA, pp. 75–79.

Decker, D.J. and Gavin, T.A. (1987) Public attitudes toward a suburban deer herd. *Wildlife Society Bulletin* **15**, 173–180.

Decker, D.J. and Purdy, K.G. (1988) Toward a concept of wildlife acceptance capacity in wildlife management. *Wildlife Society Bulletin* **16**, 53–57.

Decker, D.J. and Richmond, M.E. (1995). Managing people in an urban deer environment: the human dimensions challenges for managers. In J.B. McAninch (ed.), *Urban Deer: A Manageable Resource?* Proceedings of a Symposium of the 55th Midwest Fish and Wildlife Conference, The Wildlife Society, Bethesda, USA, pp. 3–10.

Decker, D.J., Lauber, T.B. and Siemer, W.F. (2002) *Human–Wildlife Conflict Management: A Practitioner's Guide.* Ithaca, NY: Northeast Wildlife Damage Management Research and Outreach Cooperative.

Decker, D.J., Raik, D.B. and Siemer, W.F. (2004) *Community-Based Suburban Deer Management: A Practitioner's Guide.* Ithaca, NY: Northeast Wildlife Damage Management Research and Outreach Cooperative.

Delahay, R.J., Smith, G.C., Barlow, A.M., Walker, N., Harris, A., Clifton-Hadley, R.S. and Cheeseman, C.L. (2007) Bovine tuberculosis infection in wild mammals in the south-west region of England: a survey of prevalence and a semi-quantitative assessment of the relative risks to cattle. *The Veterinary Journal* **173**, 287–301.

DeNicola, A.J., Weber, S.J., Bridges, C.A. and Stokes, J.L. (1997) Nontraditional techniques for management of overabundant deer populations. *Wildlife Society Bulletin* **25**, 496–499.

DeNicola, A.J., VerCauteren, K.C., Curtis, P.D. and Hyngstrom, S.E. (2000) *Managing White-tailed Deer in Suburban Environments.* Ithaca, NY: Cornell Cooperative Extension.

DeNicola, A.J. and Williams, S.C. (2008) Sharpshooting suburban white-tailed deer reduces deer–vehicle collisions. *Human–Wildlife Conflicts* **2**, 28–33.

Doerr, M.L., McAninch, J.B. and Wiggers, E.P. (2001) Comparison of four methods to reduce white-tailed deer abundance in an urban community. *Wildlife Society Bulletin* **29**, 1105–1113.

Doherr, M.G. (2007) Brief review of the epidemiology of transmissible spongiform encephalopathies. *Vaccine* **25**, 5619–5624.

Duarte, J., Farfán, M.Á. and Vargas. J.M. (2012) Ungulados en las nuevas zonas urbanas de la Costa del Sol (Málaga). III Reunión sobre Ungulados Silvestres Ibéricos. *Ungulados, biodiversidad y actividades humanas: gestión de conflictos.* Girona, SpaIn Castelló d'Empúries, p. 17. Available at: http://www.castello.cat/rusi3/documents/Ponencias.pdf.

East, M.L., Bassano, B. and Ytrehus, B. (2011) The role of pathogens in the population dynamics of European ungulates. In R.J. Putman, M. Apollonio and R. Andersen (eds), *Ungulate Management in Europe: Problems and Practices.* Cambridge, UK: Cambridge University Press, pp. 319–348.

El 9 Nou. (2012) L'Ametlla programa batudes puntuals per matar senglars pels danys a horts i jardins. El 9 Nou.cat. L'Ametlla del Vallès, Catalonia, Spain. Available at: http://www.el9nou.cat/noticies_v_0_/22754/l%C3%82%E2%80%99ametlla_programa_batudes_puntuals_per_matar_senglars_pels.

ELO (European Landowners' Organisation) and Polytechnic University of Madrid (2012) *L'explosion démographique du sanglier en Europe Enjeux et Défis.* European Landowners Organisation, Brussels. Available at: http://www.europeanlandowners.org/files/pdf/2012/Etude%20explosion%20demogaphique%20sanglier%20ELO%2002%2007%202012%20FINAL-2.pdf.

Fagerstone, K.A., Miller, L.A., Eismann, G.S., O'Hare, J.R. and Gionfriddo, J.P. (2008) Registration of wildlife contraceptives in the USA with OVoControl and GonaCon immunocontraceptive vaccines as examples. *Wildlife Research* **35**, 586–592.

Ferroglio, E., Gortazar, C. and Vicente, J. (2011) Wild ungulate diseases and the risk for livestock and public health. In R.J. Putman, M. Apollonio and R. Andersen (eds), *Ungulate Management in Europe: Problems and Practices.* Cambridge, UK: Cambridge University Press, pp. 192–214.

Fraker, M.A., Brown, R.G., Gaunt, G.E., Kerr, J.A. and Pohajdak, W. (2002) Long-lasting single dose immunocontraception of feral fallow deer in British Columbia. *Journal of Wildlife Management* **66**, 1141–1147.

Franco, G. and Mathieu, V. (2012) Surpopulation de sangliers en Europe. Parliamentary Question E-009973-12 for the European Commission. Available at: http://www.europarl.europa.eu/sides/getDoc.do?pubRef=-//EP//TEXT+WQ+E-2012-009973+0+DOC+XML+V0//FR.

Froliche, K., Thiede, T., Kozikowski, T. and Jakob, W. (2002) A review of mutual transmission of important infectious diseases between livestock and wildlife in Europe. *Annals of the New York Academy of Sciences* **969**, 4–13.

Georgii, B., Dinter, U. and Meierjürgen, U. (1991) Wild boar (*Sus scrofa* L.) in an urban forest. *Abstracts of the XXth Congress of the International Union of Game Biologists,* University of Agricultural Sciences, Gödöllõ, Hungary, pp. 31.

Gortazar, C., Lopez-Olvera, J., Paz Martin-Hernando, M. and Falconi, C. (2009) *The response of red deer to bluetongue infection.* A report to the Scottish Executive, Edinburgh, UK.

Goulding, M.J. and Roper, T.J. (2002) Press responses to the presence of free-living wild boar (*Sus scrofa*) in southern England. *Mammal Review* **32**, 272–282.

Green, P. (2007) *Can contraception control deer populations in the UK?* Report for the Deer Initiative, Wrexham UK. Available at www.thedeerinitiative.co.uk.

Green, P. (2008) Disease, health and other animal welfare issues arising: presentation at the British Deer Society Urban Deer Seminar, 21 October 2008. Available at www.thedeer initiative.co.uk.

Groot Bruinderink, G.W.T.A. and Hazebroek, E. (1996) Ungulate traffic collisions in Europe. *Conservation Biology* **10**, 1059–67.

Harder, J.D. (1970) Evaluating winter deer use of orchards in Western Colorado. *Transactions of the North American Wildlife and Natural Resources Conference* **35**, 35–47.

Harris, S. and Rayner, J.M.V. (1986) Urban fox (*Vulpes vulpes*) populations estimates and habitat requirements in several British cities. *Journal of Animal Ecology* **55**, 575–591.

Hesse, G. (2010). *British Columbia Urban Ungulate Conflict Analysis.* British Columbia, Canada: Ministry of the Environment. Available at: http://www.env.gov.bc.ca/cos/info/ wildlife_human_interaction/UrbanUngulatesConflictAnalysisFINALJuly5-2010.pdf.

Ishmael, W.E. and Rongstad, O.J. (1984) Economics of an urban deer-removal program. *Wildlife Society Bulletin* **12**, 394–398.

Jaenson, T.G.T. and Tällerklint, L. (1992) Incompetence of roe deer as reservoirs of the Lyme borreliosis spirochete. *Journal of Medical Entomology* **29**, 813–817.

Jansen, A., Schöneberg, I., Frank, C., Alpers, K., Schneider, T. and Stark, K. (2005) Leptospirosis in Germany, 1962–2003. *Emerging Infectious Diseases* **11**, 1048–1054.

Jansen, A., Luge, E., Guerra, B., Wittschen, P., Gruber, A., Loddenkemper, C., Schneider, T., Lierz, M., Ehlert, D., Appel, B., Stark, K. and Nöckler, K. (2007) Leptospirosis in urban wild boars, Berlin, Germany. *Emerging Infectious Diseases* **13**(5), 739–742.

Jones, J.M. and Witham, J.H. (1995) Urban deer problem-solving in northeast Illinois: an overview. In J.B. McAninch (ed.). *Urban Deer: A Manageable Resource?* Proceedings of a Symposium of the 55th Midwest Fish and Wildlife Conference, The Wildlife Society, Bethesda, USA, pp. 58–65.

Jordan, P.A., Moen, R.A., Degayner, E.J. and Pitt, W.C. (1995) Trap-and-shoot and sharpshooting methods for control of urban deer: The case history of North Oaks, Minnesota. In J.B. McAninch (ed.), *Urban Deer: A Manageable Resource?* Proceedings of a Symposium of the 55th Midwest Fish and Wildlife Conference, The Wildlife Society, Bethesda, USA, pp. 97–104.

Kellert, S.J. (1988) Public view of deer management. *Proceedings of a Symposium on Deer Management in an Urbanising Region: Problems and Alternatives to Traditional Management.* Washington DC: The Humane Society of the United States, pp. 8–11.

Kilpatrick, H.J. and LaBonte, A.M. (2003) Deer hunting in a residential community: the community's perspective. *Wildlife Society Bulletin* **31**, 340–348.

Kilpatrick, H.J. and LaBonte, A.M. (2007) *Managing Urban Deer in Connecticut: A Guide for Residents and Communities.* Hartford, CT: Connecticut Department of Environmental Protection, Wildlife Division.

Kilpatrick, H.J. and Walter, W.D. (1999) A controlled archery deer hunt in a residential community: cost, effectiveness and recovery rates. *Wildlife Society Bulletin* **27**, 115–123.

Kilpatrick, H.J., Eccleston, K.A. and Ellingwood, M.R. (1996) Attitudes and perceptions of a suburban community experiencing deer/human conflicts. *Transactions of the Northeast Fish Wildlife Conference* **52**, 19.

Kirkpatrick, J.F.,Turner, J.W., Liu, I.K.M. and Fayrer-Hosken, R. (1996) Applications of pig *zona pellucida* immunocontraception to wildlife fertility control. *Journal of Reproduction and Fertility, Supplement* **50**, 183–89.

Kirkpatrick, J.F., Turner, J.W., Liu, I.K.M., Fayrer-Hosken, R. and Rutberg, A.T. (1997) Case studies in wildlife immunocontraception: wild and feral equids and white-tailed deer. *Reproduction, Fertility and Development* **9**, 105–110.

Kilpatrick, H.J., LaBonte, A.M. and Barclay, J.S. (2007) Acceptance of deer management strategies by suburban homeowners and bowhunters. *Journal of Wildlife Management* **71**, 2095–2101.

König, A. (2008) Fears, attitudes and opinions of suburban residents with regards to their urban foxes.A case study in the community of Grünwald: a suburb of Munich. *European Journal of Wildlife Research* **54**, 101–109.

Kotulski, Y. and König, A. (2008) Conflicts, crises and challenges: wild boar in the Berlin City– a social, empirical and statistical survey. *Natura Croatica* **17**(4), 233–246.

Kurtenbach, K., Sewell, H.S., Ogden, N.H., Randolph, S.E. and Nuttal, P.A. (1998) Serum complement sensitivity as a Key factor in Lyme disease ecology. *Infections and Immunology* **66**, 1248–1251

Kurtenbach, K., Haninková, K., Tsao, J.I., Margos, G., Fish, D. and Ogden, N.H. (2006) Fundamental process in the evolutionary ecology of *Lyme borreliosis*. *National Review of Microbiology* **4**, 660–669.

Lagos, L., Picos, J. and Valero, E. (2012) Temporal pattern of wild ungulate-related traffic accidents in northwest Spain. *European Journal of Wildlife Research* **58**, 661–668.

Langbein, J. (2008) Deer vehicle collisions in peri-urban areas: a risky life for deer. Presentation to British Deer Society Urban Deer Conference at Linnean Society, London. Available at: http://www.bds.org.uk/c2/uploads/langbeinurbandeervc1.pdf.

Langbein, J., Putman, R.J. and Pokorny, B. (2011) Road traffic accidents involving ungulates and available measures for mitigation. In R.J. Putman, M. Apollonio and R. Andersen (eds), *Ungulate Management in Europe: Problems and Practices.* Cambridge: Cambridge University Press, pp. 215–259.

Levi, T., Kilpatrick, A.M., Mangel, M. and Wilmers, C.C. (2012) Deer, predators, and the emergence of Lyme disease. *Proceedings of the National Academy of Sciences* **109**, 10942–10947.

Lewis, J.C. (1966) Observations of pen-reared European hogs released for stocking. *Journal of Wildlife Management* **30**, 832.

Licoppe, A., Prévot, C., Heymans, M., Bovy, C., Casaer, J. and Cahill, S. (2013) *Wild boar/ feral pigs in (peri-)urban areas.* International survey report, Department for Natural and Agricultural Environment Study, Gembloux, Haute Ecole Lucia de Brouckère, Université de Liège, Research Institute for Nature and Forest, Geraardsbergen (INBO), Consorci del Parc Natural de la Serra de Collserola and Barcelona, Brussels, Belgium. Available at: http://www.iugb2013.org/docs/Urban%20wild%20boar%20survey.pdf.

Llimona, F., Cahill, S., Tenés, A. and Cabañeros, L. (2005) Relationships between wildlife and periurban Mediterranean areas in the Barcelona metropolitan region (Spain). *Extended Abstracts of the 27th Congress of the International Union of Game Biologists,* 143–144.

Loker, C.A., Decker, D.J. and Schwager, S.J. (1999) Social acceptability of wildlife management actions in suburban areas: three case studies from New York. *Wildlife Society Bulletin* **27**, 152–159.

López, R., López, J., Gavela, J., Bosch, J. and Ballesteros, C. (2010) Wild boar capture methodology (*Sus scrofa,* Linnaeus 1758) in a suburban area: the case of Las Rozas de Madrid (central Spain). *Abstracts of the 8th International Symposium on Wild Boar and other Suids,* 62. Available at: https://secure.fera.defra.gov.uk/wildboar2010/documents/book OfAbstractsWildBoar-Nov10.pdf.

Maes, E., Lecomte, P. and Ray, N. (1998) A cost-of-illness study of Lyme disease in the United States. *Clinical Therapeutics* **20**, 993–1008.

Manrique, J. (2012) Los jabalíes dañan los cultivos de las huertas de la Mejana de Tudela. *Diario de Navarra,* **32**.

Martín-Hernando, M. P., González, L. M., Ruiz–Fons, F., Garate, T. and Gortazar, C. (2008) Massive presence of *Echinococcus granulosus* (Cestoda, Taeniidae) cysts in a wild boar (*Sus scrofa*) from Spain. *Parasitology Research* **103**, 705–707.

Massei, G. and Genov, P.V. (2004) The environmental impact of wild boar. *Galemys* **16**, 135–145.

Matschke, G.K, de Calesta, D.S. and Harder, J.D. (1984) Crop damage and control. In L.K. Halls (ed.), *White-tailed Deer: Ecology and Management.* Mechanicsburg, PA: Stackpole Books.

Matthews, N., Paul-Murphy, J. and Frank, E. (2005) *Evaluation of a trap-sterilize-release program for white-tailed deer management in Highland Park, Illinois.* Report prepared for the Highland Park City Council.

McAninch, J.B. (ed. 1995) *Urban deer: a manageable resource?* Proceedings of a Symposium of the 55th Midwest Fish and Wildlife Conference, North Central Section, 12–14 December 1993, St. Louis, MI, USA.The Wildlife Society.

McCarthy, A.J. and Rotherham, I.D. (1994) Deer in the Sheffield region including the Eastern Peak District. *The Naturalist* **119**, 103–110.

McCarthy, A.J., Baker, A. and Rotherham, I.D. (1996) Urban-fringe deer management issues: a South Yorkshire case study. *British Wildlife* **8**, 12–19.

McCarthy, A.J., Howes, C.A. and Baker, A. (1999) The Sheffield and Doncaster Deer Surveys 1990–1996. In M. Jones, I.D. Rotherham and A.J. McCarthy (eds), Deer or the New Woodlands? *Journal of Practical Ecology and Conservation,* special edition.

Meiner, A., Georgi, B., Petersen, J.-E. and Uhel, R. (2010) *The European Environment. State and Outlook 2010: Land Use.* Copenhagen, Denmark: European Environment Agency. Available at: http://www.eea.europa.eu/soer/europe/land-use.

Meng, X.J., Lindsay, D.S. and Sriranganathan, N. (2009) Wild boars as sources for infectious diseases in livestock and humans. *Philosophical Transactions of the Royal Society B* **364**, 2697–2707.

Merrill, J.A., Cooch, E.G. and Curtis, P.D. (2007) Managing an over-abundant deer population by sterilization: effects of immigration, stochasticity and the capture process. *Journal of Wildlife Management* **70**, 268–277.

Messmer, T.A., Cornicelli, L., Decker, D.J. and Hewitt, D.G. (1997a) Stakeholder acceptance of urban deer management techniques. *Wildlife Society Bulletin* **25**, 360–366.

Messmer, T.A., George, S.M. and Cornicelli, L. (1997b) Legal considerations regarding lethal and nonlethal approaches to managing urban deer. *Wildlife Society Bulletin* **25**, 424–429.

Morel, P. (2011) Chasser le sanglier, un droit risqué. Le Soir (online). Belgium. Available at: http://www.lesoir.be/partners.

Navarro-Gonzalez, N., Casas-Díaz, E., Porrero, C.M., Mateos, A., Domínguez, L., Lavín, S. and Serrano, E. (2013a) Food-borne zoonotic pathogens and antimicrobial resistance of indicator bacteria in urban wild boars in Barcelona, Spain. *Veterinary Microbiology* **167**, 686–689.

Navarro-Gonzalez, N., Porrero, M.C., Mentaberre, G., Serrano, E., Mateos, A., Domínguez, L. and Lavín, S. (2013b) Antimicrobial resistance in indicator *Escherichia coli* from free-ranging livestock and sympatric wild ungulates in a natural environment (NE Spain). *Applied and Environmental Microbiology* **79**, 6184–6186.

Nidaira, M., Taira, K., Itokazu, K., Kudaka, J., Nakamura, M., Ohno, A. and Takasaki, T. (2007) Survey of the antibody against Japanese encephalitis virus in Ryukyu wild boars (*Sus scrofa riukiuanus*) in Okinawa, Japan. *Japanese Journal of Infectious Diseases* **60**, 309–311.

O'Bryan, M.K. and McCullough, D.R. (1985) Survival of black- tailed deer following relocation in California. *Journal of Wildlife Management* **49**, 115–119.

Onac, D., Győrke, A., Oltean, M., Gavrea, R. and Cozma,V. (2013) First detection of *Echinococcus granulosus* G1 and G7 in wild boars (*Sus scrofa*) and red deer (*Cervus elaphus*) in Romania using PCR and PCR-RFLP techniques. *Veterinary Parasitology* **193**, 289–291.

Opsteegh, M., Swart, A., Fonville, M., Dekkers, L. and van der Giessen, J. (2011) Age-related *Toxoplasma gondii* seroprevalence in Dutch wild boar inconsistent with lifelong persistence of antibodies. *PLoS ONE* **6**: e16240.

Ortuño, A., Quesada, M., López-Claessens, S., Castellà, J., Sanfeliu, I., Antón E. and Segura-Porta, F. (2007) The role of wild boar (Sus scrofa) in the eco-epidemiology of *R. slovaca* in northeastern Spain. *Vector-borne and Zoonotic Diseases* **7**, 59–64.

Ostfeld, R.S. (2012) Effects of host and vector diversity on disease. *Annual Review of Ecology, Evolution, and Systematics* **43**, 157–182.

Pannwitz, G., Freuling, C., Denzin, N., Schaarschmidt, U., Nieper, H., Hlinak, A., Burkhardt, S., Klopries, M., Dedek, J., Hoffmann, L., Kramer, M., Selhorst, T., Conraths, F.J., Mettenleiter, T. and Müller, T. (2012) A long-term serological survey on Aujeszky's disease virus infections in wild boar in East Germany. *Epidemiology and Infection* **140**, 348–358.

Penfold, L.M., Jochle, W., Trigg, T.E. and Asa, C.S. (2007) Inter and intra species variability in response to GnRH agonists. *Proceedings of the 6th International Symposium on Fertility Control in Wildlife,* York, UK.

Philip, L.J. and Macmillan, D. (2003) *Public Perceptions of, and Attitudes Towards the Control of Wild Animal Species in Scotland.* Stirling, UK: Scotecon, University of Stirling.

Piorr, A., Ravetz, J. and Tosics, I. (2011) *Peri-urbanisation in Europe: Towards a European Policy to Sustain Urban–Rural Futures.* Frederiksberg, Denmark: Academic Books, Life Sciences. Available at: http://www.plurel.net/images/Peri_Urbanisation_in_Europe_printversion.pdf.

Podgórski, T., Baś, G., Jędrzejewska, B., Sönnichsen, L., Śnieżko, S., Jędrzejewski, W. and Okarma, H. (2013) Spatiotemporal behavioral plasticity of wild boar (*Sus scrofa*) under contrasting conditions of human pressure: primeval forest and metropolitan area. *Journal of Mammalogy* **94**, 109–119.

Prior, R. (2000) *Roe Deer; Management and Stalking.* Shrewsbury UK: Swan Hill Press.

Putman, R.J. (1996) *Guidelines for the management of deer in and around agricultural, urban fringe, woodland and conservation habitats.* Report for the Ministry of Agriculture, Fisheries and Foods, London, UK.

Putman, R.J. (1997) *Chemical and immunological methods in the control of reproduction in deer and other wildlife: potential for population control and welfare implications.* Technical Bulletin, Royal Society for the Prevention of Cruelty to Animals, Horsham, UK.

Putman, R.J. (2004) *The Deer Manager's Companion: A Guide to Deer Management in the Wild and in Parks.* Shrewsbury, UK: Swan Hill Press.

Putman, R.J. (2011) A review of the various legal and administrative systems governing management of large herbivores in Europe. In R.J. Putman, M. Apollonio and R. Andersen (eds), *Ungulate Management in Europe: Problems and Practices.* Cambridge, UK: Cambridge University Press, pp. 54–79.

Putman, R.J. and Kjellander, P. (2003) Deer damage to cereals: economic significance and predisposing factors. In F. Tattersall and W. Manley (eds), *Conservation and Conflict: Mammals and Farming in Britain.* London: Linnean Society Occasional Publications, pp. 186–197.

Putman, R.J. and Moore, N. P. (1998) Impact of deer in lowland Britain on agriculture, forestry and conservation habitats. *Mammal Review* **28**, 141–164.

Quarrell, D. (2012) *Controlling urban deer: roe deer.* Available at: www.thestalkingdirectory.co.uk.

Radeloff, V., Hammer, R., Stewart, S., Fried, J., Holcomb, S. and McKeefry, J. (2008) The wildland–urban interface in the United States. *Ecological Applications* **15**(3), 799–805.

Raik, D.B., Decker, D.J. and Siemer, W.F. (2006) Capacity building: A new focus for collaborative approaches to community-based suburban deer management? *Wildlife Society Bulletin* **34**, 525–530.

Raoult, D., Lakos, A., Fenollar, F., Beytout, J., Brouqui P. and Fournier, P.E. (2002) Spotless rickettsiosis caused by *Rickettsia slovaca* and associated with dermacentor ticks. *Clinical Infectious Diseases* **34**, 1331–1336.

Reimoser, F. and Putman, R.J. (2011) Impact of large ungulates on agriculture, forestry and conservation habitats in Europe. In R.J. Putman, M. Apollonio and R. Andersen (eds), *Ungulate Management in Europe: Problems and Practices.* Cambridge, UK: Cambridge University Press, pp. 144–191.

Richomme, C., Afonso, E., Tolon, V., Ducrot, C., Halos, L., Alliot, A., Perret, C., Thomas, M., Boireau, P. and Gilot–Fromont, E. (2010) Seroprevalence andfactors associated with *Toxoplasma gondii* infection in wild boar (*Sus scrofa*) in a Mediterranean island. *Epidemiology and Infection* **138**, 1257–1266.

Rudolph, B.A., Porter, W.F. and Underwood, H.B. (2000) Evaluating immunocontraception for managing suburban white-tailed deer in Irondequoit, New York. *Journal of Wildlife Management* **64**, 463–473.

Rutberg, A.T. and Naugle, R.E. (2008a) Population effects of immunocontraception in white-tailed deer (*Odocoileus virginianus*). *Wildlife Research* **35**, 494–501.

Rutberg, A.T. and Naugle, R.E. (2008b) Deer–vehicle collision trends at a suburban immuno-contraception site. *Human–Wildlife Conflicts* **2**, 60–67.

Rutberg, A.T., Naugle, R.E., Thiele, L.A. and Liu, I.K.M. (2004) Effects of immuno-contraception on a suburban population of white-tailed deer *Odocoileus virginianus. Biological Conservation* **116**, 243–250.

Sáez-Royuela, C. and Tellería, J.L. (1986) The increased population of the wild boar (*Sus scrofa* L.) in Europe. *Mammal Review* **16**, 97–101.

Schielke, A., Sachs, K., Lierz, M., Appel, B., Jansen, A. and Johne, R. (2009) Detection of hepatitis E virus in wild boars of rural and urban regions in Germany and whole genome characterization of an endemic strain. *Virology Journal* **6**, 58.

Schley, L. and Roper, T.J. (2003) Diet of wild boar *Sus scrofa* in Western Europe, with particular reference to consumption of agricultural crops. *Mammal Review* **33**, 43–56.

Scott, J.D. and Townsend, T.W. (1985) Characteristics of deer damage to commercial tree industries of Ohio. *Wildlife Society Bulletin* **13**, 135–143.

Seagle, S.W. and Close, J.D. (1996) Modelling white-tailed deer *Odocoileus virginianus* population control by contraception. *Biological Conservation* **76**, 87–91.

Siemer, W.F., Decker, D.J., Lowery, M.D. and Shanahan, J.E. (2000) The Islip Deer Initiative: A collaborative approach to suburban deer management. *Proceedings of the Ninth Wildlife Damage Management Conference,* 247–264.

Siemer, W.F., Lauber, T.B., Chase, L.C., Decker, D.J., McPeake, R.J. and Jacobson, C.A. (2004) Deer/elk management actions in suburban environments: what will stakeholders accept? *Proceedings of the 4th International Urban Wildlife Symposium on Urban Wildlife Conservation,* 228–237.

Sigurdson, C.J. and Aguzzi, A. (2007) Chronic wasting disease. *Biochemica et Biophysica Acta* **1772**, 610–618.

Simpson, V. (2002) Wild animals as reservoirs of infectious diseases in the UK. *The Veterinary Journal* **163**, 128–146.

Sims, N.K.E. (2005) *The ecological impacts of wild boar rooting in East Sussex.* PhD thesis, University of Sussex, School of Biological Sciences, Sussex, England.

Stout, R.J., Stedman, R.C., Decker, D.J. and Knuth, B.A. (1993) Perceptions of risk from deer-related vehicle accidents: implications for public preferences for deer herd size. *Wildlife Society Bulletin* **21**, 237–249.

Stout, R.J., Knuth, B.A. and Curtis, P.D. (1997) Preferences of suburban landowners for deer management techniques: A step towards better communication. *Wildlife Society Bulletin* **25**, 348–359.

Stradtmann, M.L., McAninch, J.B., Wiggers, E.P. and Parker, J.M. (1995) Police sharpshooting as a method to reduce urban deer populations. In J.B. McAninch (ed.), *Urban Deer: A Manageable Resource?* Proceedings of a Symposium of the 55th Midwest Fish and Wildlife Conference, The Wildlife Society, Bethesda, USA, pp. 117–122.

Telford, S.R., Mather, T.N., Moore, S.I., Wilson, M.L. and Spielman, A. (1988) Incompetence of deer as reservoirs of the Lyme disease spirochete. *American Journal of Tropical Medicine and Hygiene* **39**, 105–109.

ver Steeg, J.M., Witham, J.H. and Beissel, T.J. (1995) Use of bowhunting to control deer in a suburban park in Illinois. In J.B. McAninch (ed.), *Urban Deer: A Manageable Resource?* Proceedings of a Symposium of the 55th Midwest Fish and Wildlife Conference, The Wildlife Society, Bethesda, USA, pp. 110–116.

Walter, W.D., Lavelle, M.J., Fischer, J.W., Johnson, T.L., Hygnstrom, S.E. and VerCauteren, K.C. (2010) Management of damage by elk (*Cervus elaphus*) in North America: a review. *Wildlife Research* **37**, 630–646.

Ward, A.I. and Smith, G.C. (2012) Predicting the status of wild deer as hosts of *Mycobacterium bovis* infection in Britain. *European Journal of Wildlife Research* **58**, 127–135.

Ward, A.I., Etherington, T.R. and Smith, G.C. (2008) *Exposure of cattle to* Mycobacterium bovis *excreted by deer in southwest England: a quantitative risk assessment.* Consultancy report to TB Programme, Food and Farming Group, Department of Environment, Food and Rural Affairs, London.

Watson, P., Putman, R.J., Green, P. and Langbein, J. (2009) *Methods for control of wild deer appropriate for use in the urban environment in England.* Report for Defra (Department of the Environment, Food and Rural Affairs), London. Available at: www.thedeerinitiative.co.uk.

West, B.C. and Parkhurst, J.A. (2002) Interactions between deer damage, deer density, and stakeholder attitudes in Virginia. *Wildlife Society Bulletin* **30**, 139–147.

Williams, E.S., Kirkwood, J.K. and Miller, M.W. (2001) Transmissible spongiform encephalopathies. In E.S.Williams and I.K. Barker (eds), *Infectious Diseases of Wild Mammals.* Ames IA: Blackwell Publishing, pp. 292–301.

Wilson, C.J. (2003) *Current and future deer management options.* Report on behalf of the Wildlife Division of the Department for the Environment, Food and Rural Affairs, UK.

Wilson, C.J. (2005) *Feral wild boar in England Status, impact and management.* A report on behalf of Defra European Wildlife Division, Department for Environment, Food and Rural Affairs (DEFRA) London, UK. Available at: http://www.naturalengland.org.uk/Images/wildboarstatusImpactmanagement_tcm6-4512.pdf.

Witham, J.H. and Jones, J.M. (1990) Post-translocation survival and movements of metropolitan white-tailed deer. *Wildlife Society Bulletin* **18**, 434–441.

Wright, R.G. (1993) Wildlife management in parks and suburbs: alternatives to sport hunting. *Renewable Resources Journal* **8**, 18–22.

Chapter 8

The Management of Ungulates in Protected Areas

Stefano Grignolio, Marco Heurich, Nikica Šprem
and Marco Apollonio

"Protected areas are essential for biodiversity conservation. They are the cornerstones of virtually all national and international conservation strategies, set aside to maintain functioning natural ecosystems, to act as refuges for species and to maintain ecological processes that cannot survive in most intensely managed landscapes and seascapes".

This is the opening sentence of the IUCN Guidelines for Applying Protected Areas Management Categories (Dudley, 2008). From the beginning of the document it is clear that the IUCN accords to protected areas a fundamental role for the conservation of species and ecosystems in the past, in the present and in the future. First of all, the IUCN defines a protected area ('A clearly defined geographical space, recognised, dedicated and managed, through legal or other effective means, to achieve the long-term conservation of nature with associated ecosystem services and cultural values') and identifies six categories:
I Strict protection (Ia: Strict nature reserve; Ib: Wilderness area); II Ecosystem conservation and protection (i.e. National park); III Conservation of natural features (i.e. Natural monument); IV Conservation through active management (i.e. Habitat/species management area); V Landscape/seascape conservation and recreation (i.e. Protected landscape/seascape); VI Sustainable use of natural resources (i.e. Managed resource protected area).

Before continuing, it is important to note that there is no consistent nomenclature for these various protected areas and there is marked disparity between the names chosen by the stakeholders and agencies in different countries, which may result in considerable confusion. Many European countries use terms such as 'National Parks' and 'Nature Reserves' in their national legislation for protected areas, but these frequently mean different things in the different countries and may well not accord to the categories with such names within IUCN's nomenclature, particularly because many of these areas were established well before the

IUCN directive. Throughout this chapter, therefore, it is important to recognise that different governments and different agencies have their own definitions of what constitutes a National Park – or any other protected area, which may not fit with this more recent IUCN terminology. Thus if a government has called, or wants to call, an area a National Park that does not mean that it has to be managed, or that it will be managed, according to the guidelines under category II.

Awareness of the need for specific, and active, management of protected areas has rapidly increased through the years. In their definition, the IUCN imply that management assumes some active steps to conserve the natural (and possibly other) values for which the protected area was established; note that 'managed' can include a decision to leave the area untouched if this is the best conservation strategy. However, protected areas have often been established without a careful analysis of the skills and abilities needed to maintain them, thus causing several difficulties in ongoing management or even endangering the existence of the protected area itself.

The aim of this chapter is to describe the role of protected areas in the conservation of ungulates, but also the problems that these animals cause to the conservation of threatened habitats. For our examples we focus on areas defined as National Parks by national governments, in order to emphasise the heterogeneity of the objective and the consequent ungulate conservation management plans inside the main important protected areas of different European countries.

8.1 Defining the goal of management within protected areas

As we have noted, a significant problem within Europe is that every country has its own definition of a National Park or other protected area. Consequently, protected areas, which apparently belong to the same category, may adopt different management objectives and follow different management practices. Indeed, even within a country, different areas, even of the same designation status, may pursue different management objectives and follow different management practices.

Ideally, rather than focusing on artificial preservation of individual species, management should be aimed at directing the natural ecological dynamics of the protected system towards protection of the target species or ecosystem, taking into account also the socioeconomics of underlying land-use practice and local peoples, in so far as this will not adversely affect the primary management aim. But for effective management, the conservation (or other) objectives for which the area has been designated have to be clearly defined. Apollonio *et al.* (2010) point to lack of clearly defined objectives as one of the major obstacles hampering effective management of ungulate populations, whether for control, exploitation or conservation. In reality, however, it is often very difficult to understand from the official documentation what is the main target (habitat and/or species) of the protected areas. We made a brief analysis of the legislation underpinning the establishment of National Parks in some European countries and in general found an anomalous situation, with, in the majority of cases, the aims being rather general in nature and not clearly defined.

For example, in Finland the overall objective and conservation goals are described in a rather general way in the Establishment Act for any given protected area. For each National Park however, a more detailed management plan is prepared; only the actual management plan for the area defines the conservation goals in more detail and, for example, defines whether or not hunting may or may not be permitted within the designated area. In Italy, legal declaration of a protected area is undertaken, by definition, to protect areas rich in biodiversity, taking into account the need to integrate human land-use patterns and land-use interests with the natural environment; indeed declaration of a site may be directed towards preservation of anthropological, archaeological, historical and architectural heritage as well as agricultural and forestry activities. As in Finland, however, the conservation target is not necessarily particularly explicit in the Acts establishing the National Park even if a hunting ban is a nationwide characteristic of these protected areas. In the same way, the German Nature Conservation Act states that a National Park is a significant area of land, in which it is assured that natural processes, in their natural dynamics, can take place in the most undisturbed manner possible. Provided this is compatible with the purpose of protection, National Parks may also serve the purposes of scientific environmental monitoring, nature education, and enabling the general public to experience nature. As a consequence National Parks have to protect the natural ecosystems including their natural dynamics within a given area. The same conservation philosophy is adopted in Austrian and Swiss National Parks. Basically in all cases, the laws instituting the protected areas establish a rather generic objective of conservation of fauna, flora and landscape.

While the identification of the more specific goal of a protected area is often difficult using only the official documents, in some cases the main objective can be deduced from the management history and the conservation initiatives which have been implemented within the area. In the next section we will describe some examples of protected areas where ungulates are among the target species for active conservation, before investigating some other instances where presence of wild ungulates may have caused problems in achieving other conservation objectives.

8.2 Protected areas devoted to the protection of ungulates

In Europe there are few species or subspecies of ungulates threatened with extinction, but nevertheless there are a few examples of National Parks or other protected areas which were established in order to protect taxa considered under threat. This may be primarily for conservation of the threatened species/subspecies or due to their importance as 'umbrella species' and 'flagship species' recognised by scientific community. The number of protected areas explicitly devoted to the protection of ungulate species or their subspecies is however low (and in some cases it is possible that the *lack* of such areas could have resulted in the loss of particular taxa: for example the lack of any large protected areas in Spain with the main objective of conserving the last populations of *Capra pyrenaica pyrenaica* has resulted in this subspecies recently becoming extinct; Perez *et al.*, 2002).

In Italy there are 22 National Parks that cover about 15,000 km², but the conservation of a specific ungulate taxon is one of the main aims only in three parks covering an overall area of approximately 2,700 km² (Gran Paradiso National Park; Abruzzo, Lazio and Molise National Park; and the Gran Sasso-Monti della Laga National Park). The Abruzzo, Lazio and Molise National Park was created in 1923 to protect the last population of Apennine chamois (*Rupicapra pyrenaica ornata*), an endangered subspecies. Currently, it is part of an ecological network together with other National Parks (Gran Sasso-Monti della Laga NP; Majella NP; Monti Sibillini NP) and local protected areas where the chamois have been reintroduced (Lovari *et al.*, 2010). This conservation strategy, also supported by European Community by means of the LIFE programs (LIFE02 NAT/IT/8538 and LIFE09 NAT/IT/000183), has allowed a significant increase in the size of the populations improving the future outlook for Apennine chamois so that its status was changed in 2008 from 'threatened' to 'vulnerable' (Herrero *et al.*, 2008).

Another example, this time from Eastern Europe, is the Białowieża National Park that was established on the former hunting estate of the Białowieża Forest, Poland (see also Chapter 9). Belonging at different times to the Duke of Lithuania, the King of Poland and the Russian Tzar, this area was the last refuge of the European bison that became extinct in the wild there in 1919; a reserve was established in 1921 and then turned into a National Park in 1932. The main purpose of its establishment was both the preservation of the only native lowland forest of Europe and as an area for reintroduction of bison to occur. A population of captive bred lowland bison were kept within the park in reserves of Białowieża so that they were simply released from their corrals in 1952 without any translocation (Krazinska and Krazinski, 2007). Notably the emblem of the park is a bison, a recurrent characteristic of protected areas devoted to the conservation of a particular species. (For further discussion of this reintroduction of bison to the Białowieża Park, see Chapters 3 and 9.)

In many instances, the conservation of ungulates is commonly threatened by hunting and so, the banning of hunting within protected areas has an obvious positive effect on target species of the protected area, but also on the other species that are no longer hunted. The Gran Paradiso National Park, for example, first established in Italy in 1922, was created initially with the specific objective of protecting the only Alpine ibex population left in the world. The history of Alpine ibex and the role of Gran Paradiso National Park in its protection and reintroduction offer one of the most striking examples of the importance of a protected area in the conservation of a single taxon (Stüwe and Nievergelt, 1991; see Chapter 3, where this example is considered in more detail). But while the establishment of the park was primarily targeted at conservation of the ibex, the protection of this ungulate and the total ban on hunting have also allowed an increase in the populations of other species such as Alpine chamois (Figure 8.1), marmots

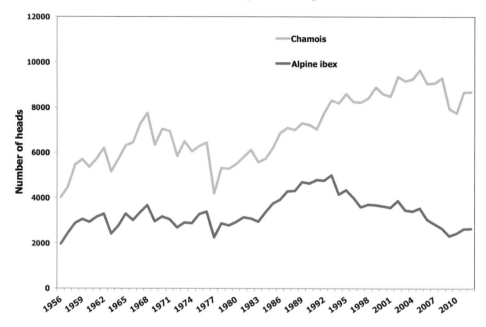

Figure 8.1 Chamois and Alpine ibex populations in the Gran Paradiso National Park (Italy). Count data provided by the Sanitary and Scientific Research Service of the Gran Paradiso National Park.

(*Marmota marmota*), red fox (*Vulpes vulpes*), golden eagles (*Aquila chrysaetos*). More generally, the protection afforded to the area has enabled the conservation of high mountain environments and avoidance of development such as construction of roads or ski resorts. The Alpine ibex has served in this as a flagship species attracting public and private funds (and in a similar way has also acted as a flagship in the Vanoise National Park in France).

However conservation initiatives of this sort must often find a compromise with social pressures and the socioeconomic interests of people living within protected areas. In Finland, for example, the Nature Conservation Act clearly distinguishes between National Parks of Northern and Southern Finland. In the National Parks of Northern Finland, hunting is generally permitted, but is firmly restricted to local people only. Moose is practically the only wild ungulate in these protected areas and they are hunted normally according to normal provisions of the Hunting Act. In Southern Finland hunting is generally banned in National Parks, but in 2011 a revision in the legislation made it possible to allow driving of moose and harvesting the introduced white tailed deer (Rautiainen Mikko, *pers. comm.*).

8.3 Areas designated for conservation of other species or habitats: Consequences for ungulate populations

In many other protected areas, the presence of ungulates may be a coincidental consequence of conservation strategies in general, or their presence may have little meaning for the biodiversity of the area. In other cases, however, the presence of large herbivores, particularly if present at high density could be a serious threat to the conservation of some target species. In the first case, i.e. when a protected area favours the expansion of an ungulate species without negative impacts to environment, the protected area has reached an important goal contributing to a natural increase of biodiversity. However, if stringently protected, ungulate populations can, over time, became overabundant in those same protected areas and resulted in a threat to ecosystems. This can present one of the most difficult situations for managers: to observe and to record, in other words, to study the environmental changes, or take action to reduce the impacts caused by ungulates?

In fact, the choice should be easy if the designated area has clearly defined and explicit aims and objectives. But, as explained above, there are few situations in which the conservation aims are so explicit. In all other cases, the choice is delegated to the managing authorities of the parks, taking into account overall conservation aims and the lobbying of interested stakeholders. In some situations it is mandatory to take action to manage ungulates. Below we will offer some examples of National Parks where management of overabundant ungulate populations, belonging to species of least conservation concern, had become problematic in relation to their impact on different components of the ecosystem. This occurs, for example, when human activities have resulted in unnatural density and where the animals are unable to move away from the area.

The Brijuni National Park in Croatia is an island in the Brijuni archipelago; it was declared a National Park in 1983, but there had already been a zoo and acclimatisation station on the site for wild animals from the tropical belt since 1901. Nowadays, three ungulate species are present, fallow deer, axis deer and mouflon. The high density (more than 1,000 ungulates on an area of only 5.55 km^2) makes clear the need for significant management of these populations. The available natural food on the island is not enough to satisfy the needs of the animals and therefore there is a need to provide supplementary feed. In addition to this mandatory provision of feed for animals, every year a certain number of animals are removed, through culling or by live capture and translocation, in order to keep numbers under control. All population management measures are carried out in the colder part of the year when there are no tourists. Brijuni National Park is an isolated area, its island nature preventing significant natural movement of the ungulate species present; this form of more interventive management is both justified and necessary in this strictly protected landscape; as such, it is a unique example of a strictly protected area where game management (i.e. hunting, feeding) is permitted in Croatia.

In France, the hunting of ungulates is banned in National Parks, with the exception of two protected areas (Parc National des Cévennes since 1970 and the new Parc National des Calanques, created in 2012). The Cevennes National Park (913 km^2) was created in 1970 after a long negotiation over many years, partly in relation to regulation of hunting. Hunting is carried out by a large Hunting Association across the majority of the Park, with hunting banned only in the core areas (some 16% of the total area). More recently however (in the 1980s), because of the high impacts apparent to vegetation, managers decided to permit hunting even in the currently restricted areas. The decision to hunt deer and wild boar was taken in 1993–94; initially only local hunters were involved, but later, hunting rights were extended to hunters from outside in order to increase the effectiveness of control. Each year a ministerial decree determines the hunting bag and regulations (Poinsot and Saldaqui, 2012). Hunting with dogs is forbidden to reduce harassment. In the same way, hunting is also permitted in the new Calanques National Park because of the increase of wild boar which might otherwise result.

By comparison, the only National Park of Slovenia (Triglav National Park) is divided into three conservation zones. The central area is primarily devoted to conservation and protection of natural values, natural primeval wilderness areas, flora and fauna, the natural evolution of ecosystems and natural processes without human care, maintenance and other interventions. Traditional grazing on pasture landscapes in the high mountains is permitted, both for conservation management and in the maintenance of related cultural heritage. Within the second zone, the traditional use of natural resources is allowed in the form of sustainable agriculture and forestry and sustainable management of wildlife and fish. The third conservation zone is intended to preserve and protect biodiversity, natural and cultural heritage and both ecological and cultural landscape quality, and sustainable development in alignment with the objectives of the National Park (Triglav National Park Law, Official Gazette of Slovenia, 2010). In accordance with the intention of this zonation, game management and hunting are permitted in the outer zones, according to a specified management plan. The management plan is valid for 10 years, and is written by staff from Department of Forestry (but not the park staff themselves). No drive hunts are allowed within the Park and the introduction of non-native species is strictly forbidden. No hunting is permitted in the central area.

The Swiss National Park (170 km²) was established in 1914 and is the oldest National Park in the Alps. The aims of this protected area are to allow the unhindered development of nature without human interference, to research the ensuing natural processes and to inform visitors. To do this, no management is undertaken at all within the park and visitors are restricted to a network of well-defined paths to reduce the animal harassment to a minimum. In the middle of the 19th century the red deer had been eradicated from this region; the species reappeared within the National Park in last part of this century, migrating from north. Because there was a hunting ban in the park and adjacent areas, the red deer population increased considerably. While in 1960 about 800 red deer spent the summer in the park, their number reached 1500 in 1971 and almost 2500 animals in 1975. Also the red deer changed their migratory behaviour and left the park in winter. As a consequence of these changes, damage to vegetation in the surrounding areas of the park increased from 50,000 Swiss francs in 1970 to 200,000 in 1976 (presently equivalent to about 160,000 euro). There was also extremely high mortality of deer during the winter, with 600 red deer dying in in 1970 alone. As a further consequence of the expansion of red deer numbers within the Park, roe deer populations declined almost to extinction (Schloeth, 1972), one of the most well known cases of within-guild competition among ungulates in Europe.

To solve these problems the hunting quota in the surrounding of the park was increased; also – after long discussions – the killing of 100 sick and weak animals in the park was ordered. In 1974 a programme of winter feeding was established. With these measures it was possible to reduce the rate of increase of the red deer population. In the 1980s, however, a new controversy arose not only about the damage caused by red deer outside the park area, but also in relation to impacts occurring within the park; it was suggested that the deer could severely damage the vegetation of the park vegetation and inhibit forest regeneration. The culling of 150 deer within the Park was ordered in 1987. In the middle of the 1990s, research showed that the high red deer densities could slow down forest development but would not inevitably endanger the existence of the forests nor the reforestation of old pastures (Krüsi *et al.*, 1998; Schütz, 2005) and the killing of deer in the park was stopped. But most crucial for the situation in and outside of the Park was the introduction in the middle of the 1980s of the 'kantonale Jagdplanung' (a regional game management plan for the surrounding area). With this wider plan in place, it was possible to reduce the number of deer to about 1,500 and to reduce costs of damage to less than 25,000 Swiss francs, without further need for culling within the park area (Kupper, 2012).

Another example involving mainly red deer is the Stelvio National Park in Italy (1,307 km^2). This protected area was established in 1935 with the declared objective of offering full protection of alpine flora and fauna, preserving the particular geological formations of the site, and through time to encourage the development of tourism. At the time of the Park's establishment, there were small populations of red deer that utilised some of the areas within the park and by the end of the 1960s resident populations had established there. Current spring populations have fluctuated between 6,000 and 7,000 individuals in the last 5 years (over 10,000 if we include the surrounding areas) and indeed nowadays, about 30% of estimated population of red deer in the entire Italian Alps is located in the zone inside the Stelvio National Park and surrounding areas (thus with 30% of the estimated population occurring within only 7% of the distribution range; Carnevali *et al.*, 2008). Impacts on agricultural pastures and forest regeneration have increased significantly over the years. During the early 2000s, a study estimated a reduction of the production of hay in the pastures of approximately 180 euros per hectare (Pedrotti *et al.*, 2011). Within the confines of the park, at least 60% of the trees in the patches of deciduous forests showed clear signs of browsing by deer. In addition, the high density of red deer within the area may be having negative effects on roe deer, capercaillie (*Tetrao urogallus*) and chamois. Red deer and chamois populations are negatively affected by population density, by climate, and by the interaction between these two variables. But the chamois population also suffers as the result of the competition with red deer inside the park (Figure 8.2). High red deer densities caused trophic competition and displacement of chamois from areas where large deer herds have been grazing for the summer season (Bonardi, 2009).

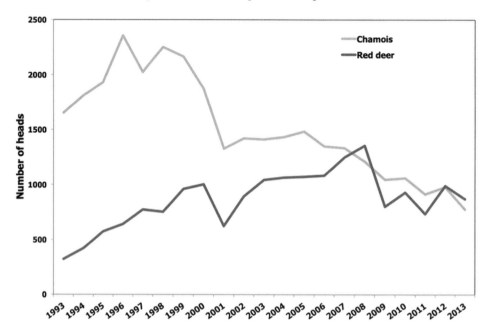

Figure 8.2 Chamois and red deer populations in the Stelvio National Park (Italy). Count data provided by the Scientific Research Service of the Stelvio National Park.

The strong hunting pressure around the Park is contributing to making the situation worse. Stelvio National Park spans a number of different provincial areas (Bolzano, Trento and Lombardy): within the Province of Trento, deer densities in 2008 inside and outside the park were 23.8 and 3.4 deer per km², respectively. Since there was no culling within the Park itself, a huge 'reserve' or 'sanctuary' effect developed in response to the heavy hunting pressure beyond the Park boundary: hinds did not leave the area of the park even in the winter, despite the fact that areas outside the reserve are more suitable for wintering.

It is well know that 'reserve effects' such as this may modify the behaviour of ungulates and, consequently, their distribution (see, for examples, Tolon *et al.*, 2009; Grignolio *et al.*, 2011). For some years now, the Stelvio National Park has been implementing a planned reduction of the population within its boundaries in the Bolzano Province. Control plans for the other two areas of the park (Trento Province and Lombardy Region) have been prepared, but have not yet been implemented. However, these interventions provide only a temporary response and on their own cannot solve the problem of deer in the Stelvio National Park, which will remain an effective sanctuary attracting immigration from the surrounding areas unless hunting pressure outside the park is reduced or alternative refuge areas for deer are also established outside of the protected area.

The competition described here between red deer and alpine chamois is similar to the problems caused by this cervid to Apennine chamois (*Rupicapra pyrenaica ornata*), a vulnerable subspecies of Pyrenean chamois. Since 1972 a total of

81 red deer have been reintroduced within the core area of the chamois range in the Abruzzo, Lazio, Molise National Park (Apollonio and Lovari, 2001). As already mentioned, at the time this Park protected the only population of the Apennine chamois. Since 2005 the abundance of chamois has declined significantly (about 30%), so that now Majella National Park, where chamois were reintroduced in 1990, harbours a larger population than that in the original Abruzzo, Lazio, Molise National Park. A recent study showed a negative effect of red deer on the availability of nutritious plant species, an interspecific overlap in resource use, and through competition for a scarce food resource, a negative influence on Apennine chamois population size (Lovari *et al.*, 2014). This example emphasises that any reintroduction project should consider both short- and long-term consequences beforehand, in order to avoid alteration of key habitats inside protected areas.

Finally, we consider the issues facing management of ungulates within the Bavarian Forest National Park. This Park was founded 1970 as the first German National Park in a forested low mountain range at the border to the Czech Republic; in 1997, the park was extended to its current size of 242 km^2. The forests in the area contain some virgin forest reserves, but most parts were managed as a forest enterprise since the 19th century. Before establishment of the National Park, harvesting trophy animals was the objective of wildlife management for the forest service, which resulted in the maintenance of high population densities and intensive feeding during winter. As a consequence more than 20 km^2 of Norway spruce stands were bark-stripped and forest regeneration was extensively browsed at this time. To address these issues the number of red and roe deer shot was increased considerably to reduce overall population density to about one animal per 100 ha. In addition, all open feeding stations were closed although managers erected four winter enclosures with a central feeding place for red deer. After the end of the rut when the first snow falls, parts of the red deer herd move to the enclosures and spend the winter there. At the beginning of May, after the flush of ground vegetation, the enclosures are opened.

With this approach it was possible to reduce the browsing and bark-stripping pressure considerably; however these measures resulted in a dramatic change in space use of the deer herd. Telemetry revealed that the animals had abandoned large parts of the National Park and retreated to an area between the border with the Czech Republic and the border fence (Iron Curtain), which ran several kilometres behind the actual border (Heurich *et al.*, 2007). There they were out of reach of both German and Czech hunters. After the first snow they moved back to the winter enclosures, where hunting was not permitted. To address this problem culling was permitted inside the enclosures from 1986.

As a further measure to control populations of roe deer, the first reintroduction of Eurasian lynx (*Lynx lynx*) in Central Europe was performed in the Bavarian Forest National Park in 1970. At that time about 5–7 lynx from the Carpathian Mountains were released (the exact number not known). As a consequence of the low number of reintroduced animals and illegal killings, the population declined. Therefore between 1982 and 1987, 17 additional animals were released in an adjacent area that is now part of the Šumava National Park in Czech Republic to support the lynx population. The lynx population increased in the following years and since the middle of the 1990s the park area is permanently settled by lynx (Wölfl *et al.*, 2001). Populations of roe have declined and the natural migration pattern of roe deer was also re-established in the area. About half of the population leave the park in winter and migrate to the valleys in adjacent areas (Cagnacci *et al.*, 2011).

As the result of all these measures, browsing pressure on white fir was reduced from about 40% in the 1980s to less than 10% – a level which is regarded sufficient to allow natural regeneration (Eiberle and Nigg, 1987). This made it possible to establish a large core area of 170 km^2 within the Park where hunting is not permitted, but where regeneration of rare tree species such as white fir is possible. After the Park administration achieved this main objective, roe deer control was stopped in 2007 in the Rachel-Lusen-area and, from 2012, the roe deer population was left to natural regulation within the whole park to proceed one step further to achieve the management objective of nonintervention in the park.

8.4 Final conclusions and comments on the management of ungulates in protected areas

In this chapter we have shown that, even if we restrict consideration only to so-called National Parks within the wide range of protected areas in Europe, we face a diversity of different ungulate management issues and practices, both among and within countries. Areas that are crucial for conservation of endangered or rare ungulate taxa are generally scarce, while by contrast there are quite commonly significant problems which may be associated with overabundance of ungulate populations within the boundaries of National Parks.

However, it is still possible to draw some general conclusions. We would suggest that:

- Only in a few protected areas can ecological processes genuinely be left to evolve in a natural way without the need for any management intervention. These would be the protected areas conforming to Classes I and II of the IUCN classification. In these situations, managers can implement a conservation strategy that could be defined as 'sit and watch'. However, we would suggest that this strategy is applicable only to areas that are sufficiently large that they are not significantly affected by management decisions outside the reserve or where management measures can be shifted to the surrounding of the park.
- A further difficulty in such 'natural-process oriented' management philosophy is that large predators are either absent or rare from such systems, or ineffective in limiting ungulate populations in very productive environments (Melis *et al.*, 2009). Due to their setting, the vast majority of National Parks (and other protected areas within Europe) encompass only one part of the total annual range of larger wildlife species, mostly the summer ranges. Protected areas in Europe are

generally too small to accommodate the natural dynamics of predator – ungulate systems.

• It is not inevitable that these situations should be dealt with by reducing the number of ungulates by means of hunting inside the protected area. Many of the examples in this chapter highlight the importance of an integrated management of ungulate populations across their wider range, both inside and outside the parks. Where this is not the case, it is very likely that ungulates will be concentrated in the protected areas (sanctuary effect), causing considerable ecological and economic damage inside and outside the park, resulting on significant social discontent. Our suggestion is that such holistic management approaches across wider geographic areas should be encouraged because it helps to reduce social problems and also contributes to maintaining a more natural ecological dynamic inside the protected area itself – facilitating a greater ability to maximise natural processes.

• Damage caused by ungulates outside the protected area is only tolerated to a limited degree. For this reason and because in most countries hunting remains a deeply engrained social tradition (see for example Putman *et al.*, 2011a), this landscape level approach to managing overall density should be the primary focus in most of the European protected areas, including National Parks, rather than isolated attempts to address localised issues of overabundance (and see for example arguments of Putman *et al.*, 2011b). We do not necessarily think that hunting should be prohibited in all protected areas. However, we argue that all protected areas define *a priori* their conservation objectives and assess whether and when to allow or ban certain practices of hunting. In these cases the hunting practices should be chosen to minimise the impact on the behaviour and to respect the natural structure of the populations – such hunting should be integrated with management over a wider geographic area more closely approximating to the true biological range of the populations of concern.

• In order conform with the IUCN Guidelines, human disturbances should be reduced as much as possible. Control measures for ungulates should thus by preference be: (i) practiced preferably outside of the limits of the national park; (ii) limited to the lowest necessary scale; (iii) carried out with the least possible amount of disturbance, with a species-specific approach, (iv) accompanied by a monitoring system.

Finally, a key element for the success of any improvement in wildlife management beyond the boundaries of the protected areas themselves is a sufficient level of acceptance within the various interest groups. Much as in the sense of an 'ecosystem approach', it is necessary to involve the local population in the various projects. In addition to the usual educational and informational efforts ('No one is likely to protect what they are not familiar with'), it is also desirable to change our inner, emotional attitude. In order to replace the anthropocentric 'useful versus harmful' way of thinking with a more enduring attitude based on respect for the inherent values of nature, a new creed must become the basis for working towards a sensitive and lasting acceptance: 'No one is likely to protect what they do not appreciate'.

Acknowledgements

We thank our colleagues Philippe Ballon (France), Bruno Bassano (Italy), Juan Carranza (Spain), Luca Pedrotti (Italy), Mikko Rautiainen (Finland), Achaz von Hardenberg (Italy), for the information on the situation in their respective countries.

References

Apollonio, M. and Lovari, S. (2001) Reintroduzioni di cervi e caprioli nei parchi nazionali, con note sulle immigrazioni naturali. In S. Lovari and A. Sforzi. (eds), *Progetto di monitoraggio dello stato di conservazione di alcuni Mammiferi particolarmente a rischio della fauna Italiana*. Rome: Ministero dell'Ambiente, pp. 462–475.

Apollonio, M., Andersen, R. and Putman, R. (2010) Present status and future challenges for European ungulate management. In M. Apollonio, R. Andersen and R. Putman (eds), *European Ungulates and their Management in the 21st Century*. Cambridge, UK: Cambridge University Press, pp. 578–604.

Bonardi, A. (2009) *Provisional models for management and conservation of Alpine fauna: the case of the red deer* (Cervus elaphus) *in the Stelvio National Park*. PhD Thesis, University of Insubria, Italy.

Cagnacci, F., Focardi, S., Heurich, M., Stache, A., Hewison, A.J.M., Morellet, N., Kjellander, M.P., Linnell, J.D.C., Mysterud, A., Neteler, M., Delucchi, L., Ossi, F. and Urbano, F. (2011) Partial migration in roe deer: migratory and resident tactics are end points of a behavioural gradient determined by ecological factors. *Oikos* **120**, 1790–1802.

Carnevali, L., Pedrotti, L., Riga, F. and Toso, S. (2008) *Banca Dati Ungulati. Status, distribuzione, consistenza, gestione, prelievo venatorio*. Rapporto INFS 2001-2005, Italy.

Ciuti, S., Muhly, T.B., Paton, D.G., McDevitt, A.D., Musiani, M. and Boyce, M.S. (2012) Human selection of elk behavioural traits in a landscape of fear. *Proceedings of the Royal Society B: Biological Sciences* **279**, 4407–4416.

Dudley, N. (2008) *Guidelines for Applying Protected Area Management Categories*. Gland, Switzerland: International Union for the Conservation of Nature (IUCN).

Eiberle, K. and Nigg, H. (1987). Basis for assessing game browsing in montane forest. *Schweizerische Zeitschrift für Forstwesen* **138**, 747–785.

Grignolio, S., Merli, E., Bongi, P., Ciuti, S. and Apollonio, M. (2011) Effects of hunting with hounds on a non-target species living on the edge of a protected area. *Biological Conservation* **144**, 641–649.

Herrero, J., Lovari, S. and Berducou, C. (2008) *Rupicapra pyrenaica*. IUCN Red List of Threatened Species. Version 2010.3. Available at: http://www.iucnredlist.org.

Heurich, M., Stache, A. and Horn, M. (2007) Habitat utilisation by red deer in relation to wildlife management practices and forest development in the Bavarian Forest National Park. *Proceedings of the XXVIII IUGB Congress*, 32.

Krasinska, M. and Krasinski, Z. (2007) *European Bison: The Nature Monograph*. Białowieża, Poland: Mammal Research Institute, Polish Academy of Sciences.

Krüsi, B.O., Schütz, M. Bigler, C. Grämiger H. and Achermann G. (1998) Huftiere und Vegetation im Schweizerischen Nationalpark von 1917–1997. Teil 1: Zum Einfluss der Huftiere auf die botanische Vielfalt der subalpinen Weiden. Teil 2: Zum Einfluss der

Huftiere auf das Weild-Freilandverhältnis. In R. Cornelius and R. Hofmann (eds), *Extensive Haltung robuster Haustierrassen, Wildtiermanagement, Multi-Species-Projekte. Neue Wege in der Landschaftspflege*. Berlin, Germany: Institut fur Zoo- und Wildtierforsch, pp. 62–74.

Kupper, P. (2012) *Wildnis schaffen. Eine transnationale Geschichte des Schweizerischen Nationalparks*. Bern, Stuttgart, Wien, Austria: Haupt Verlag.

Lovari, S., Artese, C., Damiani, G. and Mari, F. (2010) Re-introduction of Apennine chamois to the Gran Sasso-Laga National Park, Abruzzo, Italy. In P.S. Soorae (ed.), *Global re-introduction Perspectives: Additional Case-studies from Around the Globe*. Abu Dhabi, United Arab Emirates: IUCN/SSC Re-introduction Specialist Group, pp. 281–284.

Lovari, S., Ferretti, F., Corazza, M., Minder, I., Troiani, N., Ferrari, C. and Saddi A. (2014) Unexpected consequences of reintroductions: competition between increasing red deer and threatened Apennine chamois. *Animal Conservation* doi: 10.1111/acv. 12103.

Melis, C., Jędrzejewska, B., Apollonio, A., Bartoń, K.A., Jędrzejewski, W., Linnell, J.D.C., Kojola, I., Kusak, J., Adamic, M., Ciuti, S., Delehan, I., Dykyy, I., Krapinec, K., Mattioli, L., Sagaydak, A., Samchuk, N., Schmidt, K., Shkvyrya, M., Sidorovich, V.E., Zawadzka, B. and Zhyla, S. (2009) Predation has a greater impact in less productive environments: variation in roe deer, *Capreolus capreolus*, population density across Europe. *Global Ecology and Biogeography* **18**, 724–734.

Pedrotti, L., Bonardi, A., Gugiatti, A., Bragalanti, N., Carmignola, G., Gunsch, H. and Platter, W. (2011) Effetto del controllo su popolazioni di ungulati all'interno di un'area protetta–il caso del cervo nel Parco Nazionale dello Stelvio. In E. Raganella Pelliccioni, F. Riga and S. Toso (eds), *Linee guida per la gestione degli ungulati*. Rome, Italy: ISPRA, pp. 192–209.

Perez, J.M., Granados, J.E., Soriguer, R.C., Fandos, P., Marquez, F.J. and Crampe, J.P. (2002) Distribution, status and conservation problems of the Spanish Ibex, *Capra pyrenaica* (Mammalia: Artiodactyla). *Mammal Review* **32**, 26–39.

Poisont, Y. and Saldaqui, F. (2012) La maîtrise des populations de grands ongulés dans les espaces naturels protégés : comment gérer la spatialité animale par des territoires humains? *Cybergeo : European Journal of Geography* **596**. Available at: http://cybergeo.revues.org/25226#ftn1.

Putman, R., Andersen, R. and Apollonio, M. (2011a) Introduction. In R. Putman, R. Andersen and M. Apollonio (eds), *Ungulate Management in Europe: Problems and Practices*. Cambridge, UK: Cambridge University Press, pp. 1–11.

Putman, R.J., Watson, P. and Langbein, J. (2011b) Assessing deer densities and impacts at the appropriate level for management: a review of methodologies for use beyond the site scale. *Mammal Review* **41**, 197–219.

Schloeth, R. (1972) Die Entwicklung des Schalenwildbestandes im Schweizerischen Nationalpark von 1918 bis 1971. *Schweizer Zeitung Forstwesen* **123**, 565–571.

Schütz, M. (2005) Huftiere als 'driving forces' der Vegetationsentwicklung. *Forum für Wissen* **2005**, 27–30.

Stüwe, M. and Nievergelt, B. (1991) Recovery of Alpine ibex from near extinction: the results of effective protection, captive breeding and reintroduction. *Applied Animal Behaviour Science* **29**, 379–387.

Tolon, V., Dray, S., Loison, A., Zeileis, A., Fischer, C. and Baubet, E. (2009). Responding to spatial and temporal variations in predation risk: space use of a game species in a changing landscape of fear. *Canadian Journal of Zoology* **87**, 1129–1137.

Wölfl, M., Bufka, L., Červený, J., Koubek, P., Heurich, M., Habel, H., Huber, T. and Poost, W. (2001) Distribution and status of the lynx in the border region between Czech Republic, Germany and Austria. *Acta Theriologica* **46**, 181–194.

Chapter 9

Challenges in the Management of Cross-border Populations of Ungulates

*Carlos Fonseca, Rita Torres, João P.V. Santos,
José Vingada and Marco Apollonio*

"Borders separate, Nature unites": European Green Belt

In 1999, the Council of Europe held the first Symposium of the Pan European Ecological Network with the theme 'Nature does not have any borders: towards transfrontier ecological networks'. This Symposium highlighted the importance of transborder cooperation as a pivotal aspect for the management and conservation of biological and landscape diversity.From a worldwide perspective, border areas house some of the most pristine ecosystems in the world, many of which are located in isolated and inhospitable areas (Westing, 1998).

However, while transboundary cooperation could help to protect localised cross-border populations of threatened taxa or could promote real ecological corridors enhancing population connectivity and shared monitoring, different management objectives and different management policies in neighbouring areas can also create problems and conflicts, particularly because borders exist for political, economic, cultural and social purposes (Meidinger, 1998) and administrative limits are rarely defined in accordance with natural systems. Problems become particularly acute for highly mobile and migratory wildlife species whose ecological requirements transcend jurisdictional boundaries (Young, 1997; Zbicz and Green, 1999). If adjoining administrative authorities apply different jurisdictional, planning and management regulations, policies and goals for the wildlife resources, such discontinuities will eventually threaten ecological refuges (Westing, 1998) and jeopardise the viability of the shared wildlife populations (Schonewald-Cox and Bayless, 1986; Landres *et al.*, 1998). In the most extreme scenario, borders may even be fenced, which will have an obvious direct negative effect on wildlife movement and ecology (Griffin *et al.*, 1999). As a result of all this, many important areas may be subject to different management approaches, objectives and land-use

192

practices across borders; sometimes these practices are incompatible, risking the ecological processes and promoting wildlife management conflicts.

To avoid such contradictions, it is therefore essential that the planning of conservation and management policies should take into account the ecological concerns of local, regional, national and international boundaries. The ultimate goal of any cross border project ought be to link local, national and transboundary levels of the management of species occurring on both sides of any border. The target should be the development of shared management objectives and compatible management practices for each species, which can then be communicated to the local management as well as be shared with neighbouring areas as part of a transboundary vision. One good example of how such a transboundary approach can enhance local and regional ecological benefits and economic development (Spenceley, 2006) is the Great Limpopo Transfrontier Park, which comprises the South African Kruger National Park, Zimbabwe's Gonarezhou National Park and Mozambique's adjacent Limpopo National Park and Coutada 16 hunting reserves, where the shared ecotourism strategy is seen as a major vehicle to promote the local and regional development (Knight *et al.*, 2011).

In addition, cross border management can have an important role in controlling wildlife trade by developing a coordinated approach between authorities of the neighbouring countries to tackle cross border illegal wildlife activities. In some countries (e.g. China) heavy cross-border poaching has lead to wildlife resources being greatly reduced, with some species even facing extinction (Zhang *et al.*, 2008). With open international borders, wildlife poachers and traders have easy access to neighbouring areas – as well as easy options for escape – and generally authorities on both sides of the border are powerless, due to diplomatic restraints and lack of capacity, leading to wildlife criminals being able to operate without restrictions in many cases. In such situations, some agreement of a cross border management strategy may serve to develop a long-term wildlife trade control and monitoring in the border regions. Transboundary agreements of this kind can even facilitate international peace in some cases (e.g. the Condor Mountain Range between Peru and Ecuador facilitated the end of a regional conflict between these two countries; Braack *et al.*, 2006).

While initial efforts at such cross border cooperation were based on a simple agreement between two adjacent areas (in most instances, protected areas), experience has shown a need to think in a broader context and in terms of wider wildlife networks, ecological corridors or joint adaptive management. Under the flag of cross border, regional/provincial, or even local cooperation, large areas can be managed in a uniform and consistent manner and thereby make a substantial contribution to conservation biodiversity and wildlife management. Unfortunately, to set up a cross border networking, more is needed than simply the political will. Different economic and cultural scenarios, political tensions and traditions can hinder cross border cooperation or at least jeopardise its practical value (Apollonio *et al.*, 2010a).

Any attempt at cross border planning must take into account the ecological requirements of species and aim to maintain access for migratory and ranging species to crucial sites and resources across borders, including those needed in critical seasons (such as winter in Central and Northern Europe or summer in the Mediterranean region), or during years with extreme climatic events. Land-use plans on both sides of a border should include viable ecological corridors linking important resources needed for the entire life cycle of the species. This is especially true as human-mediated disturbance has had a profound, and ongoing, effect on habitats around the world (Loreau *et al.*, 2001) and ecosystems are increasingly fragmented as a consequence of the human population increase, urbanisation, expansion of transport infrastructure, habitat transformation and agriculture intensification (e.g. Prugh *et al.*, 2008). To manage such a developing world, where areas of wilderness are rapidly decreasing, the management and mitigation of border effects on ecological processes must involve cooperative and integrated approaches to ensure ecological connectivity (Bennett, 2003), especially in relation to the development and implementation of transnational management plans and strategies. Transboundary conservation and management practices are particularly useful for socio-cultural-political reasons but also for wildlife conservation and management, highlighted here by ungulate populations because of their ecological, economical, social and cultural importance (Apollonio *et al.*, 2010a).

9.1 Cross border cooperation in Europe

Generally, cross border cooperation in Europe is far from ideal (Wolf *et al.*, 2006). Currently, most management of ungulate populations – whatever the objective of that management – is undertaken in isolation within individual countries or regions, without wider consideration of how this may conform to, or conflict with management of those same ungulate populations in neighbouring areas. Apollonio *et al.* (2010a) have emphasised this lack of coordination of management at sufficiently large geographical scale as being one of the major problems constraining effective management of European ungulate populations. This is not purely a problem affecting appropriate cooperation between separate nation states, but may also be apparent in lack of consistency of management between different provinces within a given region, or even between adjacent game management units or hunting blocks (see below).

However, there have been some attempts to establish some integration of management activities, most commonly in regard to management of conservation areas. Under a banner of cross border cooperation, a number of 'Euroregions' or 'Euregios' were established in 1953 within which, at least in principle, local and regional authorities have joined together across one or several national borders. These 'Euregios' provide the testing ground for pilot projects to examine practical solutions of cooperation (Perkmann, 2007). Despite some good examples of collaborations, the good relations that are needed depend almost solely on private personal relations and contacts. Only 25% of the transborder areas base their collaboration on legal and official agreements with the vast majority operating in this way on the basis of private or informal contacts. Some promising examples of European cooperation are presented below, highlighting their virtues and problems and the degree to which they have been effective in terms of wildlife conservation and management.

9.2 European initiatives of cross border cooperation

9.2.1 European Green Belt

The European Green Belt (Figure 9.1) was established in 2003 and follows the route of the former Iron Curtain from the Barents Sea at the Russian–Norwegian border, along the Baltic Coast through Central Europe and the Balkans to the Black Sea. The scale of this remarkable ecological network is notable: 40 National Parks are situated

*in accordance with UNSCR 1244 and opinion of ICJ.

Figure 9.1 General localisation of the European Green Belt (© European Green Belt Initiative/Coordination Group; reproduced with permission).

along 12,500 km of the European Green Belt, more than 3,200 protected areas can be found within a 50 km buffer on either side of the Green Belt and it crosses nearly all European biogeographical regions. Connecting 24 countries, this project symbolises cross border cooperation and the global effort for joint nature conservation and sustainable development. The major aim of this initiative is to conserve and enhance the development of this ecological network. Apart from the underlying ecological values, this project also has a historical, political and social dimension. A feature that once divided nations can now serve as a symbol of Europe growing together.

However, there are some threats to the European Green Belt due to ever-increasing urbanisation and fragmentation with subsequent loss of habitats and migration corridors. Despite its remarkable scale, connectivity within the European Green Belt may be at risk. New road infrastructures are planned on the border regions of Bavaria, Czech Republic, Austria and Slovakia as well as hydropower industries in the border rivers (Mohl *et al.*, 2009). Additionally, the general increasing trend of tourism and construction of new buildings are hazards for the natural and semi-natural landscapes of the Baltic, Mediterranean and Black Sea coasts of the European Green Belt (Zmelik *et al.*, 2011).

9.2.2 East Carpathian Biosphere Reserve

The East Carpathian Biosphere Reserve is a mountain area covering a total of 40,601 ha across Poland, Slovakia and the Ukraine (Figure 9.2). This biosphere reserve was designated in 1998, uniting an existing Polish-Slovak reserve (designated in 1992) with the Ukrainian part. The Foundation for Eastern Carpathians Biodiversity Conservation (ECBC) was established in 1995, specifically aimed at supporting transboundary cooperation. This biosphere reserve is an example of transboundary cooperation under difficult political and economic circumstances (Guziova, 1996).

This Biosphere Reserve includes three biogeographical zones and its remoteness and restricted accessibility is reflected in the ecological diversity and its significant value for conservation because of the many endemic species. The Carpathian forests play an important role in water conservation, and hence in the agriculture and industry of the Central European countries. A number of more specific projects have been implemented within the area as a whole, chiefly focused on the management of nature reserves, mountain meadows, river corridors and water ecosystem protection, lowering the impact of tourism on the core area, conservation of old monumental trees and restoration of historical buildings. Important conservation programmes are currently ongoing such as the reintroduction of the Hutzul horse, with positive impacts on local sustainable tourism development. In order to unify wildlife inventory methodologies and databases and to facilitate common decisions, a common GIS (Geographic Information System) database has been developed.

European bison were reintroduced in the Carpathian region in the 1960s and 1970s; currently, around 350 European bison live in the Carpathians in five free-ranging herds even though, as the result of a small founder population, genetic

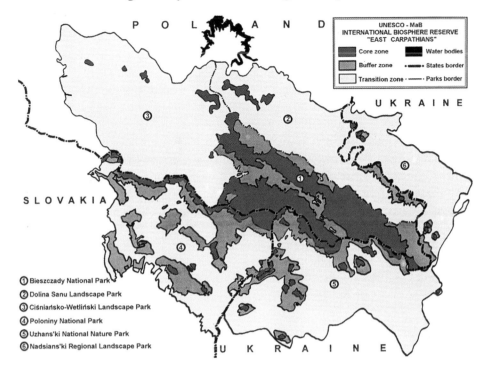

Figure 9.2 East Carpathian Biosphere Reserve location (© Z. Niewiadomski and J. Wolski; with permission).

diversity is low (Ziółkowska *et al.*, 2012; see also Chapter 3). Habitat suitability analyses showed that this region offers favourable conditions for bison (e.g. large areas of suitable habitat and relatively low human pressure) and this area is perhaps one of the best in which it would be possible to create a large and connected population (Pucek, 2004; Kuemmerle *et al.*, 2007). However, current bison herds only occupy a small portion of the available high-quality habitat and are partly isolated from each other (Ziółkowska *et al.*, 2012). The establishment of a large and more continuous bison population in the area depends on the establishment of linkages among the existing herds in the cross-border region of Poland, Ukraine and Slovakia and coordinated efforts to expand the area occupied by bison. And for that, the transboundary region covered by the East Carpathian Biosphere Reserve should be effective and proactive among countries.

9.2.3 Białowieża National Park and Belovezhskaya Pushcha Biosphere Reserve

Our third example is the Białowieża forest, a primeval forest that extends along the borders of Belarus and Poland. The Białowieża National Park (BNP) in Poland was first established as a park in 1932 and restored in 1947. The park was declared a Biosphere Reserve in 1977 and a World Heritage Site in 1979 (Agrawal, 2000). It is also a Transboundary World Heritage site in combination with the

Belovezhskaya Pushcha Biosphere Reserve (BPBR), the neighbouring park on the Belarus side of the border. BNP covers 10,500 ha and its core area (4,747 ha) is under strict protection, while BPBR covers 96,223 ha and its strictly protected area (15,677 ha) is the only one in Belarus.

Different conservation and management regimes characterise the two sides of the border: from strict protection to managed forestry. Deciduous trees dominate the strictly protected area on the Polish side while the Belarusian side is dominated by tree stands of mixed conifers. Moreover, the Polish agencies have permitted a much higher intensity of exploitation and plantation in comparison with the Belarusian forests (e.g. timber harvest on the Polish side is nearly four times higher then in the Belarusian area; Okarma *et al.*, 1996); the BNP also supports much higher densities of ungulates (red deer, moose, roe deer and wild boar) and their hunting is currently forbidden.

The symbol of the park is the European bison. In 1952, the first bisons were released into the wild in the Polish part and in 1953 into the Belarusian part (Daleszczyk and Bunevich, 2009). As the herds expanded, some individuals from Poland, mainly adult males, naturally dispersed and started to cross the border (Bunevich, 2004). However, in the early 1980s, a fence was built along the Belarusian border, now acting as a barrier not only for bison but also for movement of other ungulates and carnivores between the two countries (Daleszczyk and Bunevich, 2009). This artificial barrier (as well as different data collection or management strategies on both sides of the border) is reflected in different wildlife densities, and in the type of interactions among the different animal and plant species.

9.3 Cross border management of wide-ranging species

Despite several general political and nature conservation initiatives that aim to promote the coordination of common conservation strategies for the whole of any given ecosystem shared by different countries, administrative institutions or managers at different scales, there are some constraints to defining and implementing common strategies for border ungulate populations. Apollonio *et al.* (2010a) emphasise that most European countries show a lack of coordination of management effort over a sufficient large area for effective management from ungulate population monitoring to harvesting strategies. Effective population management needs administrative structures that are able to define and manage real biological population units, based on the ecological requirements of the species of concern; the majority of ungulate species have large home ranges and their territories may extend over many landscape units/hunting grounds, provinces or even several countries, requiring a global coordination of their management. Coordination of strategy will thus ensure the proper integration of objective, conservation and management efforts over an appropriate spatial scale.

Independent of scale, the practical difficulties and solutions for boundary ungulate population management are, in most of the cases, very similar, as is apparent in the examples below.

9.3.1 Management of red deer in the Iberian Peninsula

The red deer is a common ungulate widespread throughout the Iberian Peninsula (Carranza, 2010; Vingada *et al.*, 2010). In Spain, red deer populations in the north of the country almost disappeared in the 20th century and were restocked with individuals from central and southern regions. In the Hunting Reserve of Sierra de la Culebra (HRSC), northwest of Spain, red deer became extinct in the late 19th century and the current populations are the result of a series of reintroductions since 1972 totalling 57 males and 131 females. These reintroduced populations rapidly spread, colonising other areas, including the Montesinho Natural Park (MNP), northeast of Portugal, which borders the HRSC.

In Portugal, by the end of the 19th century, the species had also come close to extinction; however, in the three last decades, through reintroduction programmes and natural dispersion, populations have recovered and the species is now comparatively common throughout much of the country (Vingada *et al.*, 2010). The most representative red deer populations are located in border areas and that includes the red deer population of Montesinho Natural Park (MNP), above.

The red deer population has been consistently monitored in the HRSC since the '90s mainly by the rangers, but monitoring only began on the Portuguese side in 2007 closely associated with research projects (Santos, 2009; Carvalho, 2013). Up until 2011, monitoring on the Portuguese side was only based on the collection of demographic data (i.e. density, productivity and sex-ratio) (see Table 9.1) but, after that, a cross-border monitoring plan also included the evaluation of individuals' physical condition.

Table 9.1 Red deer demographic data from both sides of the border.

Area	Hunting season	Density (deer/km²)	Sex ratio (M:F)	Productivity (calves:female)
MNP	2007/2008 (autumn–spring)	1.17	0.74	0.44–0.36
	2008/2009 (winter)	3.26	0.83	0.39
	2010/2011 (autumn)	0.95	1.08	0.41
	2010/2011 (winter)	1.75	0.78	0.36
HRSC	2007/2008 (autumn)	1.21	0.81	0.40
	2008/2009 (winter)	1.61	0.93	0.40
Transboundary	2011/2012 (autumn)	3.02	0.61	0.45
	2012/2013 (autumn)	2.93	0.81	0.45

In MNP and HRSC different harvesting regimes and quotas are operated: in MNP males (seven males as maximum quota) are hunted by stalking while in HRSC deer are hunted both by stalking (17 males every year on average) and by female selective hunting.

Management of this continuous population of red deer on either side of the national border clearly has different goals. This is mainly because in Portugal the deer are contained within a protected area whereas in Spain, they fall within the boundaries of a hunting reserve (HRSC). Inevitably, this leads to conflicts between hunting associations and conservationists. However, since 2011, a great effort has been made to strengthen cross-border cooperation in the management of red deer, promoting the exchange of information between the administrative bodies of the two countries. Standardisation of monitoring methodologies has facilitated (and will continue to facilitate) comparisons of the results from the two sides of the border and the development of a framework guided by common objectives reconciling hunting activity and nature conservation.

9.3.2 Red deer management in the Alps

Another border that offers good illustration of some of the difficulties that may arise between countries with different cultures, management traditions and attitude towards wildlife is represented by the Alps. The south side of this mountain chain belongs to Italy but the north side falls within a number of different countries including France, Switzerland, Germany, Austria and Slovenia. There are a significant number of differences in management objective and practice on different sides of this mountain chain and those differences have had a profound impact on ungulate populations in recent centuries. A first example of this is given by different regimes of winter/summer feeding in Austria and Italy and its influence of spatial behaviour of red deer populations. Ungulate feeding has been compulsory by law in Austria for a long time (Reimoser and Reimoser, 2010); only in very recent times has the situation changed such that in some regions managers of hunting districts are no longer required to offer such supplementation. By contrast, across the border in Italy, there was no legal requirement to feed and there were (and remain) legal restrictions on the provision of winter feeding. The red deer is a highly mobile species, with alpine populations that are composed both by resident and by migrating individuals (Georgii, 1980; Georgii and Schröeder, 1983). In one particular border area between Italy and Austria (the area of Tarvis in north east Italy), red deer populations were known to migrate in summer reaching Austria, and taking advantage of the large amount of food offered in the feeding stations of the local hunting reserves (Luccarini *et al.*, 2006). These migrating individuals contributed to the high level of damage suffered by Austrian forests just close to the border; over winter, however, they withdrew to the Italian side of the border making them inaccessible to managers from the Austrian side. The need of some common management plan was obvious - but effectively impracticable, at least in the '90s, since most of the Italian side was inside a protected area.

9.3.3 The management of alpine chamois scabies on the Alps

A different example may be given for the same area in relation to the different strategies adopted to deal with sarcoptic mange in alpine chamois. Sarcoptic mange is a contagious infestation of the burrowing mite *Sarcoptes scabiei*, first observed in the alpine chamois (Bavaria and Styria) at the beginning of the 19th century (Miller, 1986). Among pathogens, sarcoptic mange has had the most severe impact on populations of chamois and ibex, and therefore represents a major threat for conservation. The majority of scabies-induced mortality events have been reported for the alpine chamois in Austria, Slovenia and Italy as well as for the Cantabrian chamois (Rossi *et al.*, 2007). Scabies in Alpine chamois may cause very high local mortality (up to 80%, Rossi *et al.*, 1995), particularly at the end of the winter (Rossi *et al.*, 2007); moreover in alpine ibex the impact can be even greater: the ibex colony of the Ljubelj area in Slovenia was completely extirpated by an outbreak of this disease (Adamic and Jerina, 2010).

In chamois both sexes have higher survival rates in areas formerly affected by scabies than in areas never affected, possibly because of the lower post-epidemic population density but also as a result of both acquired immunity and greater prevalence within the population of individuals with some genetic resistance to the disease. High MHC heterozygosity may help combat a wide range of pathogens, eventually resulting in higher survival rates. Schaschl *et al.* (2009) showed that males of several chamois populations on the eastern Alps tend to survive longer, if they are heterozygous at the MHC class II *DRB* locus, whereas females do not: interestingly, in the case of sarcoptic mange, males tend to carry higher parasite loads than females (Rossi *et al.*, 1995). At the same time density is a recognised factor influencing the probability of disease transmission; remarkably, whilst pre-epidemic densities in alpine chamois populations can be quite different from each other (range 2.0–22.7 individuals/100 ha), post-epidemic ones are similar (range 1.1–1.7 chamois/100 ha). This suggests that scabies will not spread further into a naive chamois population once a 'threshold' density below 2 individuals/100 ha has been attained (Rossi *et al.*, 2007).

Against such a context and in face of a steadily expanding disease, neighbouring countries adopted quite different policies: while in Austria the policy is to shoot any infested animal (Miller, 1986), in most neighbouring Italian provinces, the policy is diametrically opposite: that is hunters are discouraged from shooting any chamois with signs of infestation, instead allowing the population's own dynamics to respond to this natural challenge. The rationale behind this latter policy is to preserve individuals that may have innate resistance to the disease and thus to engineer a surviving population which is less susceptible to future infection. By converse, the policy in Austria is based on an objective of removing any potential source of infection in combination with a reduction of overall population density, to reduce rates of transmission of the parasite.

However, Ferroglio *et al.* (2011) have noted that while selective culling is often proposed as a method to contain infections in wildlife, it is only usually effective when affected individuals are readily identifiable (Wobeser, 2002). Since the status of infection of wild animals cannot easily be determined and many infected individuals cannot be detected, the proportion of animals killed by selective culling may have little or no effect on the spread of the disease, unless population densities are indeed reduced below levels where onward transmission is inhibited. Field experience, such as experience with chamois mange in the Cantabrian Mountains, Spain (Fernandez-Moran *et al.*, 1997), or infectious keratoconjuntivitis in chamois (Ferroglio *et al.*, 2011), clearly shows that selective culling of obviously infected individuals is of little or no value in disease management of free ranging wildlife and may even prove counterproductive (see also Corlatti *et al.*, 2011). In addition, it is obvious that, since the population is in fact shared between two countries, lack of coordination of control strategies is hardly well-designed to ensure effective control of the disease, since each approach, while potentially effective on its own, is countered by application of the contrasting policy in other parts of the distributional range.

9.3.4 Alpine ibex management

Another example of inconsistencies associated with transboundary management is given by alpine ibex management across the Swiss–Italian borders (Figure 9.3) although in this case the story has a more positive twist. Nowadays more than 100 populations of ibex live in the Alps, amounting to around 47,000 individuals (Tosi *et al.*, 2009). All of them have their ancestry in a remnant nucleus of around

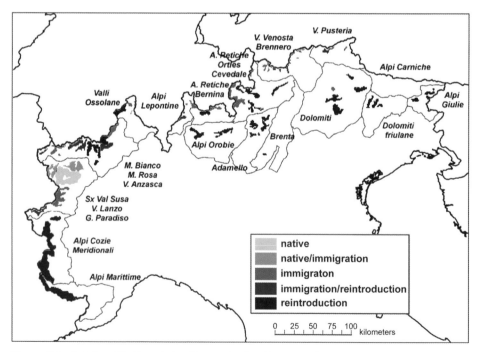

Figure 9.3 Origin of Italian alpine ibex colonies (from Tosi *et al.*, 2009).

100 ibex which survived in the 19th century, after over-hunting had driven the species to extinction in the rest of its former range. This surviving population, occupying the present Gran Paradiso National Park, was the source for captive breeding programmes that started in Swiss parks in the beginning of the 20th century. In turn, these parks, along with the Gran Paradiso population, provided animals for the first reintroductions in the wild, which were followed by a mass of additional translocations that were carried out in Switzerland, Italy, France, Austria, Germany and Slovenia (see also Chapter 3).

In Switzerland, ibex were reintroduced the beginning of the 20th century; populations expanded such that legal hunting has been permitted since the 1960s (Imesch-Bebié *et al.*, 2010) and in the time span between 1995–2005 a number of 900–1800 ibex were shot each year (Swiss Federal Hunting Office, http://www. wild.uzh.ch/jagdst/). In Austria, ibex were reintroduced from 1951 and have been hunted from 1978, with about 400–500 ibex shot each year.

By contrast, in Italy (and France) the species is still protected and no hunting is allowed with minor exception (such as in Bolzano province with less than 40–50 ibex shot per year) (Apollonio *et al.*, 2010b). In consequence of this lack of hunting, populations not only expanded through reproductive recruitment but were also enhanced by immigration from the (hunted) Swiss and Austrian population: in fact, some 48% of the colonies on the south side of the Alps originated from immigration, including some late colonies established by immigration from France and Austria. In such a case, it is clear that differences between countries may not always bring problems with them as there may be unpredictable outcomes from such disjunction in management policy.

9.4 Different hunting seasons and rules within a country

Even within a country, issues may arise of inconsistencies in management in adjacent administrative areas. By way of illustration, in Italy, for example, all hunting, with a few exceptions, is in the charge of each individual provincial government. This implies that different (even bordering) provinces may adopt different hunting rules for the same species.

In different provinces, wild boar may be hunted either from October to December or from November to January. One month shift may mean that border groups may easily escape hunting moving to one or another part of their home range, contributing to making control of boar numbers extremely difficult. This is a serious issue in that wild boar are responsible for more than 90% of agricultural damage by ungulates (Apollonio *et al.*, 2010b). An even more extreme case can be represented by north-eastern Italy where each municipality has its own hunting ground and can choose, for instance, whether or not dogs may be used in pursuit of game. The outcome is a patchwork of markedly uneven densities of roe deer, due to inconsistencies in the way hunting is managed.

Where the rights to hunt game are leased to different Hunter's Associations or Hunter's Clubs such discontinuities may become apparent at even more local levels, and perhaps reach their extreme in the UK, where a very strong concept of private property implies a total freedom of the owners of land to state their own objectives for deer management within their own property (irrespective of the choices of their neighbours) such that they may elect to manage for sport, for control of damage to agriculture and forestry, or to undertake no effective management at all, without need for consultation (Putman, 2004, 2010).

Even at very local levels therefore issues may still arise as the result of lack of coordination of management across boundaries, where conflicting management objectives and actions may be applied at geographic scales which do not encompass the entire biological range of a single population.

9.5 Management with no boundaries: opportunities and challenges

Understanding of the dynamics and movement patterns of animal populations implies that management should be adapted to those ecological and behavioural characteristics and that management, to be effective, must be at the lansdscape level, or at least at the scale of the entire range of a given biological population. In the last chapter of the volume *European Ungulates and their Management in the 21st Century,* Apollonio *et al.* highlighted the lack of coordination of management over a sufficiently large geographical scale and the inadequate scale of most management units as one of the major challenges of ensuring effective management of ungulate populations (Apollonio *et al.*, 2010a). This chapter clearly poses the question of the need for management with no boundaries.

Problems of inconsistency in management objective and management practice indeed persist even within countries (above). Efforts to coordinate and integrate the conservation and management of ungulates in trans-national areas within Europe perhaps face even greater obstacles, due to conflicts between national interests and differences in existing legislation in different countries. Inevitably therefore, the initiatives already in place in the majority of instances of developing collaboration, are restricted to shared monitoring of population size and movements across the borders. Even where these studies indicate an obvious need for integrated trans-national management action, such international management plans are rarely implemented in practice. Transboundary cooperation is simpler when the target species has some kind of protected status on both sides of the border; cooperation in also better developed when such a species lives in protected areas which cross administrative borders. But when dealing with species that are harvested in one part of the border and that are not hunted in the other side of the border, cross border cooperation poses extra challenges.

However the existence of the European Community represents a good chance to implement coordination within the 28 countries that belongs to the EU. The possibility of legislative actions that, through the setting of appropriate directives can coordinate actions over the whole Europe, offers the possibility to overcome some of these limitations. In the context of environmental management, Directive 92/43/EEC on the Conservation of Natural Habitats and Directive 79/409/EEC on the Conservation of Wild Birds give good examples of this; perhaps a similar initiative attempting to coordinate ungulate management should be considered.

9.5.1 Future opportunities

Plachter (2005) claimed 'the idea to cross national borders by joint protected area programmes is one of the noblest and convincing ones in current days'. Currently, a major challenge for conservation and wildlife management is to determine species responses to environmental gradients in order to anticipate the consequences of global change. Scientists believe that as climate change progresses, the importance of continuous cross border cooperation will increase. In response to climate change, species can shift their distribution, which has been already documented for many species (Root *et al.*, 2005; Parmesan, 2006; Rosenzweig *et al.*, 2008). Shifts can be either pole-wards (e.g. Parmesan *et al.*, 1999; Hickling *et al.*, 2006) or upslope (e.g. Colwell *et al.*, 2008; Merrill *et al.*, 2008). Mountain ungulates on the Alps and on other mountain chains in Europe are presently threatened by the last kind of distributional shift: the change of growing phase of vegetation (Pettorelli *et al.*, 2007) and the limited tolerance to high temperature (Aublet *et al.*, 2009) imposes a common approach: it is obviously not appropriate to apply different management to the same population on the two sides of a mountain.

Therefore, the maintenance of logical, effective and functional transboundary areas that may also contribute to the definition of effective migratory ecological and management units, will be crucial to allow maintaining viable transboundary populations, promoting the movement of species and adaptation of the natural systems in response to changing ecosystem conditions. Thus the long-term persistence of viable wildlife populations and habitats depends upon effective protection that transcends administrative boundaries.

References

Adamic, M. and Jerina, K. (2010) Ungulates and their management in Slovenia. In M. Apollonio, R. Andersen and R.J. Putman (eds), *European Ungulates and their Management in the 21st Century.* Cambridge, UK: Cambridge University Press, pp. 507–526.

Agrawal, A. (2000) Adaptive management in transboundary protected areas: The Białowieża National Park and Biosphere Reserve as a case study. *Environmental Conservation* 27(4), 326–333.

Apollonio, M., Andersen, R. and Putman, R. (2010a) Present status and future challenges for European ungulate management. In M. Apollonio, R. Andersen and R.J. Putman (eds), *European Ungulates and their Management in the 21st Century.* Cambridge, UK: Cambridge University Press, pp. 578–604.

Apollonio, M., Ciuti, S., Pedrotti, L. and Banti, P. (2010b) Ungulates and their management in Italy. In M. Apollonio, R. Andersen and R.J. Putman (eds), *European Ungulates and their Management in the 21st Century.* Cambridge, UK: Cambridge University Press, pp. 475–506.

Aublet, J.F., Festa-Bianchet, M., Bergero, D., Bassano, B. (2009) Temperature constraints on foraging behaviour of male Alpine ibex (*Capra ibex*) in summer. *Oecologia* **159**, 237–247.

Bennett, A.F. (2003) *Linkages in the Landscape: The role of Corridors and Connectivity in Wildlife Conservation.* Gland, Switzerland: IUCN.

Braack, L., Sandwith, T., Peddle, D. and Petermann, T. (2006) *Security considerations in the planning and management of transboundary conservation areas.* Gland, Switzerland: IUCN.

Bunevich, A.N. (2004) Spacial structure and movements of European bison in the Belarusian part of Białowieża Forest (Belavezhskaya Pushcha). In M. Krasińska, K. Daleszczyk (eds), *Proceedings of the Conference European Bison Conservation,* 30 September–2 October 2004. Białowieża, Poland: Mammal Research Institute Polish Academy of Sciences, pp. 23–26.

Carranza, J. (2010) Ungulates and their management in Spain. In M. Apollonio, R. Andersen and R.J. Putman (eds), *European Ungulates and their Management in the 21st Century.* Cambridge, UK: Cambridge University Press, pp. 419–440.

Carvalho, J. (2013) *O veado: análise ecológica e especial em três populações.* MSc thesis. University of Aveiro, Portugal.

Colwell, R.K., Brehm, G., Cardelús, C.L., Gilman, A.C. and Longino, J.T. (2008) Global warming, elevational range shifts, and lowland biotic attrition in the wet tropics. *Science* **322**(5899), 258–261.

Corlatti, L., Lorenzini, R. and Lovari, S. (2011) The conservation of the chamois *Rupicapra* spp. *Mammal Review* **41**, 163–174.

Daleszczyk, K. and Bunevich, A.N. (2009) Population viability analysis of European bison populations in Polish and Belarusian parts of Białowieża Forest with and without gene exchange. *Biological Conservation* **142**, 3068–3075.

Fernandez-Moran, J., Gomez, S., Ballesteros, F., Quiros, P., Benito, J.L., Feliu, C. and Nieto, J.M. (1997) Epizootiology of sarcoptic mange in a population of Cantabrian chamois (*Rupicapra pyrenaica parva*) in Northwestern Spain. *Veterinary Parasitology* **73**, 163–171.

Ferroglio, E., Gortazar, C. and Vicente, J. (2011) Wild ungulate diseases and the risk for livestock and public health. In R.J. Putman, M. Apollonio and R. Andersen (eds), *Ungulate Management in Europe: Problems and Practices.* Cambridge, UK: Cambridge University Press, pp. 192–214.

Georgii, B. (1980) Home range patterns of female red deer (*Cervus elaphus*) in the Alps. *Oecologia* **47**, 278–285.

Georgii, B. and Schröeder, W. (1983) Home range and activity patterns of male red deer (*Cervus elaphus*) in the Alps. *Oecologia* **58**, 238–248.

Griffin, J., Metcalfe, S., Singh, J. and Cumming, D.H.M. (1999) *Study on the development of transboundary natural resource management areas in southern Africa.* Main Report. Biodiversity Support Program. Washington DC.

Guziova, Z. (1996) East Carpathian Biosphere Reserve. In A. Breymeyer and R. Noble (eds), *Biodiversity Conservation in Transboundary Protected Areas.* Washington DC: National Academy Press, pp. 135–140.

Hickling, R., Roy, D.B., Hill, J.K., Fox, R. and Thomas, C.D. (2006) The distributions of a wide range of taxonomic groups are expanding polewards. *Global Change Biology* **12**, 450–455.

Imesch-Bebié, N., Gander, H. and Schnidrig-Petrig, R. (2010) Ungulates and their management in Switzerland. In M. Apollonio, R. Andersen and R.J. Putman (eds), *European Ungulates and their Management in the 21st Century.* Cambridge, UK: Cambridge University Press, pp. 357–391.

Knight, M.H., Seddon, P.J. and Midfa, A.A. (2011) Transboundary conservation initiatives and opportunities in the Arabian Peninsula. *Zoology in the Middle East* **54**, 183–195.

Kuemmerle, T., Hostert, P., Radeloff, V.C., Perzanowski, K. and Kruhlov, I. (2007) Post-socialist forest disturbance in the Carpathian border region of Poland, Slovakia, and Ukraine. *Ecological Applications* **17**, 1279–1295.

Landres, P.B., White, P.S., Aplet, G. and Zimmermann, A. (1998) Naturalness and natural variability: definitions, concepts, and strategies for wilderness management. In D.L. Kulhavy and M.H. Legg (eds), *Wilderness and Natural Areas in Eastern North America: Research, Management and Planning.* Nacogdoches, TX: Stephen F. Austin State University, Arthur Temple College of Forestry, Center for Applied Studies, pp. 41–50.

Loreau, M., Naeem, S., Inchausti, P., Bengtsson, J., Grime, J.P., Hector, A., Hooper, D.U., Huston, M.A., Raffaelli, D., Schmid, B., Tilman, D. and Wardle, D.A. (2001) Biodiversity and eco-system functioning: current knowledge and future challenges. *Science* **294**, 804–808.

Luccarini, S., Mauri, L., Ciuti, S., Lamberti, P. and Apollonio, M. (2006) Red deer (*Cervus elaphus*) spatial use in the Italian Alps: home range patterns, seasonal migrations, and effect of snow and winter feeding. *Ethology, Ecology, Evolution* **18**, 127–145.

Meidinger, E. (1998) Laws and institutions in cross boundary stewardship. In R.L. Knight and P.B. Landres (eds), *Stewardship Across Boundaries.* Washington DC: Island Press, pp. 87–110.

Merrill, R.M., Gutiérrez, D., Lewis, O.T., Gutiérrez, J., Díez, S.B. and Wilson, R.J. (2008) Combined effects of climate and biotic interactions on the elevational range of a phytophagous insect. *Journal of Animal Ecology* **77**, 145–155.

Miller, C., (1986) Die Gamsräude in den Alpen. *Zeitschrift für Jagdwissenschaft,* **32**, 42–46.

Mohl, A., Egger, G. and Schneider-Jacoby, M. (2009) Flowing boundaries: tensions between conservation and use along border rivers. *Natur und Landschaft* **84**, 431–435.

Okarma, H., Jedrzejewski, W. and Jedrzejewska, B. (1996) Bialowieza Primeval Forest: habitat and wildlife management. In A. Breymeyer and R. Noble (eds), *Biodiversity Conservation in Transboundary Protected Areas: Proceedings of an International Workshop,* May 1994, Poland, pp. 167–77.

Parmesan, C. (2006) Ecological and evolutionary responses to recent climate change. *Annual Review of Ecology, Evolution, and Systematics* **37**, 637–669.

Parmesan, C., Ryrholm, N., Stefanescu, C., Hill, J.K., Thomas, C.D., Descimon, H., Huntley, B., Kaila, L., Kullberg, J., Tammaru, T., Tennant, W.J., Thomas, J.A. and Warren, M. (1999) Poleward shifts in geographical ranges of butterfly species associated with regional warming. *Nature* **399**, 579–583.

Perkmann, M. (2007) Construction of new territorial scales: A framework and case study of the EUREGIO cross-border region. *Regional Studies* **41**, 253–266.

Pettorelli, N., Pelletier, F., von Hardenberg, A., Festa-Bianchet, M., Coté, S. (2007) Early onset of vegetation growth vs. rapid green-up: impacts on juvenile mountain ungulates. *Ecology* **88**, 381–390

Plachter, H. (2005) The World Heritage Convention of UNESCO: A flagship of the Global Nature Conservation Strategy, UNU Global. Seminar Series, Inaugural Shimane-Yamaguchi Session, Yamaguchi, Japan.

Prugh, L.R., Hodges, K.E., Sinclair, A.R. and Brashares, J.S. (2008) Effect of habitat area and isolation on fragmented animal populations. *Proceedings of the National Academy of Sciences* **105**, 20770–20775.

Pucek, Z. (2004) *European Bison. Status survey and Conservation Action Plan.* IUCN/SSC Bison Specialist Group. Gland, Switzerland: IUCN.

Putman, R.J. (2004) *The Deer Manager's Companion: A Guide to Deer Management in the Wild and in Parks.* Shrewsbury, UK: Swan Hill Press.

Putman, R.J. (2010) Ungulates and their management in Great Britain and Ireland. In M. Apollonio, R. Andersen and R.J. Putman (eds), *European Ungulates and their Management in the 21st Century.* Cambridge, UK: Cambridge University Press, pp. 129–164.

Reimoser, F. and Reimoser, S. (2010) Ungulates and their management in Austria. In M. Apollonio, R. Andersen and R. Putman (eds), *European Ungulates and their Management in the 21st Century.* Cambridge, UK: Cambridge University Press, pp. 338–356.

Root, T.L., MacMynowski, D.P., Mastrandrea, M.D. and Schneider, S.H. (2005). Human-modified temperatures induce species changes: joint attribution. *Proceedings of the National Academy of Sciences of the United States of America* **102**, 7465–7469.

Rosenzweig, C., Karoly, D., Vicarelli, M., Neofotis, P., Wu, Q., Casassa, G., Menzel, A., Root, T.L., Estrella, N., Seguin, B., Rryjanowski, P., Liu, C., Rawlins, S. and Imeson, A. (2008) Attributing physical and biological impacts to anthropogenic climate change. *Nature* **453**, 353–357.

Rossi, L., Meneguz, P.G., De Martin, P. and Rodolfi, M. (1995) The epizootiology of sarcoptic mange in chamois, *Rupicapra rupicapra,* from the Italian Eastern Alps. *Parasitologia* **37**, 233–240

Rossi, L., Fraquelli, C., Vesco, U., Permunian, R., Sommavilla, G.M., Carmignola, G., Da Pozzo, R. and Meneguz, P.G. (2007) Descriptive epidemiology of scabies epidemic in chamois in the Dolomite Alps, Italy. *European Journal of Wildlife Research* **53**, 131–141.

Santos, J. (2009) *Estudo populacional do veado (Cervus elaphus L.) no Nordeste Transmontano.* MSc thesis. University of Aveiro, Portugal.

Schaschl, H., Suchentrunk, F., Morris, D.L., Ben Slimen H., Smith, S. and Arnold, W. (2009) Sex-specific selection for MHC class II heterozygosity in Alpine chamois. *Book of Abstracts of the Fifth World Conference on Mountain Ungulates,* November, Granada, Spain, pp. 250–251.

Schonewald-Cox, C.M. and Bayless, J.W. (1986) The boundary model: a geographical analysis of design and conservation of nature reserves. *Biological Conservation* **38**, 305–322.

Spenceley, A. (2006) Tourism in the Great Limpopo Transfrontier Park. *Development in Southern Africa* **23**, 649–667.

Tosi, G., Apollonio, M., Giacometti, M., Lanfranchi, P., Lovari, S Meneguz, P.G., Molinari,P., Pedrotti, L., Perco, F., Toso, S. and Vigorita, V. (2009) *Piano di conservazione, diffusione e gestione dello stambecco sull'arco alpino italiano.* Provincia di Sondrio, IT: Settore Agricoltura e Risorse Ambientali.

Vingada, J., Fonseca, C., Cancela, J., Ferreira, J. and Eira, C. (2010) Ungulates and their Management in Portugal. In M. Apollonio, R. Andersen and R.J. Putman (eds), *European Ungulates and their Management in the 21st Century.* Cambridge, UK: Cambridge University Press, pp. 392–418.

Westing, A.H. (1998) A transfrontier reserve for peace and nature on the Korean peninsula. *International Environmental Affairs* **10**, 8–17.

Wobeser, G.A. (2002) Disease management strategies for wildlife. *Revue Scientifique et Technique de l'Office International des Epizooties* **21**, 159–178.

Wolf, U., Hollederer, A. and Brand, H. (2006) Cross-border cooperation in Europe: what are Euregios? *Das Gesundheitswesen* **68**, 667–673.

Young, O.R. (ed.), (1997) *Global Governance: Drawing Insights from the Environmental Experience.* Cambridge, MA: MIT Press.

Zbicz, D.C. and Green, M.J.B. (1999) Status of the world's transfrontier protected areas. Paper presented at the International Conference on Transboundary Protected Areas as a Vehicle for International Co-operation, 16–18 September 1997, Somerset West, South Africa.

Zhang, L., Hua, N. and Sun, S. (2008) Wildlife trade, consumption and conservation awareness in southwest China. *Biodiversity and Conservation* **17**, 1493–1516.

Ziółkowska, E., Ostapowicz, K., Kuemmerle, T., Perzanowski, K., Radeloff, V.C. and Kozak, J. (2012) Potential habitat connectivity of European bison (*Bison bonasus*) in the Carpathians. *Biological Conservation* **146**, 188–196.

Zmelik, K., Schindler, S. and Wrbka, T. (2011) The European Green Belt: international collaboration in biodiversity research and nature conservation along the former Iron Curtain. *Innovation: The European Journal of Social Science Research* **24**, 273–294.

Chapter 10

Novel Management Methods: Immunocontraception and Other Fertility Control Tools

Giovanna Massei, Dave Cowan and Douglas Eckery

Impacts of overabundant ungulate populations on human activities and conservation include crop and forestry losses, collisions with vehicles, disease transmission, nuisance behaviour, damage to infrastructures, predation on livestock and native species, and reduction of biodiversity in plant and animal communities (e.g. Curtis *et al.*, 2002; Massei *et al.*, 2011; Reimoser and Putman, 2011; Ferroglio *et al.*, 2011; Langbein *et al.*, 2011).

Current trends in human population growth and landscape development indicate that human–ungulate conflicts in Europe, as well as in the United States, are likely to increase in parallel with increased expansion in numbers and range of many of these species (Rutberg and Naugle, 2008; Brainerd and Kaltenborn, 2010; Gionfriddo *et al.*, 2011a). Many of these conflicts have been traditionally managed by lethal methods. However, current trends in distribution and numbers of wild boar, feral pigs and deer in Europe and in the United States (e.g. Saez-Royuela and Telleria, 1986; Waithman *et al.*, 1999; Ward, 2005; Apollonio *et al.*, 2010) suggest that recreational hunting is not sufficient to control ungulate densities. In addition, ethical considerations regarding humane treatment of animals are increasingly shaping public attitudes about what are considered acceptable methods of mitigating human–wildlife conflicts, and lethal control is often opposed (Beringer *et al.*, 2002; Wilson, 2003; Barfield *et al.*, 2006; McShea, 2012).

Public antipathy towards lethal methods increasingly constrains the options available for ungulate management, particularly in urban and suburban areas and in protected areas where culling is often opposed on ethical, legal or safety grounds (Kirkpatrick *et al.*, 2011; Boulanger *et al.*, 2012; Rutberg *et al.*, 2013). Consequently, interest in non-lethal methods, such as translocation or fertility control, has increased (Fagerstone *et al.*, 2010).

Reviews of translocations of problem wildlife as a mechanism for reducing human–ungulate conflicts concluded that this method may cause significant stress, increase mortality and traffic accidents, is relatively expensive and has the potential to spread diseases and pathogens (Daszak *et al.*, 2000; Corn and Nettles, 2001; Conover, 2002; Beringer *et al.*, 2002; Massei *et al.*, 2010a). Examples of translocations of pathogens and hosts include the spread of bovine brucellosis and bovine tuberculosis following the translocation of bison (*Bison bison*) in Canada (Nishi *et al.*, 2006), the potential spread and dissemination of diseases such as the Aujeszky's disease virus following the translocation of wild boar between hunting estates in Spain (Ruiz-Fons *et al.*, 2008) and warble and nostril flies spread to conspecifics by caribou (*Rangifer tarandus*) after translocation of animals from Norway to Greenland (in Kock *et al.*, 2010).

Fertility control is often advocated as a safe, humane alternative to culling for managing overabundant wildlife (Fagerstone *et al.*, 2010; McLaughlin and Aitken, 2011; Kirkpatrick *et al.*, 2011). Early attempts to use fertility control to manage ungulates failed for reasons that included toxicity of the drugs used, transfer of these drugs to the food chain, manufacturing costs and the fact that repeated applications of contraceptives were required to induce long-term infertility (Gray and Cameron, 2010; Kirkpatrick *et al.*, 2011). In the last two decades, a reawakened interest in alternatives to surgical sterilisation for companion animals and livestock has led to the development of novel fertility control agents (Herbert and Trigg, 2005; Naz *et al.*, 2005; Massei *et al.*, 2010b). In parallel, several fertility control agents have emerged for wildlife applications.

In this chapter we provide a comprehensive, critical overview of fertility control to mitigate human–ungulate conflicts. In particular, we discuss the availability and use of fertility control agents in ungulates, we review delivery methods for these agents, we provide a synthesis of the conclusions of empirical and theoretical studies of fertility control applied to populations and we offer suggestions to guide decisions regarding the suitability of fertility control to mitigate human–ungulate conflicts.

10.1 Fertility inhibitors for ungulates

10.1.1 Fertility control and reproduction

Chemical fertility control can be achieved through contraception or sterilisation. Contraception prevents the birth of offspring but maintains fertility, whilst sterilisation renders animals infertile (Kutzler and Wood, 2006). In mammals, the series of events that leads to ovulation and spermatogenesis begins in the brain, where gonadotropin-releasing hormone (GnRH) is produced in the hypothalamus. GnRH is transported through small blood vessels to the anterior pituitary gland, where it binds to GnRH receptors to stimulate the release of the pituitary gonadotropins, LH (luteinizing hormone) and FSH (follicle-stimulating hormone) (Figure 10.1).

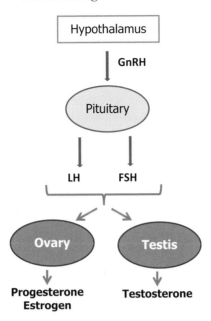

Figure 10.1 Schematic illustration of the fertility axis in male and female mammals.

These gonadotropins in turn stimulate the synthesis and secretion of sex hormones such as oestrogen, progesterone, and testosterone which are responsible for ovulation, spermatogenesis and sexual behaviour. The reproductive cycle and the production of eggs and sperm can be disrupted by administration of substances that interfere with the hypothalamic–pituitary–gonadal axis by blocking the synthesis, release or actions of hormones produced by the hypothalamus, the pituitary gland, or the testes and ovary. In females, a further target for contraception is the zona pellucida (ZP), a protein coat that surrounds the ovulated egg and allows species-specific sperm recognition and fertilization. In males, sterilisation can also be achieved by chemicals that cause testicular sclerosis and permanent sterility (Crawford *et al.*, 2011). The following section presents a brief overview of fertility control agents commercially available or widely tested on ungulates. Taking into account field applications, the review includes only those drugs that induce infertility for at least 6–12 months following administration of a single dose.

The majority of the fertility inhibitors reported in the literature target females, although some are effective for both genders and a few have been specifically developed for males. In many ungulate species the mating system is promiscuous, thus requiring extremely high levels of male sterility for fertility control to have any effect at the population level. For instance, in feral horses (*Equus caballus*) breeding still occurred even when 100% of the dominant harem stallions were sterilized (Turner and Kirkpatrick, 1991; Garrott and Siniff, 1992). In addition, some contraceptives may affect secondary sexual characteristics such as antler development (see later sections) and their use is not recommended for male deer.

A fertility control agent suitable for field applications should ideally have the following characteristics (Turner and Kirkpatrick, 1991; Fagerstone *et al.*, 2002; Massei and Miller, 2013):

1. Nil or acceptable side effects on the target animal's physiology, behaviour and welfare, including no interference with pre-existing pregnancy or lactation
2. Effective for at least one reproductive season when delivered through a single, injectable dose or implant, or when administered in one or multiple oral doses
3. Render all or the majority of treated animals infertile
4. Inhibit female reproduction but ideally prevent reproduction in both sexes
5. Relatively inexpensive to produce and deliver
6. No effect on any food chain
7. Species specificity
8. Stability under a wide range of field conditions.

Although none of the fertility control agents currently available meet all the above features, several exhibit many of these characteristics.

10.1.2 Hormonal contraceptives

Synthetic progestins such as norgestomet, melengestrol acetate (MGA), megestrol acetate (MA) and levonorgestrel have been widely used in zoo animals, livestock and wildlife. By binding to progesterone receptors, synthetic progestins disrupt ovulation and egg implantation in females and impair spermatogenesis in males (Asa and Porton, 2005). For instance, norgestomet, administered to white-tailed and black-tailed deer, caused infertility in 92–100% of the females for at least one year (Jacobsen *et al.*, 1995; DeNicola *et al.*, 1997). These drugs may cause abortion, although this effect depends on progestin type, species, dose and time of administration during pregnancy (Waddell *et al.*, 2001; Asa and Porton, 2005). MGA did not affect pregnancy in several ungulate species, but delayed or prevented parturition in treated white-tailed deer (Plotka and Seal, 1989; Asa and Porton, 2005). Progestin implants, with an estimated duration of efficacy of ≥2 years, have been widely used for suppression or synchronisation of oestrus in cattle and they have been employed as contraceptives in zoos for about 20 years. MA implants induced infertility in female mountain goats for at least 5 years, with reproduction recorded in 10% treated goats against 68% untreated controls (Hoffman and Wright, 1990).

Implants containing different concentrations of steriods such as ethinyloestradiol (EE), and progesterone (P) have been successful in preventing pregnancy in feral mares. Suppression of ovulation appeared to be inversely related to the concentration of EE used in the implant. The percentage of animals ovulating after 2 years was 12–20% for groups that had received a combination of P and EE or the highest dose of EE respectively, against 100% for control mares; pregnancy rate for the same groups was 0% for both P+EE and EE and 100% for control females. All animals that were pregnant at the time of contraceptive treatment delivered normal foals. The results demonstrated effective contraception of feral mares for up to 36 months without compromising pregnancy (Plotka *et al.*, 1992).

Another group of hormones widely used as contraceptives are the gonadotropin-releasing hormone (GnRH) agonists: these are synthetic peptides that mimic GnRH and stimulate the production and release of FSH and LH. Chronic administration of these drugs (e.g. >4 weeks) results in a downregulation of the pituitary gland and suppression of the secretion of FSH and LH. However, immediately following administration, a 'flare up' effect often occurs that can stimulate oestrus in females and cause temporary enhancement of testosterone and semen production in males (Patton *et al.*, 2007). As agonists have a higher affinity for and do not quickly dissociate from the GnRH receptors, the 'flare up' is followed by prolonged oestrus inhibition and infertility (Gobello, 2007) as long as the drug is present. The effectiveness of GnRH agonists depends on type of agonist, release system, dose rate and duration of treatment (Gobello, 2007; Patton *et al.*, 2007). The side-effects are equivalent to gonad removal but are reversible; however, GnRH agonists may cause abortion and thus their application to free-living ungulates is limited to those species that have a well-defined, relatively short breeding season (Asa and Porton, 2005).

Sustained-release subcutaneous implants containing GnRH agonists have been tested successfully in several livestock and wildlife species. For instance, implants of the GnRH agonist deslorelin (Suprelorin©) have been used to inhibit reproduction for 1–2 years in cattle and in several other wildlife species (e.g. D'Occhio *et al.*, 2002; Herbert and Trigg, 2005; Eymann *et al.*, 2007). Another GnRH agonist, leuprolide, administered in biodegradable implants was found effective at preventing pregnancy for one breeding season in 100% of female elk (wapiti) and mule deer with no effects on behaviour, body condition, haematology and blood chemistry (Baker *et al.*, 2002, 2004; Conner *et al.*, 2007). Regardless of proven efficacy, the use of hormonal contraceptives on free-ranging ungulates is still controversial because of potential welfare effects on pregnancy, environmental impact and possible transfer to consumers through the food chain (Kirkpatrick *et al.*, 1996; De Nicola *et al.*, 2000).

10.1.3 Immunocontraceptive vaccines

Most studies of fertility control applications in free-ranging ungulates have focussed on immunocontraceptive vaccines. These vaccines stimulate the immune system to produce antibodies to proteins or hormones essential for reproduction (Miller and Killian, 2002), thus rendering animals contracepted or infertile. To achieve long-term infertility, adjuvants are used, which are chemicals, large molecules or entire cells of killed pathogens, that enhance the immune response to a vaccine (Fraker *et al.*, 2002). Using liposome-based formulations has also been shown to increase the immune response of some immunocontraceptive vaccines (Fraker and Brown, 2011). The effectiveness, duration and side effects of immunocontraceptive vaccines can vary with species, sex, age, individual differences in immunocompetence, as well as the active component of the vaccine, its formulation, delivery system and the dose and type of adjuvant (Miller *et al.*, 2008a, 2009; Holland *et al.*, 2009; Kirkpatrick *et al.*, 2011). The most studied immunocontraceptives in ungulates are zona pellucida- and GnRH-based vaccines (Table 10.1).

Table 10.1 Effectiveness of single-dose immunocontraceptive vaccines to cause infertility in ungulate species in captivity and field trials. The effectiveness is expressed as proportion of infertile females in the control (C) and treatment (T) groups in the years following administration of the vaccine.

Species	Type of study	Vaccine type, adjuvant type and vaccine dose	% infertile females	References
White-tailed deer	Captive	GonaCon and AdjuVac various formulations	T GonaCon-KLH = 100% 60% 50% 50% 25% T GonaCon-Blue = 100% 100% 80% 80% 80%	Miller *et al.* (2008a)
White-tailed deer	Field	GonaCon-KLH and AdjuVac	T = 67% 43% C = 8% 17%	Gionfriddo *et al.* (2011a)
White-tailed deer	Field	GonaCon-KLH and AdjuVac	T = 88% 47% C = 15% 0%	Gionfriddo *et al.* (2009)
White-tailed deer	Field	PZP (SpayVac) and AdjuVac	T = 100% 100% C = 22%	Locke *et al.* (2007)
White-tailed deer	Field	PZP and AdjuVac	T = 100% C = 22%	Hernandez *et al.* (2006)
White-tailed deer	Captive	PZP and SpayVac, with AdjuVac or Alum	SpayVac-AdjuVac: 100% 100% 100% 80% 80% IVT-PZP-AdjuVac: 100% 80% 80% 80% 80% SpayVac-Alum: 80% NWRC-PZP-AdjuVac: 80% 0% (200 µg) NWRC-PZP-AdjuVac: 100% 20% 20% 20% 0% (500 µg); C = 0%	Miller *et al.* (2009)

Species	Type	Method	Results	Reference
Wapiti	Captive	GonaCon-B and AdjuVac	T = 90% 75% 50% 25% C = 0% 0% 0% 14%	Powers *et al.* (2011)
Wapiti	Captive	GonaCon-KLH and AdjuVac	GonaCon-KLH (1000 µg) = 92% 90% 100% GonaCon-KLH (2000 µg) = 90% 100% 100% C = 27% 25% 0%	Killian *et al.* (2009)
American Bison	Captive	GonaCon-KLH and AdjuVac	T = 100% C = 0%	Miller *et al.* (2004)
Wild boar	Captive	GonaCon-KLH and AdjuVac	T = 92% infertile for at least 4-6 years C = 0%	Massei *et al.* (2008) Massei *et al.* (2012)
Fallow deer	Field	PZP (SpayVac) and FCA	T = 100% 100% 100% C = 4% 3% 4%	Fraker *et al.* (2002)
Feral horse	Captive	GonaCon-KLH and AdjuVac	T = 93% 64% 57% 43% C = 25% 25% 12% 0%	Killian *et al.* (2008)
Feral horse	Field	GonaCon-B and AdjuVac	T = 61% 58% 69% C = 40% 31% 14%	Gray, *et al.* (2010)
Feral horse	Captive	PZP (SpayVac) and AdjuVac	T = 100% 83% 83% 83% C = 25% 25% 12% 0%	Killian *et al.* (2008)
Feral horse	Field	PZP with FCA and QS-21	T = 95% 85% 68% 54% C = 47% 42% 49% 48%	Turner *et al.* (2007)
Feral horse	Field	PZP and AdjuVac	T= 63% 50% 56% C = 40% 31% 14%	Gray *et al.* (2010) Gray *et al.* (2011)

The zona pellucida (ZP) that surrounds an ovulated egg is composed of four types of proteins, named ZP1, ZP2, ZP3 and ZP4, each with different functions in mediating structure and species-specific sperm recognition and binding. Differences in these proteins among mammals are partly responsible for the variable results obtained when using a particular ZP vaccine on different species (Kitchener *et al.*, 2009; Gupta and Bansal, 2010). For instance, porcine ZP (PZP) immunocontraceptive vaccines, derived from ZP isolated from pig ovaries, inhibit fertilisation in many wildlife species including ungulates (Table 10.1) but not rodents, cats and wild pigs (Fagerstone *et al.*, 2002; Kirkpatrick *et al.*, 2009, 2011). Likewise, differences in the results of studies using ZP-based vaccines may reflect different formulations of native, purified or recombinant ZP vaccines and different methods of extraction of PZP from pig ovaries (Miller *et al.*, 2009; Kirkpatrick *et al.*, 2011; Bechert *et al.*, 2013).

Early immunocontraceptive vaccines had to be delivered as a primer injection followed by a booster, which made field applications impractical (Putman, 1997). Initial vaccine formulations also used Freund's complete adjuvant (FCA). Some constituents of this adjuvant, namely mycobacteria (*Mycobacterium tuberculosis*) and mineral oil, were found responsible for granulomas (thickened tissue filled with fluid) at injection sites, for false-positive results in TB skin tests in deer treated with these vaccines and for potential carcinogenicity to consumers of treated animals (Kirkpatrick *et al.*, 2011). Significant progress has been made through the development of a novel adjuvant (AdjuVac™, National Wildlife Research Center, United States), containing inactivated *Mycobacterium avium* and based on a modified version of the Johne's disease vaccine.

Injectable ZP-based immunocontraceptives have been employed extensively to reduce fertility in zoo ungulates, in free-living deer, feral horses and elephants (Table 10.1). In particular, the combination of AdjuVac and PZP-vaccine made ungulates infertile for several years after a single dose (Table 10.1). In some species, such as white-tailed deer, some ZP vaccines may cause pathologies such as inflammation of the ovary (Curtis *et al.*, 2007) but in others, such as wild horses, no ovarian damage was observed after 3 years of treatment (Patton *et al.*, 2007). Following injection of ZP-based immunocontraceptives, injection site reactions such as granulomas are common, whilst the occurrence of draining abscesses is around 1% in various species (Gray *et al.*, 2010; Kirkpatrick *et al.*, 2009). As ZP-based immunocontraceptives inhibit fertilisation but not ovulation, animals treated with these vaccines tend to have multiple infertile oestrus cycles which may lead to extended breeding seasons, increased movements and potential late births (Miller *et al.*, 2000; Curtis *et al.*, 2007; Nuñez *et al.*, 2009, 2010; reviewed in Kirkpatrick *et al.*, 2009, 2011). Multiple infertile oestrus cycles following treatment with PZP vaccine were observed in white-tailed deer, wapiti and horses (Heilmann *et al.*, 1998; Killian and Miller, 2001; Curtis *et al.*, 2002; Ransom *et al.*, 2013). Other studies suggested that treatment with ZP vaccines did not affect behaviour and body condition of mares (Ransom *et al.*, 2010; Kirkpatrick *et al.*, 2011), white-tailed deer (Hernandez *et al.*, 2006) and wapiti (Heilmann *et al.*, 1998). However, an extension of the breeding season in deer treated with PZP vaccine resulted in increased energy expenditure by males (Curtis *et al.*, 2002). PZP vaccines were found safe to administer to pregnant or lactating females (Kirkpatrick and Turner, 2002; Patton *et al.*, 2007) and had no long-term effect on health of white tailed deer (Miller *et al.*, 2001). In 2012, an injectable PZP-based vaccine, ZonaStat-H, was registered by the US Humane Society and approved by the US Environment Protection Agency (EPA) as a contraceptive for population control of wild and feral horses and feral donkeys.

Vaccines based on gonadotropin-releasing hormone (GnRH) generate antibodies towards GnRH, thus preventing the hormonal cascade that leads to ovulation and sperm production. Several multi-dose GnRH-based immunocontraceptive vaccines, developed for use in livestock are unsuitable for wildlife due to the difficulty of recapturing individuals to administer booster doses and to their relatively short-term (a few months) effectiveness (reviewed in Naz *et al.*, 2005; McLaughlin and Aitken, 2010). Single-dose injectable GnRH-based vaccines, specifically formulated for wildlife applications, offer better prospects for managing ungulates. Among these vaccines, the most studied is Gonacon™, currently registered in the United States by the EPA as an immunocontraceptive for white-tailed deer, feral horses and feral donkeys. Gonacon™ consists of a synthetic GnRH coupled to a mollusc protein (Miller *et al.*, 2008b).

Formulated as an injectable, single-dose immunocontraceptive, Gonacon™ caused infertility for several years in males and females of several ungulates (e.g. Miller *et al.*, 2000; Killian *et al.*, 2008; Massei *et al.*, 2008; Gray *et al.*, 2010; Massei *et al.*, 2012) (Table 10.1). As Gonacon™ interferes with steroid production, treated females do not exhibit oestrus behaviour. Male deer treated with GonaCon™ also showed a complete lack of sexual activity and reduced testicle size, but they also exhibited abnormal antler development (Miller *et al.*, 2000, 2009; Fagerstone *et al.*, 2008). The lack of sexual activity, in species where this behaviour might increase human–wildlife conflicts such as deer collisions with vehicles during the rut, could be advantageous, although the effect on antler development suggests GonaCon™ should not be used on male deer. Similar to ZP-based contraceptives, antibodies to GnRH decrease with time and fertility may be restored in the years after treatment unless animals are administered booster vaccinations (Miller *et al.*, 2008b; Massei *et al.*, 2012).

The main side effect of Gonacon™ is the formation of a granuloma or a sterile abscess at the injection site; the severity and incidence of injection site reactions vary with species. In white-tailed deer, injection-site granulomas and sterile abscesses occurred in the deep hind-limb musculature of >85% of treated animals, although no evidence of limping or impaired mobility was observed in these deer during a 2-year study (Gionfriddo *et al.*, 2011b). These reactions are typical responses to injection of adjuvanted vaccines formulated as water-in-oil emulsions (Miller *et al.*, 2009). On the other hand, GonaCon™ had no adverse effects on major organs, body condition, fat deposits or blood chemistry in wild boar, white-tailed deer and wapiti (Massei *et al.*, 2008, 2012; Gionfriddo *et al.*, 2009, 2011b).

When given to pregnant bison and elk, GonaCon™ did not affect pregnancy (Miller *et al.*, 2004; Powers *et al.*, 2011). Other studies found that GonaCon™ did not induce infertility and did not prevent sexual development when administered to 3–4-month-old white-tailed deer (Miller *et al.*, 2008a; Gionfriddo *et al.*, 2011a). Like ZP-based vaccines, GnRH vaccines are broken down if ingested, thus they do not pose risks to predators or human consumers.

10.2 Delivery methods

Although a fertility control agent should be ideally species specific, this is rarely the case and specificity must be achieved through the delivery method. At present, fertility control agents that induce at least 1 year of infertility are administered by direct injection following capture, by implant or are delivered remotely through biobullets and syringe-darts (see below). Subcutaneous implants that release contraceptive agents into the body over a sustained period of time have been successfully employed to induce infertility for 1–5 years in a variety of wildlife species (e.g. Plotka and Seal, 1989; Nave *et al.*, 2002; Coulson *et al.*, 2008; Lohr *et al.*, 2009). However, steroid implants have the potential for transferring active ingredients to predators and scavengers.

Biobullets are biodegradable projectiles used for remote administration of veterinary products (DeNicola *et al.*, 2000). Syringe-darts, routinely employed to anaesthetise wild animals, have also been used to administer contraceptives to large ungulates at ranges of ≤40 m (Rudolph *et al.*, 2000; Aune *et al.*, 2002; Delsink *et al.*, 2006). The advantages of remote administration of contraceptives to ungulates are that delivery can be targeted to specific individuals (unlike oral delivery), and that this method minimises the welfare and economic costs of trapping (Kreeger, 1997). Potential disadvantages of these delivery systems include the inability to identify successfully vaccinated animals, cost, dose regulation and incomplete intra-muscular injection (De Nicola *et al.*, 1997; Kreeger, 1997; Aune *et al.*, 2002). The inability to identify previously vaccinated animals is important because these animals can receive multiple doses: whilst this is not expected to have welfare costs, it certainly reduces the efficiency of any fertility control programme. Another approach to a single-dose, multiple-year immunocontraceptive is to mimic the effects of booster injections by incorporating the vaccine into controlled-released polymers formulated as injectable pellets. This approach was successfully tested with wild horses by using simultaneous intramuscular injection of 1-, 3- and 12-month pellets to provide *in vivo* delivery of booster doses of the PZP vaccine (Turner *et al.*, 2007; Rutberg *et al.*, 2013).

Injectable forms of fertility control vaccines have been shown to effectively block fertility in a number of species. However, to be of further practical use in wildlife management, more efficient means of delivery are required. There is great interest in the development of mucosal (e.g. oral or intranasal) vaccines in human pharmaceuticals (reviewed in Woodrow *et. al.*, 2012) and this will aid in efforts towards wildlife applications where some research has already been conducted (Cui *et al.*, 2010). Once developed, oral fertility control agents are likely to be less expensive to administer than injectable forms, in part, because capture and handling of animals will not be necessary for the delivery of these contraceptives. However, unlike injectable vaccines, oral fertility control agents will likely require repeated applications to cause infertility (Cross *et al.*, 2011). As oral forms of fertility control might also affect non-target animals, species specificity could be achieved through targeted delivery methods. One example is the BOS (Boar-Operated System) developed as a specific delivery system for wild boar and feral pigs (Massei *et al.*, 2010c; Long *et al.*, 2010; Campbell *et al.*, 2011) (Figure 10.2).

Immunocontraceptive vaccines delivered through genetically modified, self-sustaining infectious vectors have been developed in Australia. Criticism of this approach involved concerns regarding irreversibility, the difficulty of controlling the vectors once released, possible mutations of the vectors that could affect non-target species and possible development of resistance (Barlow, 2000; Williams,

Figure 10.2 Free-living wild boar feeding on maize-based baits from a Boar-Operated System (BOS). The metal cone slides along the pole and fully encloses the base onto which the baits are placed. Several studies found that free-living wild boar and wild pigs fed regularly from the BOS and that the device successfully prevented bait uptake by non-target species. The BOS can be used to deliver vaccines, contraceptives or other pharmaceuticals employed to manage overabundant populations of wild suids.

2002). In New Zealand genetically modified transmissible organisms, such as species-specific nematode parasites, have been explored to deliver contraceptives, although no data are available for ungulates (McDowell *et al.*, 2006; Cowan *et al.*, 2008; Cross *et al.*, 2011).

10.3 Fertility control and population responses

Most recent field studies on fertility control have used immunocontraceptives, whilst modelling studies have focussed on generic contraceptives of different levels and duration of induced infertility (Table 10.2). Comparing the relative merits of fertility and lethal control to manage overabundant populations, recent research suggests that large, long-lived species are easier to manage with fertility control than smaller, shorter-lived ones because a lower proportion of the population must be targeted each year (Hone, 1999), particularly if lifelong contraceptives are employed (Hobbs *et al.*, 2000).

   Modelling the impact of fertility control versus culling for a geographically closed population of white-tailed deer, Merrill *et al.* (2003) concluded that, for instance, to achieve a 60% reduction over 4 years, culling should remove 40% of available fertile females each year. To maintain this level of reduction, only 13% of the available females should be sterilised every year. Based on this model, the authors suggest that an effective management strategy to control overabundant urban deer populations would require two steps. The first step will reduce the population to a given level: to achieve this, culling would be more efficient than sterilisation. The second step will maintain the population at a set level and sterilisation will become more efficient as the number of sterilised females increases (Hobbs *et al.*, 2000). However, in long-lived species and in populations characterised by slow turnover, the benefits of using fertility control to decrease population size will only accrue in the long term (Twigg *et al.*, 2000; Kirkpatrick and Turner, 2008; Cowan and Massei, 2008).

   The effects of fertility control on population dynamics also depend on species-specific social and reproductive behaviours, on the type of contraceptive used and on its mode of action, as well as on whether a population is isolated or open. There is general consensus that fertility control is most effective for managing relatively small (50–200 animals) isolated populations of ungulates (Rudolph *et al.*, 2000; Kirkpatrick and Turner, 2008). Avoiding disruption of behaviour is crucial, as fertility-control-induced changes in immigration and emigration might prevent fertility control achieving the required reduction in population growth (e.g. Davis and Pech, 2002; Merril *et al.*, 2006).

   On the other hand, using fertility control methods that inhibit normal sexual behaviour can potentially reduce disease transmission by decreasing contact rates between individuals (Caley and Ramsey, 2001; Ramsey, 2007). For instance, a reduction of reproductive behaviour would result in decreased transmission of venereal diseases such as pseudorabies and brucellosis (Miller *et al.*, 2004; Killian *et al.*, 2006). In this context, methods that prevent ovulation are likely to be more successful at decreasing disease transmission than those that only block fertilisation. When only fertilisation is blocked, females of many ungulate species will continue to ovulate, thus attracting males (Putman, 1997; Miller *et al.*, 2000; Curtis *et al.*, 2007; Nuñez *et al.*, 2009, 2010). This may have significant effects on prolonging the duration of the rut, enhancing and extending the period of male–male competition (and thus increasing risk of injury or male exhaustion).

The factors affecting emigration and immigration in ungulate populations managed through fertility control have received little attention. For instance, a reduction in population density due to fertility control might increase immigration rate, thus negating the benefits of using non-lethal population management. On the other hand, fertility control might also encourage emigration, particularly of males looking for mating opportunities outside their normal home range. As female white-tailed deer in urban and suburban areas have relatively small home range size and high site fidelity (Grund *et al.*, 2002), it is possible to hypothesise that fertility control will not affect the movements of these animals. Other studies found that ZP-based immunocontraceptives did not affect spatial behaviour in white-tailed deer and feral horses (Hernandez *et al.*, 2006; Ransom *et al.*, 2010).

Density-dependent regulation of population should also be taken into account: Merrill *et al.* (2003) suggested that if density-dependence was occurring, it would increase the effectiveness of sterilisation as the reproductive removal (but not the physical removal) of part of the population would intensify density-dependent feedback. Clearly, this is an area where more field studies are warranted to assess the effects of fertility control on emigration, immigration, recruitment and mortality in ungulate populations with different life-history traits.

Fertility control has been associated with increased survival and improved health condition, probably due to the reduced expenditure of energy normally required for reproduction. For example, sterilisation-induced increases in survival and total food consumption in feral Soay rams caused an increase in both animal density and impact on the plant community (Jewell, 1986). Similarly, as immunocontraceptives can significantly extend lifespan and improve body condition (Turner and Kirkpatrick, 2002; Kirkpatrick and Turner, 2007; Gionfriddo *et al.*, 2011b), the impact of increased survival on population dynamics must be taken into account when using fertility control to manage ungulate populations.

Fertility control in ungulates has been used to decrease population size or growth, reduce vertical or horizontal transmission of diseases or reduce impacts of local populations on human activities (Table 10.2). The relative merits of fertility control and culling have been much debated, with advocates of the two methods often holding opposite, irreconcilable positions (Kirkpatrick, 2007; Curtis *et al.*, 2008; Fagerstone *et al.*, 2010). Modelling studies concluded that in several instances the outcome of the two methods in reducing population size or disease transmission depends on the definition of 'efficiency'. If efficiency is defined in terms of the time taken to achieve the desired effect, then culling is

Table 10.2 Examples of empirical and theoretical applications of fertility control (FC) at population level in wildlife and in feral ungulate populations.

Aim	Species	Trial	Method	Results and conclusions	Reference
Evaluate impact of FC on population size	White-tailed deer	Field	PZP vaccine	FC feasible to maintain small (<200) suburban deer populations at 30–70% of carrying capacity	Rudolph *et al.* (2000)
	White-tailed deer	Field	PZP vaccine	FC induced a 7.9% population decline in a suburban deer population	Rutberg *et al.* (2004)
	White-tailed deer	Field and model	PZP vaccine	FC caused a 27–58% decline in population size in the 5–10 years following treatment of females	Rutberg and Naugle (2008)
	Wild horse	Field	PZP vaccine	The effort required to achieve zero population growth decreased, as 95, 83, 84, 59 and 52% of all adult mares were treated in the first 5 years. FC increased longevity and improved body condition	Turner and Kirkpatrick (2002)
	Wild horse	Field	PZP vaccine	FC prevented population growth within 2 years; by year 11, the population had declined by 22.8%. FC also increased longevity of mares	Kirkpatrick and Turner (2008)
	Wild horse	Model	PZP vaccine	FC can be used to reduce population size to the target number in 5–8 years	Ballou *et al.* (2008)
	African Elephant	Field	PZP vaccine	FC prevented population growth	Delsink (2006)

Objective	Species		Contraception	Finding	Reference
	African Elephant	Model	Immuno-contraception	'Rotational' FC can be used to increase calving interval, slow population growth rate and alter age structure	Druce *et al.* (2011)
	Wildlife	Model	Generic contraception	FC was more effective than culling in reducing population size for medium and large-size animals	Zhang (2000)
	White-tailed deer	Model	Generic contraceptive	FC was more efficient than culling in reducing population size provided >50% females are maintained infertile	Hobbs *et al.* (2000)
	Wapiti	Model	Yearlong *vs.* lifelong contraceptive	FC using lifetime contraceptives was more efficient than any other population control option	Bradford and Hobbs (2008)
Evaluate impact of removal and FC on population size	Feral horse	Model	Generic contraception	Compared to removal, FC resulted in smaller, less fluctuating population size	Gross (2000)
Evaluate factors affecting time to reduce a population through FC	White-tailed deer	Model	Permanent sterilisazion	FC could reduce a population by 30–60% in 4–10 years if 25–50% of fertile females were sterilised every year	Merrill *et al.* (2003)
Evaluate effects of immigration, stochasticity and variation in capture process on FC to manage population size	White-tailed deer	Model	Permanent sterilisazion	FC was unlikely to reduce the size of an open population. In a closed population, permanent sterilisation could reduce population size if 30–45% deer were captured each year	Merrill *et al.* (2006)

always the most efficient solution (Bradford and Hobbs, 2008). Conversely, if efficiency is defined as the proportion of the population to be targeted, fertility control can be regarded as potentially more efficient than culling (Hobbs *et al.*, 2000; Merrill *et al.*, 2003). By defining efficiency as the proportion of the population that must be treated, the time and costs required are deliberately ignored (Merrill *et al.*, 2003). In this scenario, modelling suggests that fertility control agents that render animals infertile for many years are likely to be more efficient than culling, provided that the fertility status of the treated animals is known, for instance, through ear-tags that identify animals previously treated with contraceptives.

Other advantages of fertility control over culling include:

1. Compared to fertility control, culling is more likely to cause social perturbation, increased contact rates and hence increased likelihood of disease transmission (e.g. Ramsey *et al.*, 2006; Carter *et al.*, 2007)
2. Animals in improved body condition, following treatment with contraceptives, might be less susceptible to disease and also mount a better immune response to disease vaccines
3. Infertile animals remain in the population, thus maintaining density-dependent feedback to recruitment and survival (Zhang, 2000)
4. A growing recognition that fertility control in conjunction with disease vaccination can be as effective as culling to manage disease transmission (Smith and Cheeseman, 2002).

    As animals vaccinated against a disease reproduce, new susceptible individuals enter the population and dilute the level of herd immunity provided by disease vaccination; combining disease vaccination and fertility control, to prevent the recruitment of new susceptibles can thus reduce the effort required to eliminate the disease (Smith and Wilkinson, 2003; Carroll *et al.*, 2010).

In some instances, fertility control might be required to reduce or halt population growth rather than to decrease population size. Exploring options to manage a small, isolated population of African elephants, Druce *et al.* (2011) suggested that using reversible immunocontraceptives on an individual rotational basis would increase inter-calving intervals, stabilise population structure and lower population growth to a predetermined rate.

Some authors have hypothesised that the use of immunocontraceptive vaccines to manage wildlife could result in the evolution of resistance, through selection for individuals that remain fertile because of low or no response to vaccination (e.g. Gross, 2000; Magiafoglu *et al.*, 2003; Cooper and Larsen, 2006; Holland *et al.*, 2009). These authors argue that when females only are treated with immunocontraceptives, resistance might evolve if the response to the vaccine is specific for this gender and could be inherited through the maternal line. No studies have so far demonstrated such effects although unresponsiveness to immunocontraceptive vaccines was found to have a genetic component in brushtail possums (Holland *et al.*, 2009).

10.4 Can fertility control mitigate human–ungulate conflicts?

Human–ungulate conflicts often demand immediate solutions. Stakeholders have a significant impact on management options but often hold opposite opinions. For instance, animal welfare groups tend to advocate fertility control to manage these conflicts (Curtis *et al.*, 2008), whilst many hunting groups oppose the use of fertility control because of concerns that this method will replace sport hunting (Kirkpatrick, 2007; Fagerstone *et al.*, 2010).

The studies carried out so far indicate that if fertility control is the sole method employed to manage overabundant populations, a substantial initial effort is required (Rudolph *et al.*, 2000; Walter *et al.*, 2002; Merrill *et al.*, 2003, 2006). In addition, changes in survival and immigration can reduce population-level efficacy of fertility control (Ransom *et al.*, 2013). However, as the proportion of infertile females increases, this effort will decline and remain constant once the desired density has been achieved. Successful examples are the marked reduction in sub-urban white-tailed deer obtained over a 10-year timescale (Rutberg and Naugle, 2004, 2008), the zero-population growth of an isolated population of elephants achieved within 2 years (Delsink *et al.*, 2006) and of an island population of wild horses obtained within 2 years (Kirkpatrick and Turner, 2008). For closed populations, Merril *et al.* (2006) suggested that, at least in white-tailed deer, contraception of 30–45% of the animals would decrease population size after 2–3 years and that a population reduction of 60% would be achieved in 10 years.

Depending on how urgent the resolution of the conflict is, fertility control can be used alone or once the population size has been reduced through other methods (Barlow, 1997; Hobbs *et al.*, 2000). When fertility control is chosen to mitigate human–ungulate conflicts, a number of issues should be considered before field applications are implemented. These issues cover humaneness, efficacy, feasibility, cost, timeframe and sustainability as well as alternative methods for population control. As humaneness is one of the primary public concerns regarding any type of wildlife management, defining this term is crucial to obtaining and maintaining public support in relation to specific, well-defined objectives. For instance, humaneness can be defined as (i) the level of stress experienced by treated animals, (ii) the severity and type of side effects, (iii) the proportion of animals likely to experience negative side effects following treatment with a contraceptive, (iv) the proportion of animals that will suffer from capture, handling and anaesthesia associated with administering the contraceptives, or (v) a combination of all these definitions.

When lethal control is illegal, unacceptable or unfeasible, fertility control might be the only option available for managing overabundant populations of ungulates. In these instances, key issues to be discussed at the planning stage include assessing the overall proportion of the population that must be rendered infertile to mitigate the conflict, estimating the relative effort and time required to achieve the target population size and evaluating the feasibility of field application of contraceptives

(Hobbs *et al.*, 2000; Bradford and Hobbs, 2008). This feasibility in turn is likely to depend on factors such as animal density, approachability of individual animals, access to private and public land, and efficacy of the contraceptive treatment (Rudolph *et al.*, 2000; Walter *et al.*, 2002; Rutberg and Naugle, 2008; Boulanger *et al.*, 2012). In the early planning stages, modelling the impact of fertility control on population dynamics can assist determining whether the application of this method will meet specific management goals (e.g. Jacob *et al.*, 2008).

The economic cost of reducing ungulate population growth through fertility control agents that require capture and handling of the animals is expected to be high. For instance, Rutberg (2005) estimated that the cost of rendering infertile a medium-to-large size individual mammal varied between US$25 and US$500. Delsink *et al.* (2007) calculated that in 2005 the average cost of managing elephants through aerial vaccination with immunocontraceptives was US$98–110 per animal, inclusive of darts, vaccine, helicopter and veterinary assistance. Walter *et al.* (2002) reported that the cost of trapping and injecting 30 white-tailed deer with immunocontraceptives for 2 years (with a spring capture and vaccination followed by two boosters in autumn of year 1 and year 2) was US$1128/deer. Labour accounted for 64% of the total cost and equipment, supplies, lodging and travel accounted for the remaining 36% of the total cost. However, after the initial year, the cost per deer dropped to US$270 (Walter *et al.*, 2002). Boulanger *et al.* (2012) found that the cost of capture, handling and administering contraceptives to white-tailed deer in various studies was about US$1,000 but that 75% of this cost was due to drugs, including anaesthetics, and a veterinarian's time. It is conceivable that costs would drop significantly if immunocontraceptives were delivered by trained staff (i.e. by wildlife managers instead of veterinarians) and ungulate capture was organised with the assistance of volunteers donating their time and skills to the project. Hobbs *et al.* (2000) suggested that fertility control of deer will only be cost-effective, compared to culling, where professionals are employed to cull deer instead of recreational hunters.

Identifying who should bear the costs of population management might raise awareness of the economics of available options amongst stakeholders and add a different perspective to ungulate management. This awareness would be further enhanced if the full costs, including negative environmental and welfare consequences, associated with each option are included.

In addition to the practical challenges of using fertility control on ungulate populations, regulatory and legal requirements for field applications of contraceptives must be met. For products that have not been registered in a country, trials can often be carried out under experimental permits and on a case-by-case basis (Humphrys and Lapidge, 2008).

In summary, this review highlighted that safe, effective contraceptives are now available allowing field applications aimed at reducing population growth in ungulates. Although many challenges still exist, we believe the next decade will witness a large number of field studies carried out to manage ungulate populations through fertility control. We recommend that, for each context, the use of fertility control, alone or in conjunction with other methods, is evaluated and compared with alternative options for population control. Only then can the costs and benefits of different methods be fully established and the optimum options selected to mitigate the conflicts between human interests and ungulate populations.

Acknowledgements

The authors would like to thank Alastair Ward for reading the manuscript and providing comments and suggestions.

References

Apollonio, M., Andersen, R. and Putman, R.J. (eds), (2010) *European Ungulates and their Management in the 21st Century.* Cambridge, UK: Cambridge University Press.

Asa, C.S. and Porton, I.J. (eds), (2005) *Wildlife Contraception: Issues, Methods and Applications.* Baltimore, MD: Johns Hopkins University Press.

Aune, K., Terry, J., Kreeger, T.J., Thomas, J. and Roffe, T.J. (2002) Overview of delivery systems for the administration of vaccines to elk and bison of the Greater Yellowstone Area. In T.J. Kreeger (ed.), *Proceedings of Brucellosis in Elk and Bison in the Greater Yellowstone Area.* Cheyenne, WY: Wyoming Game and Fish Department, pp. 66–79.

Baker, D.L., Wild, M.A., Conner, M.M., Ravivarapu, H.B., Dunn, R.L. and Nett, T.M. (2002) Effects of GnRH agonist (leuprolide) on reproduction and behaviour in female wapiti (*Cervus elaphus nelsoni*). *Reproduction* **60**, 155–167.

Baker, D.L., Wild, M.A., Conner, M.M., Ravivarapu, H.B., Dunn, R.L, and Nett, T.M. (2004) Gonadotropin-releasing hormone agonist: a new approach to reversible contraception in female deer. *Journal of Wildlife Diseases* **40**, 713–724.

Ballou, J.D., Traylor-Holzer, K., Turner, A., Malo, A.F., Powell, D., Maldonado, J. and Eggert, L. (2008) Simulation model for contraceptive management of the Assateague Island feral horse population using individual-based data. *Wildlife Research* **35**, 502–512.

Barfield, J.P., Nieschlag, E. and Cooper, T.G. (2006) Fertility control in wildlife: humans as a model. *Contraception* **73**, 6–22.

Barlow, N.D. (1997) Modelling immunocontraception in disseminating systems. *Reproduction Fertility and Development* **9**, 51–60.

Barlow, N.D. (2000) The ecological challenge in immunocontraception: editor's introduction. *Journal of Applied Ecology* **37**, 897–902.

Bechert, U., Bartell, J., Kutzler, M., Menino, A., Bildfell, R., Anderson, M. and Fraker, M. (2013) Effects of two porcine zona pellucida immunocontraceptive vaccines on ovarian activity in horses. *Journal of Wildlife Management* **77**, 1386–1400.

Beringer, J., Hansen, L. P., Demand, J.A. and Sartwell, J. (2002) Efficacy of translocation to control urban deer in Missouri: costs, efficiency, and outcome. *Wildlife Society Bulletin* **30**, 767–774.

Boulanger, J.R., Curtis, P.D., Cooch, E.G. and DeNicola, A.J. (2012) Sterilization as an alternative deer control technique: a review. *Human–Wildlife Interactions* **6**, 273–282.

Bradford, J.B. and Hobbs, N.T. (2008) Regulating overabundant ungulate populations: An example for elk in Rocky Mountain National Park, Colorado. *Journal of Environmental Management* **86**, 520–528.

Brainerd, S.M. and Kaltenborn, B.P. (2010) The Scandinavian model. A different path to wildlife management. *The Wildlife Professional* **4**, 52–56.

Caley, P. and Ramsey, D. (2001) Estimating disease transmission in wildlife, with emphasis on leptospirosis and bovine tuberculosis in possums, and effects of fertility control. *Journal of Applied Ecology* **38**, 1362–1370.

Campbell, T.A., Long, D.B. and Massei G. (2011) Efficacy of the Boar-Operated-System to deliver baits to feral swine. *Preventive Veterinary Medicine* **98**, 243–249.

Carroll, M.J., Singer, A., Smith, G.C., Cowan, D.P. and Massei, G. (2010) The use of immuno-contraception to improve rabies eradication in urban dog populations. *Wildlife Research* **37**, 676–687.

Carter, S.P., Delahay, R.J., Smith, G.C., Macdonald, D.W., Riordan, P., Etherington, T., Pimley, E. and Cheeseman, C.L. (2007) Culling-induced social perturbation in Eurasian badgers *Meles meles* and the management of TB in cattle: an analysis of a critical problem in applied ecology. *Proceedings of the Royal Society B* **274**, 2769–2777.

Conner, M.M., Baker, D.L., Wild, M.A., Powers, J.G., Hussain, M.D., Dun, R.L. and Nett, T.M. (2007) Fertility control in free-ranging elk using gonadotropin releasing hormone agonist leuprolide: effects on reproduction, behavior, and body condition. *Journal of Wildlife Management* **71**, 2346–2356.

Conover, M.R. (2002) *Resolving Human–Wildlife Conflicts: The science of wildlife damage management.* Boca Raton, FL: Lewis.

Cooper, D.W. and Larsen, E. (2006) Immunocontraception of mammalian wildlife: ecological and immunogenetic issues. *Reproduction* **132**, 821–828.

Corn, J.L. and Nettles, V.F. (2001) Health protocol for translocation of free-ranging elk. *Journal of Wildlife Diseases* **37**, 413–426.

Coulson, G., Nave, C.D., Shaw, J. and Renfree, M.B. (2008) Long-term efficacy of levonorgestrel implants for fertility control of eastern grey kangaroos (*Macropus giganteus*). *Wildlife Research* **35**, 520–524.

Cowan, D.P. and Massei, G. (2008) Wildlife contraception, individuals and populations: How much fertility control is enough? In R.M. Timm and M.B. Madon (eds), *Proceedings of the 23rd Vertebrate Pest Conference.* Davis, CA: University of California, pp. 220–228.

Cowan, P.E., Grant, W.N. and Ralston, M. (2008) Assessing the suitability of the parasitic nematode *Parastrongyloides trichosuri* as a vector for transmissible fertility control of brushtail possums in New Zealand: ecological and regulatory considerations. *Wildlife Research* **35**, 573–577.

Crawford, J.L., McLeod, B.J. and Eckery, D.C. (2011) The hypothalamic–pituitary–ovarian axis and manipulations of the oestrous cycle in the brushtail possum. *General and Comparative Endocrinology* **170**, 424–448.

Cross, M.L., Zheng, T., Duckworth, J.A. and Cowan, P.E. (2011) Could recombinant technology facilitate the realisation of a fertility-control vaccine for possums? *New Zealand Journal of Zoology* **38**, 91–111.

Cui, X., Duckworth, J.A., Lubitz, P., Molinia, F.C., Haller, C., Lubitz, W. and Cowan, P.E. (2010) Humoral immune responses in brushtail possums (*Trichosurus vulpecula*) induced by bacterial ghosts expressing possum zona pellucida 3 protein. *Vaccine* **28**, 4268–4274.

Curtis, P.D., Pooler, R.L., Richmond, M.E., Miller, L.A., Mattfeld, G.F. and Quimby, F.W. (2002) Comparative effects of GnRH and porcine zona pellucida (PZP) immunocontraceptive vaccines for controlling reproduction in white-tailed deer (*Odocoileus virginianus*). *Reproduction* **60**, 131–141.

Curtis, P.D., Richmond, M.E., Miller, L.A. and Quimby, F.W. (2007) Pathophysiology of white-tailed deer vaccinated with porcine zona pellucida immunocontraceptive. *Vaccine* **25**, 4623–4630.

Curtis, P.D., Richmond, M.E., Miller, L.A. and Quimby, F.W. (2008) Physiological effects of gonadotropin-releasing hormone immunocontraception on white-tailed deer. *Human–Wildlife Conflicts* **2**, 68–79.

Daszak, P., Cunningham, A.A. and Hyatt, A.D. (2000) Emerging infectious diseases of wildlife – Threats to biodiversity and human health. *Science* **287**, 443–449.

Davis, S.A. and Pech, R.P. (2002) Dependence of population response to fertility control on the survival of sterile animals and their role in regulation. *Reproduction* **60**, 89–103.

Delsink, A.K., van Altena, J.J., Grobler, D., Bertschinger, H., Kirkpatrick, J. and Slotow, R. (2006) Regulation of a small, discrete African elephant population through immunocontraception in the Makalali Conservancy Limpopo, South Africa. *South African Journal of Science* **102**, 403–405.

Delsink, A.K., van Altena, J.J., Grobler, D., Bertschinger, H., Kirkpatrick, J. and Slotow, R. (2007) Implementing immunocontraception in free-ranging African elephants at Makalali Conservancy. *Journal of the South African Veterinary Association* **78**, 25–30.

DeNicola, A.J., Kesler, D.J. and Swihart, R.K. (1997) Dose determination and efficacy of remotely delivered norgestomet implants on contraception of white-tailed deer. *Zoo Biology* **16**, 31–37.

DeNicola, A.J., VerCauteren, K.C., Curtis, P.D. and Hygnstrom, S.E. (2000) *Managing White-tailed Deer in Suburban Environments: A Technical Guide.* Ithaca, NY: Cornell Cooperative Extension.

D'Occhio, M.J., Fordyce, G., Whyte, T.R., Jubb, T.F., Fitzpatrick, L.A., Cooper, N.J., Aspden, W.J., Bolam, M.J. and Trigg, T.E. (2002) Use of GnRH agonist implants for long-term suppression of fertility in extensively managed heifers and cows. *Animal Reproduction Science* **74**, 151–62.

Druce, H.C., Mackey, R.L. and Slowtow, R. (2011) How immunocontraception can contribute to elephant management in small, enclosed reserves: Munyawana population. *PLoS ONE* **6**, 1–10.

Eymann, J., Herbert, C.A., Thomson, B.P., Trigg, T.E., Cooper, D.W. and Eckery, D.C. (2007) Effects of deslorelin implants on reproduction in the common brushtail possum (*Trichosurus vulpecula*). *Reproduction, Fertility and Development* **19**, 899–909.

Fagerstone, K.A., Coffey, M.A., Curtis, P.D., Dolbeer, R.A., Killian, G.J., Miller, L.A. and Wilmot, L.M. (2002) Wildlife fertility control. *Wildlife Society Technical Review* **02–2**, 1–29.

Fagerstone, K.A., Miller, L.A., Eisemann, J.D., O'Hare, J.R. and Gionfriddo, J.P. (2008) Registration of wildlife contraceptives in the United States of America, with OvoControl and GonaCon immunocontraceptive vaccines as examples. *Wildlife Research* **35**, 586–592.

Fagerstone, K.A., Miller, L.A., Killian, G.J. and Yoder, C.A. (2010) Review of issues concerning the use of reproductive inhibitors, with particular emphasis on resolving human–wildlife conflicts in North America. *Integrative Zoology* **1**, 15–30.

Ferroglio, E., Gortazar, C. and Vicente, J. (2011) Wild ungulate diseases and the risk for livestock and public health. In R.J. Putman, M. Apollonio and R. Andersen (eds), *Ungulate Management in Europe: Problems and Practices.* Cambridge, UK: Cambridge University Press, pp. 192–214.

Fraker, M.A. and Brown, R.G. (2011) Efficacy of SpayVac® is excellent: a comment on Gray *et al.* (2010). *Wildlife Research* **38**, 537–538.

Fraker, M.A., Brown, R.G., Gaunt, G.E., Kerr, J.A. and Pohajdak, B. (2002) Long-lasting, single-dose immunocontraception of feral fallow deer in British Columbia. *Journal of Wildlife Management* **66**, 1141–1147.

Garrott, R.A. and Siniff, D.B. (1992) Limitations of male-oriented contraception for controlling feral horse populations. *Journal of Wildlife Management* **56**, 456–464.

Gionfriddo, J.P., Eisemann, J.D., Sullivan, K.J., Healey, R.S. and Miller, L.A. (2009) Field test of a single-injection gonadotrophin-releasing hormone immunocontraceptive vaccine in female white-tailed deer. *Wildlife Research* **36**, 177–184.

Gionfriddo, J.P., DeNicola, A.J., Miller, L.A. and Fagerstone, K.A. (2011a) Efficacy of GnRH immunocontraception of wild white-tailed deer in New Jersey. *Wildlife Society Bulletin* **35**, 142–148.

Gionfriddo, J.P., DeNicola, A.J., Miller, L.A. and Fagerstone, K.A. (2011b) Health effects of GnRH immunocontraception of wild white-tailed deer in New Jersey. *Wildlife Society Bulletin* **35**, 149–160.

Gobello, C. (2007) New GnRH analogs in canine reproduction. *Animal Reproduction Science* **100**, 1–13.

Gray, M.E. and Cameron, E.Z. (2010) Does contraceptive treatment in wildlife result in side effects? A review of quantitative and anecdotal evidence. *Reproduction* **139**, 45–55.

Gray, M.E., Thain, D.S., Cameron, E.Z. and Miller, L.A. (2010) Multi-year fertility reduction in free-roaming feral horses with single-injection immunocontraceptive formulations. *Wildlife Research* **37**, 475–481.

Gray, M.E., Thain, D.S., Cameron, E.Z. and Miller, L.A. (2011) Corrigendum. Multi-year fertility reduction in free-roaming feral horses with single-injection immunocontraceptive formulations. *Wildlife Research* **38**, 260.

Gross, J. (2000) A dynamic simulation model for evaluating effects of removal and contraception on genetic variation and demography of Pryor Mountain wild horses. *Biological Conservation* **96**, 319–330.

Grund, M.D., McAnich, J.B. and Wiggers, E.P. (2002) Seasonal movements and habitat use of female white-tailed deer associated with an urban park. *Journal of Wildlife Management* **66**, 123–130.

Gupta, S.K. and Bansal, P. (2010) Vaccines for immunological control of fertility. *Reproductive Medicine and Biology* **9**, 61–71.

Heilmann, T.J., Garrott, R.E., Caldwell, L.L. and Tiller, B.L. (1998) Behavioral responses of free-ranging elk treated with an immunocontraceptive vaccine. *Journal of Wildlife Management* **62**, 243–250.

Herbert, C.A. and Trigg, T.E. (2005) Applications of GnRH in the control and management of fertility in female animals. *Animal Reproduction Science* **88**, 141–153.

Hernandez, S., Locke, S.L., Cook, M.W., Harveson, L.A., Davis, D.S., Lopez, R.R., Silvy, N. J. and Fraker, M.A. (2006) Effects of SpayVac® on urban female white-tailed deer movements. *Wildlife Society Bulletin* **34**, 1430–1434.

Hobbs, N.T., Bowden, D.C. and Baker, D.L. (2000) Effects of fertility control on populations of ungulates: General, stage-structured models. *Journal of Wildlife Management* **64**, 473–491.

Hoffman, R.A. and Wright, R.G. (1990) Fertility control in a non-native population of mountain goats. *Northwest Science* **64**, 1–6.

Holland, O.J., Cowan, P.E., Gleeson, D.M., Duckworth, J.A. and Chamley, L.W. (2009) MHC haplotypes and response to immunocontraceptive vaccines in the brushtail possum. *Journal of Reproductive Immunology* **82**, 57–65.

Hone, J. (1999) On rate of increase (r): patterns of variation in Australian mammals and the implications for wildlife management. *Journal of Applied Ecology* **36**, 709–718.

Humphrys, S., and Lapidge, S.J. (2008) Delivering and registering species-tailored oral antifertility products: a review. *Wildlife Research* **35**, 578–585.

Jacob, J., Singleton, G.R. and Hinds, L.A. (2008) Fertility control of rodent pests. *Wildlife Research* **35**, 487–493.

Jacobsen, N.K., Jessup, D.A. and Kesler, D.J. (1995) Contraception in captive black-tailed deer by remotely delivered norgestomet ballistic implants. *Wildlife Society Bulletin* **23**, 718–722.

Jewell, P. (1986) Survival in a feral population of primitive sheep in St Kilda, Outer Hebrides, Scotland. *National Geographic Research* **2**, 402–406.

Killian, G.J. and Miller, L.A. (2001) Behavioral observations and physiological implications for white-tailed deer treated with two different immunocontraceptives. In M. C. Brittingham, J. Kays, and R. McPeake (eds), *Proceedings of the 9th Wildlife Damage Management Conference,* Pennsylvania State University, PA, pp. 283–291.

Killian, G., Miller, L.A., Rhyan, J. and Doten, H. (2006) Immunocontraception of Florida feral swine with single-dose GnRH vaccine. *American Journal of Reproductive Immunocology* **55**, 378–384.

Killian, G., Thain, D., Diehl, N.K., Rhyan, J. and Miller, L.A. (2008) Four-year contraception rates of mares treated with single-injection porcine zona pellucida and GnRH vaccines and intrauterine devices. *Wildlife Research* **35**, 531–539.

Killian, G., Kreeger, T.J., Rhyan, J., Fagerstone, K., and Miller, L.A. (2009) Observations on the use of Gonacon™ in captive female elk (*Cervus elaphus*). *Journal of Wildlife Diseases* **45**, 184–188.

Kirkpatrick, J.F. (2007) Measuring the effects of wildlife contraception: the argument for comparing apples with oranges. *Reproduction, Fertility and Development* **19**, 548–552.

Kirkpatrick, J.F. and Turner, A. (2002) Reversibility of action and safety during pregnancy of immunization against porcine zona pellucida in wild mares (*Equus caballus*). *Reproduction* **60**, 197–202.

Kirkpatrick, J.F. and Turner, A. (2007) Immunocontraception and increased longevity in equids. *Zoo Biology* **26**, 237–244.

Kirkpatrick, J.F. and Turner, A. (2008) Achieving population goals in long-lived wildlife with contraception. *Wildlife Research* **35**, 513–519.

Kirkpatrick, J.F., Turner, J.W. and Liu, I.K.M. (1996) Contraception of wild and feral equids. In T.J. Kreeger (ed.), *Contraception in Wildlife Management.* Washington DC: US Government Printing Office, pp. 161–169.

Kirkpatrick, J.F., Rowan, A., Lamberski, N., Wallace, R., Frank, K. and Lyda, R. (2009) The practical side of immunocontraception: zona proteins and wildlife. *Journal of Reproductive Immunology* **83**, 151–157.

Kirkpatrick, J.F., Lyda, R.O., and Frank, K.M. (2011) Contraceptive vaccines for wildlife: a review. *American Journal of Reproductive Immunology* **66**, 40–50.

Kitchener, A.L., Kay, D.J.,Walters, B., Menkhorst, P., McCartney, C.A., Buist, J.A., Mate, K.E. and Rodger, J.C. (2009) The immune response and fertility of koalas (*Phascolarctos cinereus*) immunised with porcine zonae pellucidae or recombinant brushtail possum ZP3 protein. *Journal of Reproductive Immunology* **82**, 40–47.

Kock, R.A., Woodford, M.H. Rossiter, P.B. (2010) Disease risks associated with the translocation of wildlife. *Revue scientifique et technique* **29**, 329–362.

Kreeger, T.J. (1997) Overview of delivery systems for the administration of contraceptives to wildlife. In T.J. Kreeger (ed.), *Contraception in Wildlife Management.* Washington DC: US Government Printing Office, pp. 29–48.

Kutzler, M. and Wood, A. (2006) Non-surgical methods of contraception and sterilization. *Theriogenology* **66**, 514–525.

Langbein, J., Putman, R.J. and Pokorny, B. (2011) Road traffic accidents involving ungulates and available measures for mitigation. In R.J. Putman, M. Apollonio and R. Andersen (eds), *Ungulate Management in Europe: Problems and Practices.* Cambridge, UK: Cambridge University Press, pp. 215–259.

Locke, S.L., Cook, M.W., Harveson, L.A., Davis, D.S., Lopez, R.R., Silvy, N.J. and Fraker, M.A. (2007) Effectiveness of SpayVac® on reducing white-tailed deer fertility. *Journal of Wildlife Diseases* **43**, 726–730.

Lohr, C.A., Mills, H., Robertson, H. and Bencini, R. (2009) Deslorelin implants control fertility in urban brushtail possums (*Trichosurus vulpecula*) without negatively influencing their body-condition index. *Wildlife Research* **36**, 324–332.

Long, D.B., Campbell, T.A. and Massei, G. (2010) Evaluation of feral swine-specific feeder systems. *Rangelands* **32**, 8–13.

Magiafoglou, A., Schiffer, M., Hoffmann, A.A. and McKechnie, S.W. (2003) Immunocontraception for population control: will resistance evolve? *Immunology and Cell Biology* **81**, 152–159.

Massei, G. (2013) Fertility control in dogs. In N. Macpherson, F.X. Meslin and A.I.Wandeler (eds), *Dogs, Zoonoses and Public Health.* Wallingford, UK: CABI International, pp. 259–270.

Massei, G. and Miller, L.A. (2013) Non-surgical fertility control for managing free-roaming dog populations: a review of products and criteria for field applications. *Theriogenology* **80**, 829–838.

Massei, G., Cowan, D.P., Coats, J., Gladwell, F., Lane J.E. and Miller, L.A. (2008) Effect of the GnRH vaccine GonaCon™ on the fertility, physiology and behaviour of wild boar. *Wildlife Research* **35**, 1–8.

Massei, G., Quy, R., Gurney, J. and Cowan, D.P. (2010a) Can translocations be used to manage human–wildlife conflicts? *Wildlife Research* **37**, 428–439.

Massei, G., Miller, L.A. and Killian, G.J. (2010b) Immunocontraception to control rabies in dog populations. *Human–Wildlife Interactions* **4**, 155–157.

Massei, G., Coats, J., Quy, R., Storer, K. and Cowan, D.P. (2010c) The BOS (Boar-Operated-System): a novel method to deliver baits to wild boar. *Journal of Wildlife Management* **74**, 333–336.

Massei, G., Roy, S. and Bunting, R. (2011) Too many hogs? A review of methods to mitigate impact by wild boar and feral pigs. *Human–Wildlife Interactions* **5**, 79–99.

Massei, G., Cowan, D. P., Coats, J., Bellamy, F., Quy, R., Brash, M. and Miller, L.A. (2012) Long-term effects of immunocontraception on wild boar fertility, physiology and behaviour. *Wildlife Research* **39**, 378–385

McDowell, A., McLeod, B.J., Rades, T. and Tucker, I.G. (2006) Application of pharmaceutical drug delivery for biological control of the common brushtail possum in New Zealand: a review. *Wildlife Research* **33**, 679–689.

McLaughlin, E.A. and Aitken, R.J. (2011) Is there a role for immunocontraception? *Molecular and Cellular Endocrinology* **335**, 78–88.

McShea, W.J. (2012) Ecology and management of white-tailed deer in a changing world. *Annals of the New York Academy of Sciences* **1249**, 45–56.

Merrill, J.A., Cooch, E.G. and Curtis, P.D. (2003) Time to reduction: factors influencing management efficacy in sterilizing overabundant white-tailed deer. *Journal of Wildlife Management* **67**, 267–279.

Merrill, J.A., Cooch, E.G. and Curtis, P.D. (2006) Managing an over-abundant deer population by sterilization: effects of immigration, stochasticity and the capture process. *Journal of Wildlife Management* **70**, 268–277.

Miller, L.A. and Fagerstone, K.A. (2000) Induced infertility as a wildlife management tool. In T.P. Salmon and A.C. Crabb (eds), *Proceedings of the 19th Vertebrate Pest Conference.* Davis, CA: University of California, pp. 160–168.

Miller, L.A. and Killian, G.J. (2002) In search of the active PZP epitope in white-tailed deer immunocontraception. *Vaccine* **20**, 2735–2742.

Miller, L.A., Johns, B.E. and Killian, G.J. (2000) Immunocontraception of white-tailed deer with GnRH vaccine. *American Journal of Reproductive Immunology* **44**, 266–274.

Miller, L.A., Crane, K., Gaddis, S. and Killian, G.J. (2001) Porcine zona pellucida immunocontraception: long-term health effects on white-tailed deer. *Journal of Wildlife Management* **65**, 941–945.

Miller, L.A., Rhyan, J.C. and Drew, M. (2004) Contraception of bison by GnRH vaccine: a possible means of decreasing transmission of brucellosis in bison. *Journal of Wildlife Diseases* **40**, 725–730.

Miller, L.A., Gionfriddo, J.P., Fagerstone, K.A., Rhyan, J.C. and Killian, G.J. (2008a) The Single-Shot GnRH Immunocontraceptive Vaccine (GonaCon™) in White-Tailed Deer: Comparison of Several GnRH Preparations. *American Journal of Reproductive Immunology* **60**, 214–223.

Miller, L.A., Gionfriddo, J.P., Rhyan, J.C., Fagerstone, K.A., Wagner, D.C. and Killian, G.J. (2008b) GnRH immunocontraception of male and female white-tailed deer fawns. *Human–Wildlife Interactions* **2**, 93–101.

Miller, L.A., Fagerstone, K.A., Wagner, D.C. and Killina, G.J. (2009) Factors contributing to the success of a single-shot, multiyear PZP immunocontraceptive vaccine for white-tailed deer. *Human–Wildlife Conflicts* **3**, 103–115.

Nave, C.D., Coulson, G., Short, R.V., Poiani, A., Shaw, G. and Renfree, M.B. (2002) Long-term fertility control in the kangaroo and the wallaby using levonorgestrel implants. *Reproduction* **60**, 71–80.

Naz, R.K., Gupta, S.K., Gupta, J.C., Vyas, H.K. and Talwar, A.G. (2005) Recent advances in contraceptive vaccine development: a mini-review. *Human Reproduction* **20**, 3271–3283.

Nishi, J.S., Shury, T. and Elkin, B.T. (2006) Wildlife reservoirs for bovine tuberculosis (*Mycobacterium bovis*) in Canada: strategies for management and research. *Veterinary Microbiology* **112**, 325–338.

Nuñez, C.M.V., Adelman, J.S., Mason, C. and Rubenstein, D.I. (2009) Immunocontraception decreases group fidelity in a feral horse population during the non-breeding season. *Applied Animal Behaviour Science* **117**, 74–83.

Nuñez, C.M.V., Adelman, J.S. and Rubenstein, D.I. (2010) Immunocontraception in wild horses (*Equus caballus*) extends reproductive cycling beyond the normal breeding season. *PLoS ONE* **5**, e13635.

Patton, M.L., Jochle, W. and Penfold, L.M. (2007) Review of contraception in ungulate species. *Zoo Biology* **26**, 311–326.

Plotka, E.D. and Seal, U.S. (1989) Fertility control in female white-tailed deer. *Journal of Wildlife Diseases* **25**, 643–646.

Plotka, E.D., Vevea, D.N., Eagle, T.C., Tester, J.R. and Siniff, D.B. (1992) Hormonal contraception of feral mares with Silastic rods. *Journal of Wildlife Diseases* **28**, 255–262.

Powers, J.G, Baker, D.L.T.L., Conner, M.M.A.H. and Nett, T.M. (2011) Effects of gonadotropin-releasing hormone immunization on reproductive function and behavior in captive female Rocky Mountain elk (*Cervus elaphus nelsoni*). *Biology of Reproduction* **85**, 1152–1160.

Putman, R.J. (1997) *Chemical and Immunological Methods in the Control of Reproduction in Deer and Other Wildlife: potential for population control and welfare implications.* RSPCA Technical Bulletin, Royal Society for the Prevention of Cruelty to Animals, Horsham, UK.

Ramsey, D. (2007) Effects of fertility control on behavior and disease transmission in brushtail possums. *Journal of Wildlife Management* **71**, 109–116.

Ramsey, D., Coleman, J., Coleman, M. and Horton, P. (2006) The effect of fertility control on the transmission of bovine tuberculosis in wild brushtail possums. *New Zealand Veterinary Journal* **54**, 218–223.

Ransom, J.I., Cade, B.S., and Hobbs, N.T. (2010) Influences of immunocontraception on time budgets, social behavior, and body condition in feral horses. *Applied Animal Behaviour Science* **124**, 51–60.

Ransom, J.I., Hobbs, N.T. and Bruemmer, J. (2013) Contraception can lead to trophic asynchrony between birth pulse and resources. *PLoS ONE* **8**, 1–9.

Reimoser, F. and Putman, R.J. (2011) Impact of large ungulates on agriculture, forestry and conservation habitats in Europe. In R.J. Putman, M. Apollonio and R. Andersen (eds), *Ungulate Management in Europe: Problems and Practices.* Cambridge, UK: Cambridge University Press, pp. 144–191.

Rudolph, B.A., Porter, W.F. and Underwood, H.B. (2000) Evaluating immunocontraception for managing suburban white-tailed deer in Irondequoit, New York. *Journal of Wildlife Management* **64**, 463–473.

Ruiz-Fons, F., Rodríguez, O., Mateu, E., Vidal, D. and Gortázar, C. (2008) Antibody response of wild boar (*Sus scrofa*) piglets vaccinated against Aujeszky's disease virus. *Veterinary Record* **162**, 484–485.

Rutberg, A.T. (2005) Deer contraception: what we know and what we don't. In A.T. Rutberg (ed.), *Humane Wildlife Solutions: The Role of Immunocontraception.* Washington DC: Humane Society Press, pp. 23–42.

Rutberg, A.T. and Naugle, R.E. (2008) Population effects of immunocontraception in white-tailed deer (*Odocoileus virginianus*). *Wildlife Research* **35**, 494–501.

Rutberg, A.T., Naugle, R.E., Thiele, L.A., and Liu, I.K.M. (2004) Effects of immunocontraception on a suburban population of white-tailed deer *Odocoileus virginianus*. *Biological Conservation* **116**, 243–250.

Rutberg, A.T., Naugle, R.E., Turner, J.W., Fraker, M.A. and Flanagan, D.R. (2013) Field testing of single-administration porcine zona pellucida contraceptive vaccines in white-tailed deer (*Odocoileus virginianus*). *Wildlife Research* **40**, 281–288.

Saez-Royuela, C. and Telleria, J.L. (1986) The increased population of the Wild boar (*Sus scrofa* L.) in Europe. *Mammal Review* **16**, 97–101.

Smith, G.C. and Cheeseman, C.L. (2002) A mathematical model for control of diseases in wildlife populations: culling, vaccine and fertility control. *Ecological Modelling* **150**, 45–53.

Smith, G.C. and Wilkinson, D. (2003) Modeling control of rabies outbreaks in red fox populations to evaluate culling, vaccination, and vaccination combined with fertility control. *Journal of Wildlife Diseases* **39**, 278–286.

Turner, J.W. and Kirkpatrick, J.F. (1991) In My Experience: new developments in feral horse contraception and their potential application to wildlife. *Wildlife Society Bulletin* **19**, 350–359.

Turner, J.W. and Kirkpatrick, J.F. (2002) Effects of immunocontraception on population, longevity and body condition in wild mares *Equus caballus*. *Reproduction Supplement* **60**, 187–195.

Turner, J.W., Liu, I.K.M., Flanagan, D.R.J.W., Rutberg, A.T. and Kirkpatrick, J.F. (2007) Immunocontraception in wild horses: one inoculation provides two years of infertility. *Journal of Wildlife Management* **71**, 662–667.

Twigg, L.E., Lowe, T.J., Martin, G.R., Wheeler, A.G., Gray, G.S., Griffin, S.L., O'Reilly, C.M., Robinson, D.J. and Hubach, P.H. (2000) Effects of surgically imposed sterility on free-ranging rabbit populations. *Journal of Applied Ecology* **37**, 16–39.

Waddell, R.B., Osborn, D.A., Warren, R.J., Griffin, J.C. and Kesler, D.J. (2001) Prostaglandin- mediated fertility control in captive white-tailed deer. *Wildlife Society Bulletin* **29**, 1067–1074.

Waithman, J.D., Sweitzer, R.A., Van Vuren, D., Drew, J.D., Brinkhaus, A.J., Gardner, I.A. and Boyce, W.M. (1999) Range expansion, population sizes, and management of wild pigs in California. *Journal of Wildlife Management* **63**, 298–308.

Walter, W.D., Perkins, P.J., Rutberg, A.T. and Kilpatrick, H.J. (2002) Evaluation of immuno contraception in a free-ranging suburban white-tailed deer herd. *Wildlife Society Bulletin* **30**, 186–192.

Ward, A.I. (2005) Expanding ranges of wild and feral deer in Great Britain. *Mammal Review* **35**, 165–173.

Williams, C.K. (2002) Risk assessment for release of genetically modified organisms: a virus to reduce the fertility of introduced wild mice, *Mus domesticus. Reproduction Supplement* **60**, 81–88.

Wilson, C. J. (2003) *Current and Future Deer Management Options.* Report on behalf of Defra European Wildlife Division.

Woodrow, K.A., Bennett, K.M. and Lo, D.D. (2012) Mucosal vaccine design and delivery. *Annual Review of Biomedical Engineering* **14**, 17–46.

Zhang, Z. (2000) Mathematical models of wildlife management by contraception. *Ecological Modelling* **132**, 105–113.

Chapter 11

Welfare Issues in the Management of Wild Ungulates

Frauke Ohl and Rory Putman

Animal welfare issues are a matter of intense societal and political debate, but discussions have generally tended to be focused primarily on laboratory animals, companion animals or farm livestock. Welfare issues, however, do not only occur in the context of closely managed animals (whose environment is strictly controlled by human activity), but may also arise in situations in which animals are living more freely, but remain nonetheless under relatively close management (such as in nature conservation areas) or for truly 'wild' animals. There is increasing pressure from society to explore what may be our responsibilities for the welfare of free-ranging animals, especially where they are directly impacted by human activities (habitat alteration, habitat fragmentation, etc.) or are exploited as game species.

As far as we are aware the Netherlands is the only country in Europe that actually imposes a legal obligation on all citizens 'to take responsibility and provide the necessary care for animals that need help' (whether wild or managed/kept).[1] In other countries any legal responsibility is generally limited to the prohibition of actions likely to *cause* suffering (or actually calculated to cause suffering), rather than imposition of a more general responsibility for welfare – and provisions relating to game animals, in particular, tend to be contained within more specific game laws. However, from various discussions it has become apparent that the general public both in the Netherlands and in the UK strongly feel that politicians and managers are morally obliged to take all necessary measures to minimise the perceived unnecessary suffering of freely living animals, irrespective of specific explicit legal responsibilities. In the UK and especially Scotland, the welfare of wildlife has increasingly become a focal point in recent years (with for example explicit mention of 'damage to welfare' in the Deer Scotland Act as a result of the Wildlife and Natural Environment Act). However, even in legislation, welfare

---

[1]Dutch Animal Health and Welfare Act (GWWD) (1992) paragraph 36 subsection 3.

is rarely explicitly defined, nor how welfare of deer (and other wildlife) might be safeguarded or enhanced.

What is clear is that conventional principles and practices in relation to safeguarding animal welfare (established primarily for closely managed animals such as production farm animals, laboratory animals or companion animals), do not translate easily to the management of free-ranging wildlife species where the environment cannot be so easily controlled. Thus, instead of a focus on alteration of environmental conditions (the way the animals are housed, sheltered, fed) to adjust welfare status, one needs a more fundamental understanding of what welfare means in biological terms. Nature is not easily managed and 'controlled'–and nor is wildlife welfare. Thus, if one aims at developing sustainable wildlife management, one has to follow natural processes instead of trying to constrain nature to what management demands.

In this chapter we will review current thinking in relation to the biological processes which promote or define welfare and then consider how we might measure and manage welfare status of wild ungulates. In such analysis it is crucial to emphasise that our considerations in the following paper relate only to responsibilities in relation to *animal welfare;* our considerations do not in any way address the rather separate issue of *animal rights* (e.g. Haynes, 2011). Thus, quite explicitly and quite deliberately, we do not address here questions of whether or not humans have the right to exploit animals for food, to use them as laboratory models, to hunt or to keep animals as pets. Simply we consider what may be the duty of care and requirements of action to ensure acceptable welfare of wild or more closely managed animals, whatever the philosophical debate about rights and wrongs of management in the first place, and whether that management may be directed towards human-oriented goals rather than objectives directed towards improving conditions for the animals themselves.

Further, even among those who are attempting to develop frameworks for the assessment and management of welfare in wild animals, much of the current emphasis is focused an avoidance of suffering (and often specifically: suffering by that individual, or others dependent upon it, when that animal may be shot by human hunters). This strong focus to date simply on the avoidance of animal suffering is hardly surprising. Much of the early literature and most existing legislation is also based on such supposition (see below). However, such emphasis on the avoidance of negative welfare states has little to do with safeguarding welfare as more broadly defined. Considerations below will lead us to suggest that in safeguarding, or actively promoting welfare in wild ungulates in more general terms, managers might be expected to have somewhat wider responsibilities and competences.

What IS welfare and what do we mean by welfare status?

In order to scope this review, can we provide simple and robust definitions of positive and negative welfare states for any individual animal? Can we clarify a distinction between negative welfare states and suffering? Welfare is in fact relatively rarely explicitly defined; rather there is a general presumption that we implicitly understand what is meant by the term (as also the term 'well-being'). A number of definitions are summarised in Ohl and Putman, 2013a; from Findlay, 2007), to which we might add from a more academic literature:

Broom (1988) defined welfare 'an individual's state as regards its attempts to cope with its environment' while noting that 'Feelings, such as pain, fear and the various forms of pleasure, are a key part of welfare'. Duncan (1993, 1996) and Fraser and Duncan (1998) suggest that welfare is entirely to do with how animals feel. At the same time, those with a medical or veterinary background sometimes present the view that physical health is all, or almost all, of welfare.

Dawkins (1980, 1990) stated that 'the feelings of the individual are the central issue in welfare but other aspects such as the health of that individual are also important', while most recently Webster (2012) notes that: 'There is now broad agreement amongst academics and real people that the welfare of a sentient animal is defined by how well it feels; how well it is able to cope with the physical and emotional challenges to which it is exposed.'

In a similar way we may explore various definitions that have been offered in relation to suffering. Thus Fraser and Duncan (1998) denote suffering as 'strong, negative affective states such as severe hunger, pain, or fear'. Dawkins (e.g. 1990, 2008) suggests that 'suffering can result from experiencing a wide range of unpleasant emotional states such as fear, boredom, pain, and hunger'; by converse, Appleby and Sandøe (2002) note that, 'Animals should *feel* well by being free from prolonged and intense fear, pain and other negative states, and by experiencing normal pleasures.' Suffering thus describes the negative emotional experience resulting from being exposed to a negative state of welfare.

11.1 A review of the general principles of animal welfare

The report of the Brambell Committee in the UK in 1965 on (farm) animal welfare of 1965[2] has cast a long shadow. While unquestionably a major advance at the time and heralding a major change of attitude to the management and treatment of livestock animals, there have been many subsequent improvements in our understanding and scientific knowledge, which require some re-examination of the Brambell principles and the concept of the Five Freedoms.

Despite such advances – and nearly 50 years after their original formulation – the Five Freedoms continue to dominate much of welfare practice and welfare thinking both within the UK and further afield. In fact, the Five Freedoms form the basis for the so-called Welfare Quality Project, which currently forms the backbone

---

[2]Brambell Committee: a technical committee set up by the UK Government in 1965 to inquire into the welfare of animals kept under intensive livestock husbandry systems (Brambell Committee (Report), HC Deb 15 December 1965 vol 722 cc279-80W).

of European animal welfare guidelines (http://ec.europa.eu/food/animal/welfare/sum_proceed_wq_conf_en.pdf).

In all fairness, the Brambell Committee's report never set out to be a 'welfare concept', but was developed specifically to establish minimum requirements to ensure the absence of negative welfare. The committee formulated the basic idea of what was subsequently to be summarised by the Farm Animal Welfare Council in terms of the 'Five Freedoms': that, in effect, compromise to animal welfare is avoided if the animals are kept free from:

- hunger, thirst or inadequate food
- thermal and physical discomfort
- injuries or diseases
- fear and chronic stress
- and were free to display normal, species-specific behavioural patterns.

The first four of these *freedoms* were formulated from the perspective that the absence of actual negative impact assures welfare; only the fifth, albeit more indirectly, potentially implied an expectation of facilitation of more positive aspects of welfare.

There have been various reformulations of these essential principles (Webster, 1994; Anon., 2009), including an attempt by Mellor to substitute the concept of 'five domains of potential welfare compromise' (Mellor and Reid, 1994); the five domains are defined in terms of: nutrition, environment, health, behaviour and mental state.[3] But these different incarnations change the basis of the construct little and as an instant index, the Five Freedoms today remain widely used today as a guideline for welfare assessment protocols, with the actual state of welfare of an animal being characterised as unimpaired if it complies with the Five Freedoms (e.g. Rutherford, 2002; Veissier and Boissy, 2007; Knierim and Winkler, 2009; Mendl *et al.*, 2010).

However, as we have emphasised, these Five Freedoms were primarily derived in relation to the welfare of closely managed farm animals and may only have restricted utility when applied to animals whose environment is less rigorously controlled by human intervention. Except in regard to the fifth Freedom, the animal is conceived as undergoing its life somewhat passively. This was perhaps legitimate in that Brambell's freedoms were originally developed primarily for application to animals whose environment was largely controlled by human management, but in application to the assessment of welfare of wild or free-ranging, animals, then they are somewhat over-restrictive (see review by Ohl and van der Staay, 2012).

More fundamentally, a view of welfare which is dominated by an emphasis on the avoidance of negative states neglects the fact that, except in the specific instances where natural selection processes have been largely countermanded by deliberate selection by humans, animals have evolved, optimising the ability to interact with and adapt to (changing conditions within) their environment and that

---

[3] Although this latter idea has not been widely adopted (other than by Mellor himself in subsequent publications: e.g. Mellor and Stafford, 2001; Mellor *et al.*, 2009).

thus exposure to environmental challenge and short periods of 'negative welfare' may be inevitable, indeed essential, if these are understood as triggers to release from the animal's repertoire the appropriate behavioural or physiological response to adapt to those challenges.

11.1.1 A more biological approach to animal welfare

The idea that animals have generally evolved adaptations to their environment, optimising the ability to adapt to changes within that environment through the expression of a variety of physiological and/or behavioural responses, was first applied within a welfare context some three decades ago (see, for example, Dantzer and Mormede, 1983; Broom, 1988; Barnett and Helmsworth, 1990) and has more recently been championed by, for example, Duncan (1993), Fraser *et al.* (1997), Fraser and Duncan (1998), Korte *et al.* (2007), Ohl and van der Staay (2012) among others. In such a concept, the animal's welfare is not at risk as long as it is able to meet environmental challenges, i.e. 'when the regulatory range of allostatic mechanisms matches the environmental demands' (Korte *et al.*, 2007).

In recognition of this, a number of authors have advocated a more dynamic view of welfare which recognises that wild and domestic animals have adaptive responses which enable them in normal conditions to respond appropriately to address some environmental or physiological challenge, to restore a more positive welfare state (except perhaps in those cases where artificial selection may have resulted in loss of some responses from domestic stock). In such a view, assessment of welfare should therefore focus not so much on the challenges which any animal may face at a given moment but on whether or not the individual possesses the appropriate (behavioural or physiological) responses to adapt appropriately to both positive and potentially harmful (negative) stimuli and has adequate opportunity to express those responses (Ohl and van der Staay, 2012).

On this basis we may then suggest that assessment of welfare should therefore focus not so much on the challenges which any animal may face at a given moment but on whether or not the animal has the freedom and capacity to react appropriately (i.e. adaptively) to both positive and potentially harmful (negative) stimuli. By the same token, welfare should not be considered as an instantaneous construct to be assessed at some moment in time. An adaptive response may take some finite period of time; crucially therefore, our assessment of welfare should not simply consider the status of any individual at a given moment in time, but needs to be integrated over the longer time periods required to execute such change.

A further problem implicit in standard methods for objective assessment of welfare status is that such protocols inevitably reflect the observer's perspective and subjective judgement, whereas most modern commentators would now acknowledge that to some significant degree, any animal's status must be that perceived and judged by that animal itself (Duncan, 1993; Fraser and Duncan, 1998; Broom, 2006; Taylor and Mills, 2007; Nordenfelt, 2011; Webster, 2011; Ohl and van der Staay, 2012).

In effect the 'decision' by any individual animal to accept its current status or to engage in behaviour designed to bring about some change of status must in part be determined by an assessment of physiological condition (hunger, thirst, etc.) but also by an assessment of a sense of 'well-being'. It is clear that emotions play an important role in this assessment and in the performance of adaptive behaviours. There is a growing literature to suggest that much of the function of emotion or emotional status may indeed be to provide a convenient proximate surrogate to reinforce behaviours which are (or were) in some way adaptive, to make performance of these appropriate behaviours in some sense pleasurable or rewarding and thus promote their expression in appropriate circumstances (Cabanac, 1971, 1979; Broom, 1988, 1991; Mendl and Deag, 1995; Dawkins, 1998; Panksepp, 1998, 2011; Lahti, 2003; Webster, 2011).

As nicely summarised by Nesse and Ellsworth (2009), 'Emotions are modes of functioning, shaped by natural selection, that coordinate physiological, cognitive, motivational, behavioural, and subjective responses in patterns that increase the ability to meet the adaptive challenges of situations that have recurred over evolutionary time.' In other words, emotions would appear to play a pivotal role in decoding and evaluating positive and negative feedbacks, in perceiving the individuals' own internal state and, finally in regulating the execution of its behaviour (Ohl and Putman, 2013a).

What is significant in this context is that there is clear variation between individuals in sensitivity/responsiveness of central nervous circuits processing emotions and internal perceptions and thus that different individuals may 'perceive' the consequences for themselves of one and the same environmental challenge in very different ways.

There may also be considerable variation between individuals in their actual behavioural or physiological abilities to respond adaptively, or in the actual coping strategy adopted. Thus there may be quite substantial variation in the way different individuals may respond to the same stressor and the strategies which they may use to cope with environmental or social challenge – not simply in relation to differences in the adaptive repertoire available to different individuals, but also in relation to coping strategy adopted (Ohl and Putman, 2013a, b).

The logical extension of such argument is now clear: If, as seems apparent, there exists significant variability in how different individual animals may assess their own welfare status; if in addition a group of animals may consist of individuals some of whom seek to optimise welfare status while others (at that moment in time) are content to satisfice (*sensu* Krippendorf, 1986; and see Ohl and Putman, 2013a, b), then we, as external observers, may expect to observe within any group of animals a series of group members with different 'absolute' welfare status (assessed by an external observer against some fixed set of criteria) yet which perceive their own welfare state as being optimal (or at least sufficient not to require action to alter that status).

Such conclusion makes clear that purely objective functional scales for measuring the welfare status of individual animals can have little validity in that, even under identical conditions, the actual welfare status of different individuals may vary widely. Further it emphasises that in assessing the welfare status of animals in groups or populations we must expect high variation in apparent welfare and in attempting to safeguard satisfactory welfare we must insert into protocols some minimum threshold value below which no individual should be allowed to fall, instead of (or in addition to) simply determining some average welfare status to be achieved.

11.1.2 Welfare at the level of the population or group

Further, at least among social species, it is probable that individual welfare should be re-evaluated as being related to the functioning of a social group, taking into account that a variety of situations exist where (social) individuals invest into the welfare of other individuals instead of maximising their own welfare. Crucially, many management decisions for free-ranging ungulates are made at the level of the group or population, or at least have impacts at population level rather than simply that of a given target individual; once again this suggests that welfare concepts and assessment methodologies need to be extended for application at the population level.

One of the first attempts to measure welfare at the group level was offered by Kirkwood *et al.* (1994) in populations of wild mammals and birds. These authors argued that an 'Assessment of the scale and severity of harm to welfare requires consideration of several factors. We propose that at the simplest level these are: (1) The number of animals affected. (2) The cause and nature of the harm. (3) The duration of the harm. (4) The capacity of the animal to suffer.'

This approach implicitly presupposed that a given cause and type of harm with a given duration will result in an identical effect on welfare in all individuals of one group as long as all individuals do have the same capacity to suffer (Kirkwood *et al.*, 1994, and later Mathews, 2010, presume that they do). Consequently, this scenario then suggests a linear correlation between group size/animal number and the scale of harm caused to (group) overall welfare. Put in another way, this implies that the welfare status of a group would be known as soon as we know about the welfare of any one of its members, since the welfare of a group is considered to being represented simply as the sum of the identical individual welfare of its members.

This calculation in practice assumes homogeneity of all members of a group in terms of state, sensitivity, and perception of welfare at any point in time – a 'universal' individual welfare. Yet, as we have argued above, there may be significant variation in what may be perceived as optimum or satisfactory welfare for different individual animals. In addition it is apparent that different individuals may have markedly different coping styles or strategies in how they may respond to environmental challenge (Mendl and Deag, 1995; Koolhaas *et al.*, 2010).[4]

If we thus follow Broom (2006) and later authors in asserting that the welfare of each individual should be related to the adaptive capacity of that individual, such recognition makes it even more likely that distinct external conditions will affect the welfare of the members of any one group to a *different* extent. This clearly has implications for efforts to assess the welfare status of individuals within a group or population–whether of domestic animals, farm livestock or free-ranging wildlife.

---

[4]Indeed, Mendl and Deag (1995) suggest the possibility that there may even exist, within groups of animals, some frequency-dependent stability of alternative coping strategies.

In such a case then, the welfare status of a group may be optimised while the (objectively determined) welfare states of its individual members may vary over a considerable range, but nonetheless all members perceive their own welfare state as optimal – or at least satisfactory. In effect, it implies not only (as above) that the welfare status of all individuals in a given set of conditions may not all be equal in absolute terms but that, in addition, the actual effect (perception) of any compromise in welfare status welfare may also not be the same for all individuals, since they may vary in the extent to which they respond to a given (positive or negative) influence (Koolhaas *et al.*, 2008).

At the individual level we assume that positive welfare is defined by the animals' freedom to adapt to environmental conditions up to a level that it perceives as positive. But we recognise that individuals may show significant variation in their perception of a given status and their 'decision' about how to respond to that perceived status. Thus we may expect that even under identical environmental conditions, different individuals within a group or population may perceive or experience their welfare status differently. Different individuals within a group or population may thus be 'satisfied' with different levels of what an external observer would consider better or worse states. In assessing the welfare status of animals in groups or populations we must therefore expect high variation in apparent welfare

In consequence, we conclude that welfare of social groups cannot adequately be assessed by assuming that the impact on all individual group members will be identical. Instead it is important to consider the welfare impact on members of a social group which may vary considerably in their welfare 'phenotype', and in addition the possible impact on the group as a whole in terms of its capacity to adapt to prevailing environmental conditions because of a given phenotypic distribution (Ohl and Putman, 2013b). By converse, as above, in seeking to ensure positive welfare of such a group, we cannot expect to optimise welfare status of all individuals; however, in recognition of variability, we should at least add an additional constraint which ensures that the status of no individual falls below some critical minimum threshold.

11.2 Some definitions

Based on the considerations explored above, we would propose the following definitions (after Ohl and Putman 2013a, c).

Adaptive capacity

The adaptive capacity describes the set of innate (physical and mental) abilities with which an animal species is naturally endowed and which an individual develops in the course of its own existence. The species-specific abilities form a basis, which is refined and developed in each individual. The adaptive capacity of an individual is not static; it is dependent on the individual's internal state as well as on its changing environmental conditions.

We suggest that group welfare may be defined by the freedom adequately to adapt to prevailing environmental circumstance *as a group* and that group welfare may be optimised while the (objectively determined) welfare states of its individual members may vary over a considerable range, with nonetheless all members perceiving their own welfare state as being optimal. The adaptive capacity of a group describes the set of (physical and mental) abilities with which a group of animals is naturally endowed.

Welfare

Welfare describes an internal state of an individual, as experienced by that individual. This state of welfare is the result of the individual's own characteristics, as well as the environmental conditions to which the individual is exposed. Human determination of an animal's state of welfare is only as good as the observer's perception of the signals that the animal emits. A negative state of welfare is perceptible via reactions that are aimed at changing the existing situation. A positive state of welfare is perceptible via reactions aimed at keeping the existing situation as it is.

Welfare describes an internal emotional state as perceived by an individual. As such, group welfare or population welfare does not exist except as an 'envelope' of the separate welfare states of its individual members. However, the welfare state of an individual represents a function of its adaptive functioning within prevailing environmental circumstances – an environment which includes also other members of its social group or population – and it is that functioning that can be assessed at the group level as well as at the level of each individual of that group. The adaptive functioning of a group then is the result of the characteristics of that group, as well as the environmental conditions to which the group is exposed.

Positive welfare

Positive (or good) welfare describes the state in which an individual or group of individuals has the freedom adequately to react to the demands of the prevailing environmental circumstances, resulting in a state that the animals themselves perceive as positive. With a growing emphasis on the importance of positive experiences (Fraser, 1993; Fraser and Duncan, 1998; Duncan, 2005), good animal welfare is not ensured by the mere absence of negative states (Knierim *et al.*, 2001; Duncan, 2005; Broom, 2010; see also Mellor, 2012) but requires the presence of positive affective states.

Negative welfare

In our view, and as a view increasingly expressed in the wider literature (e.g. Broom, 2006; Korte *et al.*, 2007; Ohl and van der Staay, 2012), negative or bad welfare status describes a state that *the animal itself* perceives as negative. Short-term negative welfare states such as suffering from hunger and fear serve as triggers for the animal to adapt its behaviour. They therefore serve a function. By corollary, a brief state of negative welfare may fall within an animal's adaptive capacity, and would not necessarily require intervention.

Welfare status is more significantly compromised when an animal or a group of animals have insufficient opportunity (freedom) to respond appropriately to a potential welfare 'challenge' through adaptation by changes in its own behaviour (either where environmental challenges exceed the adaptive capacity of the animal or the opportunities available are inadequate to permit the animal effectively to express the appropriate adaptive responses).

The adaptive functioning of a group is compromised when a group of animals have insufficient opportunity (freedom) to respond appropriately to a potential 'challenge' through adaptation by changes in behaviour (where environmental challenges exceed the adaptive capacity of the group as a whole or the opportunities available are inadequate to permit the group effectively to express the appropriate adaptive responses). Negative 'welfare' at the group level, thus, describes a state in which distinct individuals still may perceive their own state as positive but that does not allow for adaptive functioning of the group as a whole. In such a situation, it can be expected that the number of group members experiencing 'negative welfare' or 'suffering' will progressively increase over time.

Suffering

Suffering describes the negative emotional experience resulting from being exposed to a negative state of welfare. Short-term, negative welfare states such as hunger and even momentary fear serve as triggers for the animal to adapt its behaviour. They therefore serve a function. A brief state of suffering may fall within an animal's adaptive capacity, and would not necessarily require intervention.

If an individual lacks the ability or the opportunity appropriately to react to any such negative stimulus however, (for example, by escaping from a frightening situation), a challenge is created that may exceed the adaptive capacity of the individual. Where this negative experience is profound (for instance, pain caused by an injury) we may consider the suffering acute, however short lived; where negative feelings are persistent over a protracted period, the situation may be considered to impose more prolonged suffering which may in practice constitute a more significant welfare issue.

11.3 General methods for assessment of welfare status in wildlife species

Given all the above, how may we set about actually assessing the (general) welfare status of wildlife individuals? Considerations above challenge the more functional protocols developed for assessing welfare status, even for more closely managed animals – and especially in terms of any attempt to apply such protocols to free-ranging wildlife.

It becomes clear that if we accept that welfare is defined by an ability to adapt and respond to environmental challenge in an appropriate way – and that thus both positive and negative welfare states are a function of the actual adaptive capacities of the individual animal and the opportunity it has to express those responses – then our assessment of welfare must be primarily based on detailed observation of the physiological condition and behavioural responses shown by individual animals over time.

In general, physiological approaches have focused on the concept of measuring levels of stress experienced by individuals based on the belief that if stress increases, welfare decreases (Dantzer *et al.*, 1983; Dantzer and Mormede, 1983; Moberg, 1985). However, there are a number of problems with such an approach. Short-term stress responses are an inevitable part of the process triggering an adaptive response from the animal and thus may be functional in maintaining a longer-term positive welfare status. In such analysis a more relevant measure might be evidence of chronic and 'traumatic' stress, something which is not trivial to differentiate by means of physiological measurements from acute stress (McEwen *et al.*, 1992; de Kloet *et al.*, 2008 a, b).

Physiologically, stress is characterised by an activation of the hypothalamus–pituitary–adrenocortical [HPA] axis, resulting via a complex cascade into the release of cortisol into the blood (Selye, 1950). Thus, the majority of approaches to the measurement of stress consider stress levels reflected in the identification of elevated cortisol levels, although a variety of other blood or tissue parameters have been considered as well (below). However, chronic stress can result in a blunted HPA response and, in consequence, a *reduced* release of cortisol in response to acute stress (McEwen *et al.*, 1992). In other words, based on low levels of peripheral cortisol alone it is impossible to discriminate between absence of stress or a chronically stressed status.

Further, for many of the other parameters assessed (lactic dehydrogenase: LDH-5, muscle glycogen, bilirubin etc.; e.g. Jones and Price, 1990, 1992; Price and Jones, 1992; Bateson and Bradshaw, 1997; Cockram *et al.*, 2011), there is some difficulty again in separating the effects of chronic or acute stress from those which may simply be associated with vigorous or prolonged physical exertion.

Where only some average assay is required for the welfare status of a population of animals, then various samples can of course be derived from culled individuals – as long as it is certain that the method and process of culling does not itself impose additional stress. Where assay is made of levels of cortisol in sources that reflect accumulation of cortisol excretion over time, such as hair or dung, some of the confusion between measures of acute and chronic stress may be avoided (Sheriff *et al.*, 2011).

Where such methods are to be applied to the assessment of stress within living individuals it is clear that collection of blood or tissue samples from free-ranging wild animals is technically complex and, here especially, elevated levels of stress-related products may simply be due to the acute stress associated with capture. Once again, more recent approaches are now assaying accumulated cortisol levels from hair or dung (Sheriff *et al.*, 2011), which may, in combination with analyses at the brain level, offer more reliable assessments of chronic levels of stress, but here again it would appear that for managers, behavioural observations offer a simpler and more robust approach to assessing welfare in the field.

In free-ranging animals, environmental conditions that exceed the adaptive capacities of an individual are likely to translate into a lack of expression of positive emotional states (such as comfort behaviour), as well as a lack of behaviour necessary to fulfil actual demands (such as foraging) (Fraser *et al.*, 1997; Mellor, 2012). Further, behavioural extremes may be observed, such as high levels of intra-group aggression (Koolhaas *et al.*, 2010) or changes in group structures. All such changes not only indicate that prevailing environmental condition are exceeding the animals' adaptive capacities, but are exerting chronic stress themselves and will therefore facilitate a further decrease in physiological condition as well.

Positive affects or emotional states may include pleasure, comfort, contentment, curiosity and playfulness (Mellor, 2012) which suggests that regular observation of such 'positive' behaviour-types might argue for the individual(s) concerned being in a status that it itself perceives as positive. We should note though, that the absence of positive indicators is not sufficient to prove a negative welfare state.

However, routine protocols of behavioural observations should include positive indicators such as play behaviour and notice that the regular absence of play behaviour in young/adolescent individuals is likely to indicate high environmental pressure (Held and Spinka, 2011). In adult animals, play behaviour may be observed less regularly, while, for example, active exploration, social- and self-grooming behaviour can be expected to be present at a regular basis (Kikusi *et al.*, 2006; Crofoot *et al.*, 2011).

While such observations may be practicable for individuals or groups which we may study over prolonged periods of time (and may well be applicable to closely managed animals including closely managed wildlife), such an approach is clearly not likely to be feasible in application where encounters with individuals or groups are typically occasional, fleeting and at a considerable distance. Here, inevitably, we must base assessments primarily at the group level and these will be biased in favour of physical condition scores or rather coarse behavioural indicators. Further such measures will largely be applied at group rather than individual level. This may well be appropriate, given that effective management measures (mostly 'non-specific measures' *sensu* Swart, 2005) can only be targeted at the group or population level.

11.4 Assessing the welfare of wild ungulates at individual and population levels

Against such a theoretical basis, what measures might be used to assess the welfare of wild ungulates at individual or population level?

Clearly many of the physiological measures for welfare status explored in the section above are not easily applicable to live animals (requiring invasive sampling) – and many require quite sophisticated (research) methodologies. However if we accept as above that welfare is defined by an ability to adapt and respond to environmental challenge in an appropriate way – and that thus both positive and negative welfare states are a function of the actual adaptive capacities of the individual animal and the opportunity it has to express those responses – then our assessment of welfare must remain primarily based on observation of the physiological condition and behavioural responses shown by individual animals over time.

While accepting that observation of any individual wild ungulate over a prolonged period is unlikely we would still advocate the use of behavioural cues where possible, including the identification of clearly appropriate and adaptive behaviour (or lack of appropriate response in given circumstances); the expression of appropriate appetitive behaviour, where an animal can be seen to be searching for the appropriate resources to address some perceived deficiency (seeking shelter from wind or adverse weather conditions; moving appropriately to woodland cover or to more sheltered feeding grounds and so forth).

Observation should also reveal lack of expected behaviour (even when opportunity to express that behaviour may be present) and also any clearly inappropriate or atypical behaviours (such as an individual separated at some distance from an obvious social group; animals clearly being rejected or shunned by others within the group; an animal in poor condition suffering continual displacement etc.).

As noted, however, much assessment of welfare in the field will be based more directly on observations of physical condition. Many use the classic visual condition score of Riney (1955, 1960), a five-point condition score for assessment of (physical) condition of wild deer (specifically red deer), which has subsequently been modified and used by others for assessing condition of other free-ranging ungulates (e.g. ponies: Gill, 1988, 1991; Burton, 1992). Riney's condition index is based on assessment, by relation to established standards, of the amount of flesh covering ribs and rump. In effect, as with the Body Condition Scores used in assessment of condition of domestic livestock (from which it was derived) it offers simply a method for formalising a subjective assessment of the depth of subcutaneous fat (e.g. Zulu *et al.*, 2001; Broring *et al.*, 2003).

Once again, experienced managers or hunters may utilise additional physical information. Prolonged examination may reveal if an animal is scouring, while staining of the hindquarters (combined with evidence of some emaciation; above) may indicate that this has been a chronic problem, or over a significant period. As with any other physical characteristics, care must be taken to ensure that observed changes in coat colour are not due to wet/mud/vegetation. Other obvious signs of disease include limping, coughing or obvious emaciation.

We suggest that by extension of such measures managers can consider the welfare status of a wider population. Within any population however we do expect there to be significant variation in how individuals perceive their welfare status and their adaptive capacity/response strategy. Assessment should thus also consider the variance and range of condition apparent within sampled individuals to assess population welfare minimum against some predetermined acceptable minimum threshold.

Potential indicators are summarised in Table 11.1 for both individual and population level (adapted from Ohl and Putman 2013a, c). Recognising that welfare is not an instantaneous construct, and is also a function in large part of an animal's own perception of its status and its ability to adapt to environmental challenge, the indicators necessarily emphasise observation of behaviour and behavioural response to any given challenge.

Table 11.1 Indicators of welfare at the individual and group level, respectively.

Based on the animals' adaptive capacities	Assessment of animal welfare			
	Individual level		*Group level*	
	Positive indicators	*Negative indicators*	*Positive indicators*	*Negative indicators*
The animal(s) should be free adequately to react to hunger/thirst	Appetitive and successful foraging behaviour Normal activity pattern Appropriate body condition	Unsuccessful foraging behaviour Lethargy Inappropriate body condition	Appetitive and successful foraging behaviour and activity pattern as a group Normal variation of body condition	Unsuccessful foraging as a group; successful foraging only in minority of group members (extreme variation within group)
The animal(s) should be free adequately to react to climate conditions	Seeking and finding shelter Appropriate fur condition Appropriate modulation of body condition during seasons	inability to seek find shelter Bad fur condition Body condition worse than can be expected in relation to season	Seeking and finding shelter for all group members Appropriate modulation of variation in fur and body condition during seasons	Not finding shelter or finding insufficient shelter for the group Fur and body condition bad throughout the group or in extreme variation
The animal(s) should be free adequately to react to physical injury or disease	Functional immune system (e.g. appropriate wound healing/lack of scouring)	Infection/inappropriate wound healing; persistent scoring	Functional immune system (e.g. appropriate wound healing; lack of scouring)	Signs of infection across (parts of) the group; e.g. persistent scouring

(*Continued*)

Table 11.1 (*Continued*)

Based on the animals' adaptive capacities	Assessment of animal welfare			
	Individual level		Group level	
	Positive indicators	Negative indicators	Positive indicators	Negative indicators
The animal(s) should be free to express its full non-social behavioural repertoire	Adequate behavioural responses to non-social circumstances/challenges (covering both avoidance and approach behaviours)	Persistent behavioural inhibition, lethargy, context-inadequate behaviour	Adequate behavioural responses to non-social circumstances/challenges that involve the group as a whole (covering both avoidance and approach behaviours)	Behavioural responses that do not involve the whole group
The animal(s) should be free adequately to respond to social interactions	Adequate behavioural responses to social interactions (covering both socio-positive and socio-negative behaviours)	Persistently being bullied (in social species); social isolation	Social stability within the group (as displayed by adequate socio-positive and socio-negative behaviours)	Social instability; splitting up in sub-groups
The animal(s) should be free to experience the full spectrum of emotional states and respond to those states adequately	Executing anxiety-related behaviour and stress responses as well as play- or other pleasure-related behaviour in appropriate context	Inadequate emotional responses (lethargy, hyperreactivity); absence of adequate emotional responses (e.g. lack of anxiety)	Displaying anxiety-related behaviour and stress responses as well as play- or other pleasure-related behaviour at the group-level and in appropriate context	Absence of pleasure-related behaviour; inadequate emotional responses (lethargy, hyperreactivity) at the level of the group

Where, simply, rather general information is sought about the welfare status of the population as a whole (on the average) some assessment of condition may also be derived by examination of (age-related) weight and condition (kidney fat or other physical index) of animals culled from the population. Experienced assessors may also make some estimate of the level of both external and internal parasitic load. While such assessment is inevitably restricted to assessment of physical condition (which is of course only one element contributing to overall welfare status) and cannot be directly related to any individual still alive within that population, it does at least offer managers some practical way of assessing average condition within the (remainder of the) population as whole and, as above, through assessment of the variance and range of condition apparent within sampled individuals, may help identify the population minimum against some pre-determined acceptable minimum threshold.

11.5 Welfare issues associated with killing of deer or other ungulates

In addition to an assessment of welfare of individuals or populations, managers should also consider the possible changes in welfare status which may result from management decisions. In those cases within European legislation where consideration is given explicitly to the welfare of wild ungulates, this is almost always in relation to the welfare consequences of killing them through hunting or other related activity.

In this section therefore we will start by considering what published evidence is available for:

- the impact of different forms of hunting on acute or chronic stress in targeted individuals or others within the population
- data available on wounding rates associated with the use of high velocity rifles, and
- issues associated with the orphaning of juveniles.

11.5.1 Hunting method

There have been a number of published studies which have considered the effect of hunting method on (apparent) pre-mortem stress of culled deer.

Cockram *et al.* (2011) have recently published an analysis of the behavioural and physiological effects on deer shot: (i) by a single stalker during daytime; (ii) by more than one stalker during daytime (collaborative culls); (iii) by using a helicopter for the deployment of stalkers and carcase extraction; or (iv) by a single stalker at night, and compared these with farmed red deer shot in a field or killed at a slaughterhouse.[5] Earlier analyses of the stress levels experienced by

---

[5]We will in this analysis ignore samples killed in a slaughterhouse because of complications arising from potential stress of transport and lairage in unfamiliar surroundings.

deer killed by high-velocity rifle and those hunted by dogs were presented by, for example, Bateson and Bradshaw (1997), Harris *et al.* (1999) and Bradshaw and Bateson (2000).

To a large extent all studies have used similar measures, focusing on assessing the degree of disturbance to the deer by observing their behaviour and measuring a number of physiological parameters from blood or tissue samples. Measures used have included blood lactate concentration, muscle glycogen concentration, plasma cortisol and free fatty acid concentrations, creatine kinase and aspartate aminotransferase.

It is not universally accepted that these latter measures relate incontrovertibly only to the effects of physical exertion and some authors consider that lactate levels and decline in muscle glycogen may also occur in direct response to stress (e.g. Jones and Price, 1990, 1992; Price and Jones, 1992; Bateson and Bradshaw, 1997; Bradshaw and Bateson, 2000); thus such measures may conflate direct effects of stress and physical exertion. To use them simply as estimates of physical activity may oversimplify. In their studies Bateson and Bradshaw also assayed for levels of plasma haemoglobin and bilirubin as evidence of more protracted exertion. Whatever the theoretical and interpretational limitations of some of these measures (also discussed by Mason, 1998), the general conclusions are broadly clear.

Many will continue to debate the degree to which physiological changes caused in red deer hunted with dogs are due primarily to reasonable responses to physical exercise (e.g. Harris *et al.*, 1999) and/or represent changes due to excessive exertion (and thus by implication cruel exertion) and/or are direct evidence of stress (Bateson, 1997; Bateson and Bradshaw, 1997; Bradshaw and Bateson, 2000). The reality remains that levels of all parameters measured (including lactate, lactate dehydrogenase, plasma haemoglobin, cortisol) were lower in samples from animals culled by rifle than those killed after hunting with hounds (Table 2 in Bateson and Bradshaw, 1997).

Similarly, Cockram *et al.* (2011) conclude that plasma cortisol concentrations in deer shot using helicopter assistance were similar to those in deer at the slaughterhouse, but higher than deer shot at night or during the day by a single stalker, or for farmed deer shot by rifle in a field. If we accept cortisol as an appropriate indicator in this case for (acute and chronic) stress, these results are at least suggestive that animals shot by a single stalker, whether by day or night, are less stressed than those shot by other approaches.

Deer shot using helicopter assistance and also deer culled by collaborative and single stalking during the day had lower muscle glycogen concentrations than those culled by a single stalker at night, again indicative of a lower level of disturbance. This last observation is supported by behavioural observations which showed that a significant higher proportion of animals shot during the day, whether by team of stalkers or with helicopter assistance, were recorded as trotting or running and thus having engaged in some physical activity due to disturbance, before death, although the proportion of deer moving was no different between animals shot by a single stalker by day or by night.

Interestingly, in terms of effects of shooting on other members of the populations, 78% or more of non-target animals observed were recorded as running away following a shot, with numbers similar for single stalker by day and night (80%, 78%) but higher in daytime exercises with teams of stalkers or with helicopter assistance.

11.5.2 Wounding

There has been a similar intense focus on wounding rates associated with different types of culling exercise (as above) or more generally, yet with comparatively little formal data available; those data which are available are also subject to rather different interpretation by different commentators. The majority of studies available deal with culling of red deer on the open hill and thus data may be biased to some extent, since the success rate in effectively following-up a wounded animal in woodland may be significantly lower. Based on a survey of stalkers on Exmoor and the Quantocks, Bateson and Bradshaw (1999) suggest that 3.5–4.5% of deer shot, escape wounded, although these figures are based on retrospective or current analysis of stalkers' own records.

Data available for wounding rates more generally, are based for the most part on animals wounded but later retrieved, and in effect relate to the frequency with which a culled animal has been shot more than once. Thus through detailed examination of lardered carcases, Urquhart and McKendrick (2003) assessed the distribution of entry wounds and wound tracts across the body, of 1975 discernible wound tracts in 900 animals culled between July and November 2001. They did not attempt to deduce which were or were not killing shots but did explore the frequency and incidence of secondary wound tracts – where a carcase bore evidence of multiple shot.

Interpretation of such data is however difficult. In many instances a second shot may be taken 'for assurance of death' (where in fact the first shot would indeed have proved fatal, but the stalker was simply making assurance doubly sure). Thus evidence of a second shot does *not* imply that the animal concerned was sentient, and suffering, or was not indeed mortally wounded. By converse, injured animals, if nonetheless immobilised, may be despatched by use of a knife at close quarters, rather than through use of a second shot. Thus the frequency of records of multiple shots is not necessarily a good indicator for proportions of animals hit but not killed.

Urquhart and McKendrick reported that 14.5% of carcases recorded had evidence of more than one wound tract. This was more frequent among males than in females or calves and also more frequent during the rut. Such differences might reflect that a greater proportion of males culled, and especially at this time, are taken by shooting guests who may have somewhat lesser experience than professional stalkers and also the fact that stags are more hormonally charged at this time: with higher levels of testosterone and adrenaline perhaps enabling them to withstand initial impact of the shot for apparently longer before collapsing, and thus giving the appearance of not being fatally injured.

The estimated percentage of carcasses with more than one wound tract increases slightly from 14.5% to 14.6% if figures are adjusted in recognition of the fact that some wound tracts may leave no permanent trace (and on the basis of the proportion of carcasses surveyed which showed evidence of no wound tract at all) with a 95% confidence interval of 12.7 to 16.8%, a figure in reasonable agreement with that of Bateson's earlier analysis of wound tracts in 40 carcasses culled by Forestry Commission rangers in the south of England (10.0±2.8%; Bateson, 1997).

Urquhart and McKendrick made no explicit analysis of the proportion of shots recorded within carcasses which fell within areas where they were likely to have resulted in significant damage to/destruction of vital organs and thus rapid death, reporting that they do not believe that it is possible to assess the immediacy or otherwise of death without considering much more detailed information about the nature of the wound and thus would distance themselves from any general conclusions about the implications of bullet placement alone.

During their subsequent analysis of the impacts of different approaches to culling on red deer, Cockram *et al.* (2011) also considered bullet placement and wounding rates. They note that there was a statistically significant difference between the four culling methods in the percentage of deer with an entry or exit wound in either the head, neck or lower chest/heart area ($P = 0.019$). Deer culled at night had a lower percentage (50%) compared to single-rifle stalking (89%). This difference remained in multivariate analyses ($P = 0.007$). However, if a wound in the neck is excluded and a wound in the upper chest is included, there was no statistically significant difference between the four culling methods in the percentage of deer with at least one entry or exit wound in either the head, chest (upper or lower)/heart area ($P > 0.228$) – usually regarded as killing zones.

There was also no statistically significant difference between the four culling methods in the percentage of deer with at least one entry wound not in either the head, neck or chest (univariate: $P = 0.702$, multivariate: $P = 0.780$) or in the percentages of deer that had at least one entry wound in the leg or in the abdomen, which injuries were recorded in between 7 and 19% of animals culled.

11.5.3 Time to death

Although there is as yet no objective basis for determining what is an acceptable time to death, much discussion is still concerned with this.

Rather qualitative estimates based on Bateson and Brashaw's survey of 15 stalkers on Exmoor (and a total of 443 red deer killed) suggest estimates of 4.4–7.5% of animals surviving for more than 2 min having been shot. For 54% of the wild deer in the studies of Cockram *et al.* (2011), the time between the first shot and death (apparent death) was recorded as <1 min and the median time between the first shot and death for all culling methods was <1 min. It was therefore not surprising that there were no statistically significant differences between culling methods in time from the first shot to death.

11.5.4 Effects of culling on others in the populations

As we have already noted, in terms of effects of shooting on other members of the population, Cockram *et al.*'s (2011) observations record 78% or more of non-target animals as running away following a shot, with numbers similar for single stalkers by day and night (80%, 78%) but higher in day time exercises with teams of stalkers or with helicopter assistance. But interpretation of such statistics is contentious; as a cursorial animal red deer are not ill-adapted to short periods of rapid locomotion (as long as these are not protracted) and such observations are not necessarily indicative of more than minor disturbance; there is currently no independent evidence to suggest these (non-culled) members of the population have been directly or acutely stressed.

More direct effects may be expected through disruption of social groups (loss of a lead or matriarchal female in more social species; loss of a close group member) disruption of rutting aggregations (where males are killed during the period of rut) or orphaning of dependent calves by culling of the mother. How important may be the social dependency on established lead animals among European ungulate species is largely unknown. Focus has tended to concentrate on the potential implications of disruption of the rut or orphaning of juveniles by culling of the dam during a vulnerable period of juvenile dependency (Apollonio *et al.*, 2011) This is rehearsed in considerable detail in that earlier discussion and will not therefore be expanded upon here.

11.6 Wider issues

In a review for the (then) Deer Commission for Scotland in 2008, Putman considered as potential welfare issues for upland red deer:

* the problem of wounding
* the problem of orphaned calves (both considered above)

but in addition:

* winter mortality/die-off
* need for (or alternatively problems resulting from) supplementary feeding
* need for winter shelter
* problems associated with extensive fencing, fragmenting range and disrupting traditional movement patterns (particularly when such fencing prevents animals from reaching low ground shelter over winter)
* welfare issues (and public safety issues) of deer involvement in road traffic accidents.

(The list was itself by no means considered exclusive.)

Clearly our considerations above have been restricted to the implications of shooting wild ungulates (or killing by other means) and in essence have concerned themselves with competences that may be required to avoid suffering, as a consequence, in the animal shot (wounding) or in any dependent calves (problems of orphaning).

Within a more all-embracing concept of welfare which accepts that concern for welfare management may imply responsibility not simply to minimise suffering but actively to enhance positive welfare, we may feel that there are other areas where management intervention might be appropriate, not simply restricted to shooting.

Thus there may be potential in other areas of management – in management of the resource itself – to establish practices which in themselves may promote opportunities for adaptive behaviour (e.g. diversification of natural food resources to provide adequate quality and quantity of natural forages during the year; provision of adequate shelter so that animals at least have opportunities to seek shelter in adverse weather; maintenance of movement corridors between different parts of the range, between winter and summer ranges or for dispersal of juveniles – ensuring that any new fencing proposals do not obstruct such freedom of movement, etc.).

Responsibilities here may thus be proactive. In addition, however, there is presumed some need to ensure that other management practices (not necessarily explicitly directed at wild ungulates in any direct way) do not have negative impacts, and that general management practices which may impact upon ungulate populations and their welfare (fencing, felling of forestry, enclosure of parts of existing range, etc.) are subject to some form of explicit risk assessment and proper consideration whether alternative options are available which will reduce or avoid negative impacts on welfare.

Thus we may legitimately ask: to what extent may society expect managers to assume responsibilities for safeguarding or actively promoting welfare in these areas also, and what competences might we reasonably expect of managers to assess wider welfare status of wild ungulates and to develop appropriate management solutions to perceived problems?

11.7 The ethical dimension of animal welfare

Assessment of the actual welfare status of an individual animal (or group of animals) offers simply that: an assessment of welfare status. This welfare status is, *per se*, neither morally to be judged good or bad; it simply reports a factual condition. Whether or not that status constitutes a welfare problem for the animal, or welfare *issue* as perceived by a human observer, is inevitably more of a value judgement or ethical decision. Subsequent subjective analysis may thus decide that that determined status does constitute a welfare issue (i.e. a welfare status which is perceived as a moral problem by an outside observer) which needs to be addressed (the animal or animals are in a poor welfare state, or at least a state which might be improved upon); in which case, we must explore what are the (welfare) advantages or disadvantages of different alternative possible interventions.

A different line of argument notes that any changes to existing management practice towards some unrelated objective (whether or not related to addressing any welfare concerns), or indeed the adoption of some new management objective (and associated measures required to deliver that new objective) may in itself, as a side-effect, have a potential impact on animal welfare. In such a case we may wish to assess the relative welfare implications of a proposed change in management or of the alternative management measures available to deliver the new objective.

In such situations it is clear that animal welfare issues cannot simply be addressed by means of objective (scientific) biological measures of an animal's welfare status under certain circumstances. In practice, interpretation of welfare status and its translation into the active management of perceived welfare issues are both strongly influenced by context and, especially, by cultural and societal values. In assessing whether or not a given welfare status is morally acceptable, animal welfare scientists must be aware that even scientifically based, operational definitions of animal welfare will necessarily be influenced strongly by a given society's moral understanding (Meijboom and Ohl, 2012; Ohl and van der Staay, 2012).

Formal examination of the various factors which may influence decisions in any given situation can be facilitated by use of formalised and structured approaches (see, for example, Beekman *et al.*, 2006; Mepham *et al.*, 2006; RDA, 2010) that may help to make explicit, structure and analyse moral issues in policy. Ideally such assessments should be public, transparent and based on the most recent scientific knowledge as well as broadly shared public moral views.

But what then *are* our responsibilities for the safeguarding or actively enhancing the welfare of wildlife individuals or populations? There has been much debate in the more philosophical literature about human responsibilities to animals and about the moral value of animal life. A relevant part of this literature recognises animals as having moral status: that is, to be an entity (a being) towards which we can have moral duties (Warren, 1997). However, there is a diversity of arguments that underlie the recognition of this moral standing of animals. Appleby and Sandøe (2002) and Swart and Keulartz (2011), amongst others, offer comprehensive reviews.

In the past, perhaps, many philosophers, following a Kantian approach, presumed that any responsibility to consider the welfare of animals derived only indirectly as a consequence of their relationship with humans; thus that our concern for their welfare is an extension of our 'humanity' and moral responsibility. A significant number of ethicists, however, now concede that animals have some moral value that is independent of their use by humans and based rather on some sense of an inherent value of animals (Taylor, 1986) or by virtue of their being able to feel (e.g. Rollin, 2011).

This broad acknowledgement of animals as having moral status, however, appears not to result in one broadly shared view on how we should treat them. As an example, current legislative provisions for the management of animal welfare are in general largely context-dependent, such that there may be a clear legal distinction between responsibilities defined towards farm animals, laboratory animals, companion animals, closely managed wildlife and truly wild animals experiencing little management input (for overview see, for example, Vapnek and Chapman, 2010).

Partly as a reflection of that legal distinction, some authors (as, for example, Swart, 2005, and Swart and Keulartz, 2011) suggest a similar distinction in relation to context (kept and non-kept) in terms of our responsibilities for animals and the way those responsibilities might be discharged. If we wish to try to a claim for a wider duty of care, we may argue that at least within a European environment, all animals (whatever their status as kept or non-kept) are to some degree influenced by human activity – whether to a greater or lesser extent, and whether deliberate or incidental. Closely managed animals in whatever context (farm animals, laboratory animals, companion animals) have their whole environment controlled by human agency; but even for free-ranging or apparently wild animals, their habitat, movement patterns are affected by human land use and land management, and many populations are directly controlled by culling.

Thus even in terms of those philosophical traditions which base responsibilities for animals on their relationship with humans, some duty of care or responsibility might be expected, at least to *consider* the welfare of all animals, including free-ranging wildlife, as well as the impacts on that welfare of human management decisions (Ohl and Putman 2013d).

11.8 Responsibility for action?

What *does* appear to differ with context is the perceived degree of obligation (or requirement) to take action in different situations. Swart and Keulartz (2011) link this responsibility to context in presuming that responsibility to address some compromise of welfare status is higher in animals more closely managed by humans (or free-ranging animals more heavily impacted upon by human activities), because we are more closely responsible for providing all resources for more closely managed (farm or companion) animals, which thus have less freedom and have fewer opportunities to respond 'naturally'.

We ourselves would argue (Ohl and Putman 2013a, d) that this negates the initial construct of an equal moral duty to all animals, which is universal and independent of the context of the human – animal relationship. We would however suggest that, even within that subset of cases where we may feel it appropriate to intervene,[6] the obligation to take action may be constrained by the actual practicalities of intervention and the availability of (realistic) mitigation measures to effect a change in welfare status.

Thus, and from a purely practical management point of view rather than from a philosophical one, if there are no practical mitigation options possible, then almost by definition, this must lessen the obligation to take action. However simple this last constraint may appear, it opens an additional debate about the actual relation between moral and practical responsibility, which translates, for example, into a debate about

---

[6]That is, as above, in those situations where the animal has insufficient opportunities to adapt by appropriate changes of its own behaviour.

the understanding of avoidable versus unavoidable animal suffering and further, whether our duty to consider animal welfare should be restricted to the prohibition of 'avoidable' suffering or should include the enhancement of positive welfare status.

As a principal framework, we would suggest that philosophical considerations explored so far would suggest that any suffering of animals that is technically avoidable should be considered (morally) unacceptable. However, there may be constraints on intervention that are posed purely by practicalities of intervention or mitigation. Thus effective intervention in the lives of free-ranging or wild animals may simply not be feasible; in a farm context, if livestock animals are kept outdoors, which may itself be warranted in terms of promoting positive welfare in other respects, they may be exposed to extreme weather that may cause transient suffering. Sudden changes of weather are neither predictable nor controllable and occasional suffering is *unavoidable* when animals are kept outdoors.

Of course, such consideration holds the danger of being misused, in that one could argue that efficiency and economics might always justify a limitation of practical responsibilities towards animals and, thus, in itself raises the question of what we may more explicitly consider necessary or unnecessary suffering in relation to human-centred objectives. We suggest that *necessary* suffering in this context should be understood as that required to achieve some anthropocentric objective (e.g. animal experimentation, food production or pest control) or animal-centred goal (for example in veterinary practice). In such cases, there may be mitigation options that are not implemented for reasons of efficiency or economics. Here, any suffering resulting from non-intervention may be considered avoidable (in theory) but necessary (because of distinct subjective/individual human interests). (For further exploration of this axis, see Ohl and Putman, 2013a, d.)

Cutting across these dimensions of necessary or unavoidable suffering, we must recognise that even science-based, operational definitions of animal welfare and suffering will necessarily be influenced by societal mores. Thus, in addition to determining whether considering an animal's suffering may be avoidable versus unavoidable, or as necessary versus unnecessary, it is appropriate also to evaluate along a third axis, based upon what is considered by a wider society to be (morally) *acceptable* or *unacceptable* (RDA, 2012).

Of course, moral frames differ markedly between cultures, thus differ between societies as well as showing a range of 'attitudes' even within one given society. Our assessment of what is acceptable must thus be framed within a proper understanding of what the majority of members of our own contemporary society believe and accept, although this implies an additional responsibility to inform and educate that wider society so that reactions are not simply based on untutored and unreflected intuitions (see, for example, Ohl and van der Staay, 2012; Meijboom and Ohl, 2012).

11.8.1 To Intervene or not to intervene?

Following this line of reasoning, the obligation to take action *in response to a perception of suffering,* or in an attempt in general to promote more positive welfare of an individual or population, is constrained by what may be considered necessary or acceptable suffering, or by the actual practicalities of intervention and the availability of (realistic) mitigation measures to effect a change in welfare status.

If we accept the more dynamic concept of what constitutes welfare presented above, such that a welfare issue arises only when an animal or a group of animals have insufficient opportunity (freedom) to respond appropriately to a potential welfare 'challenge' through adaptation by changes in its own behaviour, it implies that biologically, we have an obligation to take action to address potential welfare problems only in those situations where the animal cannot adapt appropriately by changes of its own behaviour (sufficient to bring about appropriate environmental adjustments required to restore its positive welfare status).

Yet, from our deliberations here, we must recognise that even within such a 'subset' of cases, if there are no practical mitigation options possible, then almost by definition, this must lessen the obligation to take action. Thus the obligation to intervene (in any context, wild or closely managed) depends primarily on whether or not there are practical (and economically feasible) options available for intervention (whether or not such intervention is in the interests of avoidance of suffering or positive, proactive enhancement of general welfare status).

One final consideration here however might be an assessment of the unintended consequences of any proposed intervention. Thus as a refinement of the responsibility for intervention we might add recognition that we should intervene only in those cases where we can be certain that our intervention will not result in greater harm (to the individual or individuals it is sought to assist and/or to other animals in the wider environment – see, for example, Donaldson and Kymlicka, 2011; Sozmen, 2012).

11.9 Conclusions

In conclusion, we would suggest that for considerations on the management of animal welfare:

- the moral duty of care for animals is absolute and independent of context
- the requirement to intervene should be based on biological assessment of whether or not sufficient opportunities exist for the animal or animals to respond appropriately to a potential compromise of welfare status through appropriate and adaptive changes in its own behaviour (sufficient to bring about appropriate environmental adjustments required not only to avoid suffering, but to restore its positive welfare status); the requirement for intervention in such cases is further constrained by the physical possibility/impossibility of any effective mitigation (avoidable versus unavoidable suffering)
- intervention is further constrained by considerations of human interest in the animals concerned (thus an evaluation of what may constitute necessary versus unnecessary suffering)
- there is in addition a moral dimension in co-determining the scale of required intervention in relation to societal norms of what may be acceptable versus unacceptable suffering.

We believe that the constructs presented here cut across the rather artificial distinctions of current legal/practical provisions between kept and non-kept animals (it obviates the need for such artificial and arbitrary distinctions) and replaces these with a more robust and functional construct applicable in more general terms, allowing proper extension to welfare considerations in relation to free-ranging wildlife species. In effect, the whole neatly defines a recognition of welfare issues (in all contexts) as lack of opportunity to respond appropriately to environmental challenges, and offers a simple concept of appropriate solutions in terms of providing the opportunities to enable the animal(s) to react appropriately to such challenge. It provides a recognition that we (humans) have an obligation to take action only if an animal is (for whatever reason) unable to respond appropriately and effectively through its own adaptive behaviour; a recognition that obligations are further constrained by actual practicalities of what is possible in terms of mitigation. It opens up, however, the separate (ethical) debate of whether or not human interest may transcend considerations of welfare and thus whether or not there may be situations in which we accept a measure of necessary suffering.

A more inclusive definition of welfare also implies that welfare assessments should not be focussed or be restricted to the avoidance of negative welfare states in general (since these may be of adaptive value in triggering appropriate adaptive responses) or 'suffering' in particular. Considerations should further extend beyond the single context of the avoidance of suffering during cull operations, which, in fact, has little or nothing to do with wider considerations of safeguarding welfare.

The assessment of acute 'negative' states in animals is insufficient when aiming at safeguarding 'positive' welfare. Short-termed negative states can be of adaptive value for the animal, while persistent negative states are likely to exceed the animals' adaptive capacities. Any welfare assessments, therefore, need to be based on repeated observations over a some prolonged time period and need to take into account the functioning of the group as a whole. While indeed it is a moral obligation not to cause unnecessary suffering in the course of any culling procedure, safeguarding welfare more generally requires the assessment of the consequences of other management measures which may impinge on the welfare status of individual ungulates or ungulate populations.

Measures which might be advocated for assessment of the welfare status of free-ranging ungulates are presented in Table 11.1. It should be stressed however that the measures proposed can only be seen as indicators and not formal objective measurements. If, as seems apparent, there exists significant variability in how different individual animals may assess their own welfare status and also in coping strategy or response to a given challenge, then we, as external observers, may expect to observe within any group of animals a series of individuals with different 'absolute' welfare status (assessed by an external observer against some fixed set of criteria) yet which perceive the own welfare state as being optimal (or at least sufficient not to require action to alter that status).

Such conclusion makes clear that purely objective functional scales for measuring the welfare status of individual animals, can have little validity in that, even under identical conditions, the actual welfare status of different individuals may vary widely.

We emphasise throughout that decisions about welfare cannot be informed simply by biological assessment of welfare status and condition, whether at the level of individual animals or the population, but require a careful integration of both biological and ethical assessments.

References

Anonymous (2009) *Farm Animal Welfare in Great Britain: Past, Present and Future.* London: Farm Animal Welfare Council.

Apollonio, M., Putman, R.J, Grignolio, S. and Bartos, L. (2011) Hunting seasons for ungulates in relation to biological breeding seasons and the implications for control of population size. In M. Apollonio, R. Andersen and R.J. Putman (eds), *European Ungulates and their Management in the 21st Century.* Cambridge, UK: Cambridge University Press, pp. 80–105.

Appleby, M.C. and Sandøe, P. (2002) Philosophical debate on the nature of well-being: implications for animal welfare. *Animal Welfare* **11**, 283–294.

Barnett, J.L. and Hemsworth, P.H. (1990) The validity of physiological and behavioural measures of animal welfare. *Applied Animal Behaviour Science* **25**, 177–187.

Bateson P. (1997) *The Behavioural and Physiological Effects of Culling Red Deer.* Report to the National Trust, London.

Bateson, P. and Bradshaw, E.L. (1997) Physiological effects of hunting red deer (*Cervus elaphus*). *Proceedings of the Royal Society of London Series B* **264**, 1707–1714.

Beekman, V., Kaiser, M., Sandøe, P., Brom, F., Millar, K. and Skorupinski, B. (2006) *The development of ethical bio-technology assessment tools for agriculture and food production.* Final Report. Available at: http://www.ethicaltools.info/.

Bradshaw, E.L. and Bateson, P. (2000) Welfare implications of culling red deer *(Cervus elaphus) Animal Welfare* **9**, 3–24.

Brambell Committee (1965) *Report of the Technical Committee to enquire into the welfare of animals kept under intensive livestock husbandry systems.* Cmnd. 2836, 3 December 1965. London: Her Majesty's Stationery Office.

Broom, D.M. (1988) The scientific assessment of animal welfare. *Applied Animal Behaviour Science* **20**, 5–19

Broom, D.M. (1991) Animal welfare: concepts and measurement. *Journal of Animal Science* **69**(10), 4167–4175

Broom, D.M. (2006) Adaptation. *Tierärztliche Wochenschrift* **119**, 1–6.

Broom, D.M. (2010) Cognitive ability and awareness in domestic animals and decisions about obligations to animals. *Applied Animal Behaviour Science* **126**, 1–11.

Broring, N., Wilton, J.W. and Colucci, P.E. (2003) Body Condition score and its relationship to ultrasound backfat measurements in beef cows. *Canadian Journal of Animal Science,* **83**, 593–596.

Burton, D. (1992) *The Effects of Parasitic Nematode Infection on Body Condition of New Forest Ponies.* PhD thesis, University of Southampton, UK.

Cabanac, M. (1971) Physiological role of pleasure. *Science* **173**, 1103–1107.

Cabanac, M. (1979) Sensory pleasure. *Quarterly Reviews in Biology* **54**(1), 1–29.

Cockram, M.S., Shaw, D.J., Milne, E., Bryce, R., McClean, C. and Daniels M.J. (2011) Comparison of effects of different methods of culling red deer (*Cervus elaphus*) by shooting on behaviour and post mortem measurements of blood chemistry, muscle glycogen and carcase characteristics *Animal Welfare* **20**, 211–224

Crofoot, M.C., Rubenstein, D.I., Maiya, A.S. and Berger-Wolf, T.Y. (2011) Aggression, grooming and group-level cooperation in white-faced capuchins (*Cebus capucinus*): insights from social networks. *American Journal of Primatolology* **73**, 821–833.

Dantzer, R. and Mormede, P. (1983) Stress in farm animals: a need for re-evaluation. *Journal of Animal Science* **57**, 6–18.

Dantzer, R., Mormede, P. and Henry, J.P. (1983) Significance of physiological criteria in assessing animal welfare. In D. Smidt (ed.) *Indicators Relevant to Farm Animal Welfare.* Leiden, the Netherlands: Martinus Nijhoff, pp. 29–37.

Dawkins, M.S. (1980) *Animal Suffering: The Science of Animal Welfare.* London: Chapman and Hall.

Dawkins, M.S. (1990) From an animal's point of view: motivation, fitness and animal welfare. *Behavioural and Brain Sciences* **13**, 1–31.

Dawkins, M.S. (1998) Evolution and animal welfare. *Quarterly Review of Biology* **73**, 305–328.

Dawkins, M.S. (2008) The science of animal suffering. *Ethology* **114**, 937–945.

deKloet, E.R., Joels, M. and Holsboer, F. (2008a) Stress and the brain: from adaptation to disease. *Nature Reviews Neuroscience* **6**, 463–475.

deKloet, E.R., Karst, H. and Joels, M. (2008b) Corticosteroid hormones in the central stress response: quick and slow. *Frontiers in Neuroendocrinology* **29**, 268–272.

Donaldson, S. and Kymlicka, W. (2011) *Zoopolis: A Political Theory of Animal Rights.* Oxford: Oxford University Press.

Duncan, I.J.H. (1993) Welfare is to do with what animals feel. *Journal of Agricultural and Environmental Ethics* **6** (suppl 2), 8–14.

Duncan, I.J.H. (1996) Animal welfare defined in terms of feelings. *Acta Agriculturae Scandinavica (Section A: Animal Science)* **27**(Suppl.), 29–35.

Duncan, I.J.H. (2005) Science-based assessment of animal welfare: farm animals. *Revue Scientifique et Technique-Office International des epizooties* **24**(2), 483–492.

Findlay, J. (2007) *A report on the current perception, legal status and expectations with regard to wild deer welfare in Scotland.* Deer Commission for Scotland. Available at: http://www.snh.gov.uk/land-and-sea/managing-wildlife/managing-deer/welfare/.

Fraser, D. (1993) Assessing animal well-being: common sense, uncommon science. In *Food Animal Well-being.* Conference Proceedings and Deliberations, West Lafayette, IN: Purdue University Office of Agricultural Research Programs, pp. 37–54.

Fraser, D. and Duncan, I.J.H. (1998) Pleasures, pains and animal welfare: toward a natural history of affect. *Animal Welfare* **7**, 383–396.

Fraser, D., Weary, D.M., Pajor, E.A. and Milligan, B.N. (1997) A scientific conception of animal welfare that reflects public values. *Animal Welfare* **6**, 187–205.

Gill, E.L. (1988) *Factors Affecting Body Condition of New Forest Ponies.* PhD thesis, University of Southampton, UK.

Gill, E.L. (1991) *Factors Affecting Body Condition in Free-ranging Ponies.* Technical Report, Royal Society for the Prevention of Cruelty to Animals, Horsham, UK.

Harris, R.C., Helliwell, T.R., Singleton, W., Stickland, N. and Naylor, J.R.J. (1999) *The physiological response of red deer (Cervus elaphus) to prolonged exercise during hunting.* Newmarket, UK: R&W Publications.

Haynes, R.P. (2011) Competing conceptions of animal welfare and their ethical implications for the treatment of non-human animals. *Acta Biotheoretica* **59**, 105–120.

Held, S.D.E. and Spinka, M. (2011) Animal play and animal welfare. *Animal Behaviour* **81**, 891–899.

Jones, A.R. and Price, S. (1990) Can stress in deer be measured? *Deer* **8**(1), 25–27.

Jones, A.R. and Price, S. (1992) measuring the responses of fallow deer to disturbance. In R. Brown (ed.), *The Biology of Deer.* Berlin: Springer Verlag, pp. 211–216.

Kikusi, T., Winslow, J.T. and Mori, Y. (2006) Social buffering: relief from stress and anxiety. *Philosophical Transactions of the Royal Society B* **361**, 2215–2228.

Kirkwood, J.K., Sainsbury, A.W. and Bennett, P.M. (1994) The welfare of free-living wild animals: methods of assessment. *Animal Welfare* **3**, 257–273.

Knierim, U. and Winckler, C. (2009) On-farm welfare assessment in cattle–validity, reliability and feasibility issues and future perspectives with special regard to the Welfare Quality® approach. *Animal Welfare* **18**, 451–458.

Knierim, U., Carter, C.S., Fraser, D., Gärtner, K., Lutgendorf, S.K., Mineka, S., Panksepp, J. and Sachser, N. (2001) Good welfare: improving quality of life. In D.M. Broom (ed.), *Coping with challenge: welfare in animals including humans.* Dahlem Workshop Report 87. Berlin: Dahlem University Press, 79–100.

Koolhaas, J.M. (2008) Coping style and immunity in animals: making sense of individual variation. *Brain Behaviour and Immunology* **22**, 662–667.

Koolhaas, J.M., de Boer, S.F., Coppens, C.M. and Buwalda, B. (2010) Neuroendocrinology of coping styles: towards understanding the biology of individual variation. *Frontiers in Neuroendocrinology* **31**, 307–321.

Korte, S.M., Olivier, B. and Koolhaas, J.M. (2007) A new animal welfare concept based on allostasis. *Physiology and Behaviour* **92**, 422–428.

Krippendorff, K. (1986) *A Dictionary of Cybernetics.* Norfolk, VA: The American Society for Cybernetics.

Lahti, D.C. (2003) Parting with illusions in evolutionary ethics. *Biology and Philosophy* **18**, 639–651.

Mason, G. (1998) Animal welfare: the physiology of the hunted deer. *Nature* **391**, 22.

Mathews, F. (2010) Wild animal conservation and welfare in agricultural systems. *Animal Welfare* **19**, 159–170.

McEwen, B.S., Angulo, J., Cameron, H., *et al.* (1992) Paradoxical effects of adrenal-steroids on the brain-protection versus degeneration. *Biological Psychiatry* **31**, 177–199.

Meijboom, F.L.B. and Ohl, F. (2012) Managing nature parks as an ethical challenge: a proposal for a practical protocol to identify fundamental questions. In T. Potthast and S. Meisch (eds), *Ethics of Non-agricultural Land-management.* Wageningen, The Netherlands: Academic Publishers.

Mellor, D.J. (2012) Animal emotions, behaviour and the promotion of positive welfare states. *New Zealand Veterinary Journal* **60**(1), 1–8.

Mellor, D. J. and Reid, C.S.W. (1994) Concepts of animal well-being and predicting the impact of procedures on experimental animals. In R.M. Baker, G. Jenkin and D.J. Mellor (eds), *Improving the Well-being of Animals in the Research Environment.* Adelaide, Australia: Australian and New Zealand Council for the Care of Animals in Research and Teaching, pp. 3–18.

Mellor, D.J. and Stafford, K.J. (2001) Integrating practical, regulatory and ethical strategies for enhancing farm animal welfare. *Australian Veterinary Journal* **79**, 762–768.

Mellor, D.J., Patterson-Kane, E., and Stafford, K.J. (2009) *The Science of Animal Welfare.* Oxford: Wiley-Blackwell, Chapters 1–5.

Mendl, M. and Deag, J.M. (1995) How useful are the concepts of alternative strategy and coping strategy in applied studies of social behaviour? *Applied Animal Behaviour Science* **44**, 119–137.

Mendl, M., Burman, O.H. and Paul, E.S. (2010) An integrative and functional framework for the study of animal emotion and mood. *Biological Science* **277**, 2895–2904.

Mepham, B., Kaiser, M., Thorstensen, E., Tomkins, S. and Millar, K. (2006) *Ethical Matrix Manual*. The Hague, The Netherlands: Landbouw Economisch Instituut (LEI). Available at: http://www.ethicaltools.info/content/ ET2 Manual EM (Binnenwerk 45p).pdf.

Moberg, G.P. (1985) Biological response to stress. In G.P. Moberg (ed.), *Animal Stress*. Bethesda, MD: American Physiological Society, pp. 27–49.

Nesse, M.R. and Ellsworth, P.C. (2009) Evolution, emotions, and emotional disorders. *American Psychology* **64**, 129–139.

Nordenfelt, L. (2011) Health and welfare in animals and humans. *Acta Biotheoretica* **59**, 139–152.

Ohl, F. and Putman, R.J. (2013a) *Applying wildlife welfare principles to individual animals*. Report for Scottish Natural Heritage, Inverness. Available at: http://www.snh.gov.uk/publications-data-and-research/publications/search-the-catalogue/publication-detail/?id=2072.

Ohl, F. and Putman, R.J. (2013b) Animal welfare at the group level: more than the sum of individual welfare? *Acta Biotheoretica* **62**, 35–45.

Ohl, F. and Putman, R.J. (2013c) *Applying wildlife welfare principles at the population level*. Report for Scottish Natural Heritage. Available at: http://www.snh.gov.uk/publications-data-and-research/publications/search-the-catalogue/publication-detail/?id=2071.

Ohl, F. and Putman, R.J. (2013d) Animal welfare considerations: should context matter? *Animal* manuscript submitted.

Ohl, F. and van der Staay, F.J. (2012) Animal welfare: at the interface between science and society. *Veterinary Journal* **192**, 13–19.

Panksepp, J. (1998) The quest for long-term health and happiness: to play or not to play, that is the question. *Psychological Inquiry* **9**, 56–66.

Panksepp, J. (2011) The basic emotional circuits of mammalian brains: do animals have affective lives? *Neuroscience & Biobehavioural Reviews* **35**(9), 1791–1804.

Price, S. and Jones, A.R. (1992) Responses of farmed red deer to being handled. In R. Brown (ed.), *The Biology of Deer*. Berlin: Springer Verlag, pp. 220.

RDA (Raad voor Dierenaangelegenheden) (2010) Moral issues and public policy on animals. Report 2010/02. Gravenhage, The Netherlands. Available at: http://www.rda.nl/pages/adviezen.aspx.

RDA (Raad voor Dierenaangelegenheden) (2012) *Duty of care: naturally considered*. Report 2012/02, Gravenhage, The Netherlands. Available at: http://www.rda.nl/pages/adviezen.aspx.

Riney, T. (1955) Evaluating the condition of free-ranging red deer (*Cervus elaphus*) with special reference to New Zealand. *New Zealand Journal of Science and Technology* **36B**, 429–463.

Riney, T. (1960) A field technique for assessing the physical condition of some ungulates. *Journal of Wildlife Management* **24**, 92–94.

Rollin, B.E. (2011) Animal Pain: What it is and why it matters, *Journal of Ethics* **15**, 425–437.

Rutherford, K.M.D. (2002) Assessing pain in animals. *Animal Welfare* **11**, 31–53.

Selye, H. (1950) Stress and the general adaptive syndrome. *British Medical Journal* **1**, 1383–1392.

Sheriff, M.J., Dantzer, B., Delehanty, B., Palme, R. and Boonstra, R. (2011) Measuring stress in wildlife: techniques for quantifying glucocorticoids. *Oecologia* **166**, 869–887.

Sozmen, B. (2012) Dismissing death: an evaluation of suffering in the wild. Presentation to conference: Animal Welfare: Building Bridges between Science, the Humanities and Ethics. Abstract available at: www.uu.nl/hum/mindinganimals. [Click through via programme and then abstracts of parallel sessions.]

Swart, J.A.A. (2005) Care for the wild. Dealing with a pluralistic practice. *Environmental Values* **14**(2), 251–263.

Swart, J.A.A and Keularz, J. (2011) Wild animals in our backyard. A contextual approach to the intrinsic value of animals. *Acta Biotheoretica* **59**, 185–200.

Taylor, K.D. and Mills, D.S. (2007) Quality of life in companion animals. *Animal Welfare* **16**, 55–65.

Taylor, P. (1986) *Respect for Nature.* Princeton, NJ: Princeton University Press.

Urquhart, K.A. and McKendrick, I.J. (2003) Survey of permanent wound tracts in the carcases of culled wild red deer in Scotland. *Veterinary Record* **152**, 497–501.

Vapnek, J. and Chapman, M. (2010) *Legislative and Regulatory Options for ANIMAL welfare.* Rome: Food and Agriculture Organization of the United Nations Legal Office.

Veissier, I. and Boissy, A. (2007) Stress and welfare: Two complementary concepts that are intrinsically related to the animals' point of view. *Physiology and Behaviour* **92**, 429–433.

Warren, M.A. (1997) *Moral Status: Obligations to Persons and Other Living Things.* Oxford: Clarendon Press.

Webster, A.J.F. (1994) *A Cool Eye towards Eden.* London: Blackwell Science Ltd.

Webster, A.J.F. (2011) Zoomorphism and anthropomorphism: fruitful fallacies? *Animal Welfare* **20**, 29–36.

Webster, A.J.F. (2012) Critical control points in the delivery of improved animal welfare. *Animal Welfare* **21**(S1), 117–123.

Zulu, V.C., Nakao, T., Moriyoshi, M., Nakada, K., Sawamukai, Y., Tanaka,Y. and Zhang, W.C. (2001) Relationship between body condition scores and ultrasonographic measurement of subcutaneous fat in dairy cows. *Asian–Australasian Journal of Animal Sciences* **14**, 816–820.

Chapter 12

Management of Ungulates in the 21st Century: How Far Have We Come?

Naomi Sykes and Rory Putman

A recurring theme in this and previous books (e.g. Apollonio *et al.*, 2010; Putman *et al.*, 2011) is that approaches to management of ungulates in different countries within Europe show an immense diversity and that there is no unique, or ideal, approach. Not only are there differences in the actual mixture of species present and their relative abundance, and the problems experienced (in relation to damaging impacts or problems experienced in delivering effective management); there are also significant differences in the legal and administrative frameworks within which management is carried out (reviewed by Putman, 2011). These differences in legislative and administrative structures indubitably reflect differences in political 'style' and recent political and socioeconomic history, but have also been significantly affected by cultural traditions within a particular country and changes in attitudes to wildlife and wildlife management during the course of that history.

Thus it is clear that attitudes to wildlife in general and to wild ungulates in particular, as well as attitudes to their exploitation and management, have changed significantly over time–right up to the present time (Cartmill, 1993). Today there still remain such differences between the cultural traditions of different countries (reviewed by Putman *et al.*, 2011). We end this book with a review of how these cultural mores have affected the extent to which, and how, humans have exploited ungulate populations, to consider how far have we come in 21st century from prehistory. Have we escaped the legacy of our ancestors? Is our management approach still shaped by the same management requirements and the same issues as influenced management decisions in the past? And to what extent is effective management practice still restricted or constrained by cultural traditions and attitudes?

This chapter sets out to highlight that ungulate management strategies are as much a reflection of cultural attitudes to the natural world, and perceptions about the place of humans within it, as they are about other issues. To make the case, we look back over the last 10,000 years to examine how the archaeological record, in particular the evidence from ancient ungulate remains, can provide information about how human attitudes to wild ungulates have changed through time and with what consequences for all concerned.

12.1 How did we get here? Highlights from the last 10,000 years

Any attempt to consider how the relationship between humans and wild ungulates has changed over such a long period must first outline the available data: in this case they are highly uneven due to temporal and geographic variations in landscape, environment, fauna and human societies. Perhaps unsurprisingly, evidence for the remotest periods in time, such as the Mesolithic (20,000 to 4000 BC with regional variations) when people lived as mobile hunter-gatherers, is the scarcest. Matters are little more improved for the early farming communities of the Neolithic (7,000 to 1,700 BC with regional variations); although archaeological data are more forthcoming, in many regions material is derived primarily from ceremonial contexts – enclosures and funerary structures – rather than domestic settlements.

As we advance through time, problems of preservation and context tend to reduce but other issues remain. For instance, zooarchaeology is not undertaken in all European countries and therefore comparable datasets are patchy. For this reason, the discussion below excludes much of Eastern Europe and Scandinavia where large trans-period datasets are unavailable. More attention is given to England and Figures 12.1 and 12.2 summarise the diachronic variation in the frequency of wild ungulates for zooarchaeological assemblages dating from the Neolithic to Post-Medieval period. These data have been synthesised from the work of Sykes (2007a), Hambleton (2008), Allen (2010), Poole (2010) and Serjeantson (2011), all of which used broadly comparable methods.

From the outset, it is important to remember that the zooarchaeological record is the result of human activity and cultural choices; it is not a direct reflection of local fauna. Where wild ungulates are absent from zooarchaeological assemblages, or present only in low frequencies, it does not necessarily mean that they were absent in the landscape, as often suggested (e.g. Jennbert, 2011). Certainly, poor representation *can* reflect low populations but it can equally reflect lack of exploitation by people. On the other hand, some animals may be represented archaeologically in areas where they were not present in life, their remains having been curated and transported to regions beyond their natural distribution: for instance aurochs horns are known to have been widely transported as high-status drinking vessels, being taken to regions where the species was extirpated (van Vuure, 2005). The presence or absence of wild ungulates can also be obscured by issues of identification, as closely related species such as wild boar (*Sus scrofa*) and domestic pigs (*Sus domesticus*), domestic and wild cattle (*Bos primagenius*) and various deer species are difficult to separate, particularly when their size ranges overlap (Lister, 1996; Albarella *et al.*, 2009; Sykes *et al.*, 2013). Despite all of these caveats, the

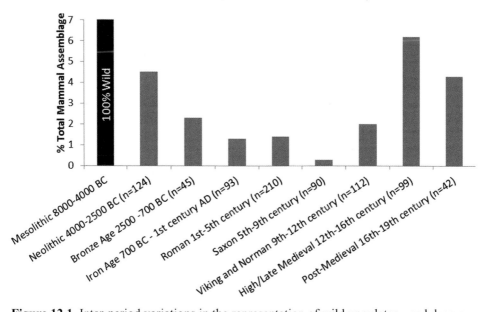

Figure 12.1 Inter-period variations in the representation of wild ungulates – red deer, roe deer, fallow deer, wild cattle and wild boar – on English archaeological sites. Bars show the average representation, calculated from all samples of each period and expressed as a percentage of the total mammal assemblage.
Sources: Sykes (2007b), Hambleton (2008), Allen (2010), Poole (2010) and Serjeantson (2011).

zooarchaeological record is an important source of information, charting the changing ways that people interacted with, and thought about, the world around them.

12.1.1 The Mesolithic period

In Mesolithic Europe, the environment and faunal spectrum were very different to those of today: woodland was far more extensive and ungulate species that are now completely extinct (e.g. aurochsen and European ass, *Equus hydruntinus*) or are locally extirpated (e.g. European elk *Alces alces*, reindeer and wild boar), would have been familiar encounters for the mobile human populations who relied on these animals for food and raw materials. As a result, the experience of Mesolithic hunter-gatherers is one entirely alien to our own, made stranger still by the complete absence of domestic animals, other than the dog. In the absence of 'the domestic' (in terms of both livestock and permanent settlements), the concept of 'wild' would not have existed either. Instead the world inhabited by the Mesolithic hunter-gatherers would have been a single integrated realm lacking the dichotomies of wild and domestic, nature and culture that are perceived by modern Western societies.

Given that the worldview of modern Europeans is so divorced from that of the Mesolithic population, it is difficult to comprehend the ideology of our ancestors. Anthropological studies of modern hunter-gatherers offer an, arguably, more relevant perspective to help understand the beliefs and practices of Mesolithic people. Whilst there is no universal 'hunter-gatherer cosmology', there are a number of traits that have been identified consistently by anthropological studies of hunter-gatherers worldwide (e.g. Ingold, 2000; Politis and Saunders, 2002; Nadadsy, 2007; Zvelebil, 2008; Conneller, 2011). The first is that modern hunter-gatherers perceive no absolute divide between humans and animals: although both are recognised as separate categories of being, it is widely accepted that animals are essentially the same as people, possessing both souls and rational thought. The intimate and indivisible connection of hunter-gatherer communities to their environment has been summarised by Ingold (2000) and Nadasdy (2007) as a kind of familial relationship, whereby hunter-gatherers perceive their environment as a nurturing parent and trust that it will share its resources with them, so long as they behave with respect: by killing no more than is necessary, by treating the animals' bodies with care, by sharing the meat amongst the community, and by disposing of the remains appropriately. Within this worldview hunting is seen as a form of inter-personal dialogue, the hunters practicing mimesis to better communicate with the animals and persuade them to give themselves up to the hunter. The death of the quarry is seen as a vital act of regeneration and, in much the same way as the phoenix rises from the ashes, the animal must die in order for its soul to be released and re-clothed with flesh so that it might return to the hunter on another occasion (Ingold, 2000; Willerslev, 2007).

Drawing upon this anthropological evidence, Conneller (2004) has proposed an interesting re-interpretation of the modified antler frontlets recovered from several Mesolithic sites in Europe, the majority coming from Star Carr in Yorkshire. Rather than invoking the usual argument that the frontlets were disguises to increase hunting success, or shamanic costumes, Conneller suggested that they may have been worn to assist the hunters' mimesis of and communication with the red deer that they hunted in such large numbers. To some extent, this ritualised behaviour can be seen as an attempt to 'manage' ungulate populations on a cosmological level. However, there is also good evidence that Mesolithic people had more direct techniques of management: selective culling and supplementary feeding, either through the provision of 'leafy hay' or through modification of the local environments to create animal-attractive browse (Tolan-Smith, 2008).

The possibility these supplementary feeding practices may have altered migration patterns is, potentially, indicated by changes in the presence of healed hunting injuries on the remains of ungulates, particularly red deer, wild boar and aurochs, from Denmark (e.g. Noe-Nygaard, 1974; Leduc, 2012). Studies have shown that, in early Mesolithic assemblages, most of the animals that displayed hunting wounds were killed outright by their injuries (70% of the injuries were fatal). By the later phases of the Mesolithic, however, a much higher proportion of remains displayed not only fatal wounds but also healed injuries: 87% of those specimens with hunting wounds had been shot (but survived) on previous occasions. It seems highly unlikely that the later Mesolithic population were just bad shots and Noe-Nygaard (1974) proposed that the rise in healed injuries reflects increasing animal sedentism that would have increased the likelihood of people and animals meeting on multiple occasions.

Certainly reduced migration has been observed in relation to modern supplementary feeding, with Mysterud (2010) and Schmidt (*in press*) arguing that, today, the impact of feeding could, in some cases, be seen as 'wildlife loss through domestication' (see also Chapter 2). The significance of this should not be underestimated because, as we see from the zooarchaeological record, domestication fundamentally changed the lives of both wild ungulate and human populations.

12.1.2 The Neolithic period

Figure 12.1 shows that, in stark contrast to the Mesolithic period where assemblages consist almost entirely of wild animals, those from Neolithic sites in England are characterised by an abundance of domestic livestock, with wild ungulate representation dropping to less than 5% (see also reviews by Serjeantson, 2011; Schulting, 2013). Where wild ungulates remains have been recovered on Neolithic sites, they consist primarily of antler from red deer (Figure 12.2), the majority of which is shed and so cannot be seen evidence for hunting (Serjeantson 2011). Indeed, Fletcher (2011) has proposed that red deer were deliberately kept alive and actively managed specifically for their antlers, which were needed to construct, and were deposited within, the many monuments and mines of the Neolithic period (Serjeantson, 2011). Fletcher (2011) has suggested that Neolithic people could have gained easy access to deer herds simply by feeding them. However, as modern studies have suggested (Schmidt, *in press*) feeding stations for deer quickly become feeding stations for large predators. It may be no coincidence that the remains of wolves and bears are fairly well represented in Neolithic assemblages (Pollard, 2006), perhaps suggesting that large predators were hunted in order to protect both domestic and wild ungulate populations. If so, this would represent an early example of human–wildlife conflict of the form seen amongst modern Saami reindeer pastoralists (Lindquist, 2000).

The idea that cervids were valued by people as living creatures could feasibly account for the rapid decline in deer hunting apparent at this time across most of northern Europe (Bartosiewicz, 2005; Gál, *in press*). However, if antler was the primary concern, it does not explain the low exploitation of non-antlered aurochsen and wild boar, which are also seldom represented in zooarchaeological assemblages of this date. It is inconceivable that the archaeological scarcity of these wild animals reflects a population crash and therefore a different explanation must be sought.

It cannot be the case that Neolithic people were simply ambivalent to wild ungulates; to the contrary, close relationships must have existed to enable the many translocations of ungulate species that occurred during the period. These population movements are easiest to detect in the case of islands, where anthropogenic transportation is the only viable explanation for the arrival of non-native animals. Recent zooarchaeological and genetic studies make clear that it was in the Neolithic that red deer and wild boar were first introduced to Ireland (Carden, 2012; Carden *et al.*, 2012) and several Scottish Isles saw the importation of red deer (Mulville, 2010). Similarly in the Neolithic of the Mediterranean, wild boar, aurochsen and Persian fallow deer (*Dama dama mesopotamica*) were introduced to

Cyprus (Vigne *et al.*, 2011; Daujat, 2013) while the European fallow deer (*Dama dama dama*) was taken to Rhodes and Crete (Masseti *et al.*, 2006; Harris, *in press*).

Some of these islands, such as Cyprus, may have been deliberately stocked with wild animals to function as self-sustaining game reserves, the animals being left to roam without fear of natural predation so that they could be hunted on occasions of human visitation. Elsewhere, particularly in northern mainland Europe, wild ungulates appear to have been conspicuously avoided. This is paralleled by evidence from stable isotope analysis of human remains: studies by Richards *et al.* (2003) and Richards and Schulting (2006) have highlighted that, across northern Europe, the beginning of the Neolithic was accompanied by a dramatic shift away from the exploitation of marine resources – a situation that endured in much of northern Europe until the arrival of the Romans. Richards and Schulting (2006) have argued that this reflects the rapid emergence of a cultural taboo over marine resources and, on the basis of the zooarchaeological data, it seems possible that the same was true for wild ungulates.

The motivation for this apparent taboo is difficult to comprehend. One possibility is that as large-scale human mobility reduced in favour of sedentism, people 'lost touch' with the areas beyond their immediate domestic surroundings and may have come to perceive them as the dwellings of ancestors and spirits, as is the case in many traditional farming societies today (Ingold, 2000). To some extent this theory finds support from the funerary practices of the period: across much of the northern and western half of Europe, for the majority of Prehistory, the principal mortuary traditions appear to have been water burial, secondary burial or excarnation, whereby corpses were left exposed to be consumed by wild animals with the fragmented remains of the body later being recovered and placed in funerary structures. If, as the anthropological studies suggest, the wilderness was seen as the spirit world or wild animals as ancestral incarnations, it is to be expected that cultural taboos may arise around hunting and the exploitation of wild resources, as Pollard (2006) has argued for the Neolithic.

Not all areas of Europe had the same funerary traditions. Across southeast Europe and Anatolia, Neolithic people were buried, often in cemeteries, suggesting that is was not deemed culturally acceptable for corpses to be eaten by wild animals. This variation in mortuary rites reflects a different attitude to the natural world and it may be no coincidence that, in these same regions, the hunting of red deer, roe deer, wild boar, aurochsen but also wild sheep and goats, and the now extinct European ass (*Equus hydruntinus*) continued at high levels. This is particularly the case in southern Germany (Stephan, *in press*), northern Italy (Boyle, *in press*), across the Balkans (Orton, *in press*) and into the Near East (Arbuckle, 2012). Towards the end of the Neolithic, however, the representation of wild ungulates starts to decline and the evidence suggests that hunting became increasingly opportunistic (e.g. Orton, *in press*).

A shift away from wild ungulate exploitation is perhaps to be expected in a period when agriculture was becoming central to many societies. Today some farming communities view wild animals with a certain reservation or actively regard them as pests because of their propensity to raid crops (Russell, 2012). Control of ungulates, or their damaging impacts, is often a primary consideration in modern management, with legislators seeking to mitigate the tensions that can arise between the perceived value of wildlife as a hunting resource (these days, admittedly, for recreational purposes) and the needs of agriculturalists (e.g. Reimoser and Putman, 2011). It may be that, in the Late Neolithic period, wild ungulates were managed via 'garden hunting', whereby animals were killed as they descended upon fields. Alternatively, hunting may have been a 'risk buffering' activity that took place only in times of crop failure or famine. This explanation has certainly been invoked in relation to Hungary where, after initially eschewing wild fauna in the early Neolithic, specialised aurochs hunting became a feature of middle and late Neolithic societies in the region (Bartosiewicz, 2005).

Although the Hungarian Late Neolithic did coincide with a period of climatic deterioration, which could have placed pressure on agricultural regimes, Bartosiewicz concludes that climate alone cannot explain the sudden interest in auroch hunting. He prefers to see the increased evidence for hunting as a reflection of social change and the emergence of a more hierarchical society that perceived the wilderness as something to be challenged through hunting. That the Late Neolithic communities of Hungary valued wild ungulates for reasons beyond protein acquisition is supported by finds of 'trophies' and jewellery made of their teeth, which have been recovered from several high-status male burials in the region, with similar examples being noted from Germany and Switzerland (Stephan, *in press*).

As in Hungary, there is some evidence that aurochsen were the main quarry of Neolithic hunters in England: Figure 12.2 shows that their remains are represented in about 20% of assemblages and, unlike the red and roe deer antler, we can be confident that these remains come from hunted animals. Work by Serjeantson (2011) has shown that the aurochs becomes one of the best-represented wild species

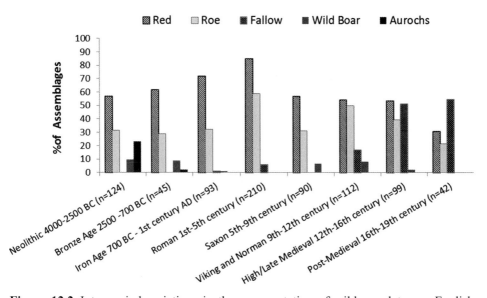

Figure 12.2 Inter-period variations in the representation of wild ungulates on English archaeological sites. Bars show the percentage of assemblages (samples sizes are shown in parentheses) in which each species is represented. Sources as for Figure 12.1.

towards the end of the Neolithic, when they are found on a comparable number of sites and in comparable relative frequencies to red deer (Serjeantson, 2011).

12.1.3 The Bronze and Iron Ages

The representation of aurochsen drops dramatically during the Bronze Age (Figure 12.2), which can be seen as evidence for the population decline that caused the species extirpation by the end of the period. However, the remains of wild cattle do continue to be found on Bronze Age sites, particularly in the burials of elite males, many of whom were also interred with archery equipment (Davis and Payne, 1993; Cotton *et al.*, 2006; Towers *et al.*, 2010).

The situation for Bronze Age England, where hunting was infrequent but concentrated on the most dangerous of wild ungulates (Figure 12.1), is consistent with that seen across much of Bronze Age Europe and the eastern Mediterranean. Generally wild ungulates are poorly represented in Bronze Age assemblages (Harding, 2000); however, the *symbolism* of hunting, in particular the hunting of wild boar and red/ fallow deer, was frequently deployed by the warrior elite as a display of power, martial ability and masculine identity (Treherne, 1995; Hamilakis, 2003; Dalix and Vila, 2007). The ideological value placed upon hunting is easier to understand through the lens of anthropology. In her study of traditional societies, Helms (1993) demonstrated that, across cultures, geographical distance is commonly equated to supernatural distance and that the hunters' ability to successfully negotiate 'outer realms' is frequently regarded as a sign of divine-bestowed authority.

There is some textual and iconographic evidence that more complex human–deer relationships began to emerge in the Bronze Age of the Mediterranean, with animals being seen as more than the quarry of male hunters. Linear B tablets recovered from Pylos on Crete list different 'kinds' of deer, including those that are wild, those that are domestic, as well as those that are used in games or for sacrifice (Yannouli and Trantalidou, 1999). In a similar way, frescoes from the site of Agia Triadha, also on Crete, appear to depict women leading fallow deer to a sacrificial altar (*ibid*). The use of seemingly 'wild' animals for sacrifice is interesting in its own right because anthropological and historical studies have demonstrated repeatedly that wild animals cannot be considered 'sacrificial' since they belong to no-one and therefore represent no personal loss (Sykes, *in press*). The evidence would therefore seem to suggest, for Crete at least, the development of either a unique attitude to wild animals or some level of deer ownership.

The zooarchaeological record for Crete suggests that the end of the Bronze Age brought a decline in wild ungulate hunting (e.g. Harris, *in press*). Frequencies of game animals drop substantially through the course of the Iron Age and the only sites where they continue to be well represented are temples and sanctuaries. The same is true across much of the Aegean and Balkans (Trantalidou, *in press*), the Near East (Arbuckle, 2012) and northern Europe; Figure 12.1, for example, indicates that the Iron Age of England saw an almost complete cessation in the hunting of wild ungulates, even on sites of high status. Figure 12.2 shows that red deer and roe deer are the best represented species in zooarchaeological assemblages from England but it is important to recognise that these remains consist almost entirely of antlers (Allen, *in press*) and most were recovered from votive deposits (Sykes, 2010).

Iron Age iconography from northern Europe hints that wild ungulates had a special status within Celtic religion. Human–animal hybrids are frequently depicted; the most famous example being on the Gunderstrup Caldron, which was recovered from a bog in Denmark (e.g. Aldhouse-Green, 2004). The caldron shows an antlered man, believed to represent Cernunnos the Celtic god of animals, who is seated amongst a variety of wild animals (Figure 12.3a; Green, 1992). The possibility that, as in the Neolithic, wild animals were viewed as sacred during the Iron Age is not without modern comparisons, although one must travel to India to find them. Today, the Bishnoi community of Rajasthan do not kill wild animals; indeed they actively manage sacred groves, some of which are more than 500 years old, as shelter for wild animals. According to Kothari and Sharma (2013) the Bishnoi villages are 'swarming' with antelopes, gazelle and other wild animals; perhaps a similar situation existed across much of Iron Age Europe where sacred groves are known to have existed (Webster, 1995).

In the absence of human hunters it must be assumed that populations of wild ungulates were kept in balance by natural processes, including predation by the wolves, bears and lynx that still inhabited much of Europe at this time. The situation began to change, however, with the rise and expansion of the Roman Empire, which took with it new cultural attitudes to the natural world.

12.1.4 The Romans

By contrast to Iron Age culture, which appears to have viewed the wilderness and wild animals as sacred, the Roman mindset saw it as their spiritual duty to dominate nature and bring it to order (Cartmill, 1993). Across the Roman Empire, the exploitation of wild animals became part of elite culture. The possession of productive estates, hounds, horses and leisure time came to be viewed as the defining traits of elite identity, with hunting serving as the nexus through which all could be communicated (Anderson, 1985; Dunbabin, 2004). Overall, the representation of wild animals does not increase dramatically on Romano-British sites but those who have studied the zooarchaeological evidence in detail have demonstrated a clear rise in in the utilisation of wild resources on large villas and military centres (Allen, *in press*). Furthermore, whereas Iron Age assemblages consist primarily of deer antler, those from Romans sites contain a much higher percentage of postcranial elements, indicating that the remains derived from hunted animals (Allen, *in press*). That it was now considered acceptable to hunt and eat wild ungulates, as well as other wild resources (stable isotope analysis of human remains indicates that marine animals once again made a considerable contribution to high-status diets; see Müldner and Richard, 2006) reflects a very different cultural attitude to the natural world. Evidence for this change in ideology is also provided by shifts in funerary practices: bodies were now either buried or cremated to ensure that they could not be consumed by wild animals, a prospect that was abhorrent to the Roman belief system (Toynbee, 1971). Iconographic representations also change

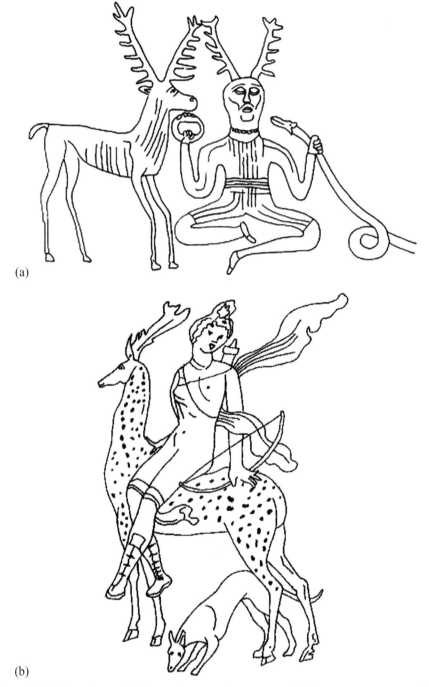

(a)

(b)

Figure 12.3 Iron Age and Roman iconography compared. There are more examples of hybridised deities in the Iron Age (such as is depicted on the Gunderstrip Cauldron, (a)) whereas animals are more often shown accompanying Roman deities, as is shown above with Diana riding a fallow deer (b).
Source: David Taylor.

somewhat, with human-animal hybrids becoming less common: rather than being mixed together with wild animals, Roman deities are depicted in their *company,* as in the case of the goddess Diana and her fallow deer familiar (Figure 12.3b).

The association of fallow deer with Diana, and her earlier Greek incarnation as Artemis, is widely known (Reinken, 1997; Yannouli and Trantalidou, 1999; Fabiš, 2003) and it seems plausible this was the very motivation for the human-assisted spread of fallow deer across the Roman Empire. A recent programme of radiocarbon dating, stable isotope analyses and ancient DNA studies has confirmed that breeding populations of European fallow deer were established in Roman Italy, Sicily, Portugal, France, Austria and certainly in southern England (Figure 12.1). These fallow deer were kept in parks, as is specified by classical authors such as Columella who was writing around AD 60 (see *De Re Rustica* IX, 1.1–9). In Italy, there is some suggestion that a park was constructed at the grand villa of Sette-finestre, built around 75 BC (Carandini, 1985). More definitive evidence for the establishment of game parks is found at another lavish villa, Fishbourne Palace on the south coast of England. This site has produced the largest collection of *D. d. dama* remains from Roman England but their remains are still very rare (n = 38) indicating that fallow deer retained an exotic status rather than becoming natural-ised (Sykes *et al.*, 2006). Fishbourne also yielded a large assemblage of red and roe deer remains, studies of which suggest that male sub-adult animals were pref-erentially targeted (Allen, *in press*). This contrasts with the picture for low-status rural sites where the remains of female animals are comparatively more abundant. These demographic profiles would seem to suggest that venison, rather than tro-phies, was the main focus of deer hunting.

12.1.5 Medieval Europe

The hunting culture transported by the Romans was not to last however. With-drawal of the Empire saw the decline of deer parks and all the evidence, including new genetic data (Baker *et al., in prep.*), suggests that any established fallow deer populations became quickly extinct (see Figure 12.2; Sykes and Carden, 2012). Zooarchaeological data are limited for this period but, for England, the early medi-eval (5th to mid-7th century) evidence indicates an almost complete abandonment of hunting traditions, with wild ungulates being less well represented than in Iron Age (Figure 12.1). Again, nearly all of the remains that are present are antler rather than post-cranial bones. It is possible that over-hunting in the Roman period, com-bined with the extensive woodland clearance that occurred during the 1st to 4th centuries, did reduce populations of wild animals: Figure 12.2 shows that whereas red deer were present on 85% of Roman sites, their frequency drops to just 57% of Anglo-Saxon sites; similarly the representation of roe deer declines from 59% to 31% of sites. The only animal that is better represented than in previous periods is the wild boar, which has been identified on almost 7% of Anglo-Saxon sites. It is clear that wild boar held a position of significance within Germanic culture, evi-denced by finds of helmets, weapons and other personal objects decorated with boar motifs (Sykes, 2011). Indeed, Pluskowski (2010a) has suggested that dangerous animals – bears, boars, wolves and, to a lesser extent, aurochsen – were particularly valued by Germanic groups, their images and body parts (e.g. teeth or skins) being used to adorn the bodies and military paraphernalia of the male warrior elite. Given that the early medieval period saw hunting focus more on predatory animals (with the exception of wild boar) it may be that the wild ungulate populations of northern Europe were able to recover from the hunting pressures of the Roman period.

The same cannot be said for the south-east half of Europe, where Kroll (2012) has shown that the exploitation of wild resources continued at comparatively high levels across Greece, the Balkans and into Anatolia: cervids and wild boar, but also leporids, make up a considerable proportion of assemblages dating to the Byzantine period (AD 4th to 14th century). Kroll (2012) suggests that the high frequencies of game may have represented a famine food, and this is a possibility. However, documentary evidence indicates that the exploitation of wild animals was a socially important pursuit across the Byzantine Empire, where the passion for hunting and the emparkment of wild ungulates appears to have persisted unbroken from the Roman period (Ševčenko, 2002).

Byzantine traditions of wild ungulate management and hunting must have come to the attention of the 8th to 9th century Frankish King Charlemagne, whose empire bordered that of the Byzantine rulers. Charlemagne is the first Early-Medieval monarch recorded as having a passion for hunting (Goldberg, 2013) and the use of hunting as a motif for kingship was promptly copied by Charlemagne's English counter-part, King Alfred: Asser's *Vita Aelfredi* (*c.* AD 893) mentioned hunting, alongside reading and warfare, as part of princely education (Marvin, 2006). Figure 12.1 shows that English assemblages dating between the 9th and 12th centuries begin to show an increased representation of wild ungulates, with considerable representation of post-cranial bones, not just antler. Red deer and roe deer are for the first time represented in similar frequencies, both being present in about 50% of assemblages (Figure 12.2), and detailed studies have revealed roe deer to be particularly abundant on sites of high status (Sykes, 2011). Indeed, it would seem that roe deer were the principal quarry for the elite of this period and it is possible that over-hunting at this time was responsible for the population decline that is indicated not only by the zooarchaeological record (Figure 12.2) but also by recent genetic studies (Baker, 2011).

Genetic studies on the fallow deer, which reappear in English assemblages during the 11th century (Figure 12.2), indicate a very different source population from that of the Roman period introductions. Preliminary results suggest that, contrary to previous suggestions of Sicily or Anatolia (e.g. Sykes and Carden, 2012), the Balkans are the most likely origin for the fallow deer that were introduced to, and remained in, north Europe (Baker, *pers. comm.*). This finding would seem to confirm that the rise in wild ungulate representation in England was inspired by Byzantine hunting traditions. It is, however, also important in terms of modern deer conservation because the north European populations can now be viewed as a relict of the now extinct Balkan fallow deer, with potential for re-introduction.

Whilst the Balkans saw the extinction of fallow deer due to over-hunting, in Britain different species succumbed to the same pressure, perhaps in combination with habitat depletion and, in the case of the wild boar, hybridisation. Figure 12.2 indicates that zooarchaeological evidence for wild boar is very scarce in assemblages post-dating the 12th century and Albarella (2010) has argued that the last native populations were probably extinct by the end of the 13th century. There is some evidence that new wild boar stock was occasionally imported from the continent, where wild boar was the most common quarry of elite hunters (Yvinec, 1993; Albarella, 2010).

The role of the elite in establishing the hunting culture of the medieval period can be seen when Figures 12.1 and 12.2 are viewed in tandem. For although Figure 12.1 highlights that wild ungulates representation reaches its highest level since the Mesolithic in the Medieval period, Figure 12.2 indicates that, with the exception of fallow deer, wild ungulates are represented in fewer assemblages in England than in the preceding period. What this means is that very large numbers of wild ungulates have been recovered from rather few sites, all of which are elite: by example fallow deer account for 28% of the mammal remains from 14th-century Okehampton Castle, Devon.

The same pattern of increased hunting can be seen across much of high medieval Europe, with Pluskowski (2007) arguing that, from the 12th century onwards, the elite of medieval western Europe engaged with wild animals in a multitude of different ways–by maintaining them alive within private parks, by hunting them, by using their body parts (furs and antlers) as objects of adornment or as trophies–to create a socially distinct but internationally recognisable 'seigneurial culture'. However, the hunting fashion of the medieval period also left a less positive legacy. To the elite hunting culture of the Middle Ages can be attributed the over-exploitation of not only wild ungulates but also their predators, which were particularly prized by the social elite: by the end of the medieval period all of Britain's large carnivores were brought to extinction and populations of bear, wolf and lynx were significantly reduced across much of mainland Europe (Hammon, 2010; Hetherington, 2010; Pluskowski, 2010b).

12.1.6 From the post-medieval to the 21st century

The last 500 years have seen perhaps the greatest change in cultural attitudes to the natural world (Thomas, 1983), arguably more dramatic than has been witnessed in the entirety of human history. Across Europe, the growth of urbanisation, industrialisation and intensive farming has seen human populations expand and become increasingly divorced from traditional rural practices and concerns. The collapse of feudalism altered both perceptions and expressions of social identity, and although ungulate hunting endured in many regions as a leisure pursuit of the elite, the level was much diminished compared to that of the medieval period.

Ironically, the elite hunting culture of the medieval period and the parks that were established to provide the aristocracy with a ready supply of game can be cited as the very two factors that led to the demise of both within English culture.

By essentially farming deer, venison became easy to acquire as it percolated into the urban black market and even became a staple of peasant feasts (Birrell, 1996; Sykes, 2007b). As soon as venison was no longer the preserve of the elite, it lost its status associations and the aristocracy began to scale down their parks and reduce their consumption of venison, as can be seen in Figures 12.1 and 12.2. As a result, many parks fell into disrepair, allowing their inmates to escape and establish feral populations without fear of predation from the now-extinct large carnivores. Of course, once the elite lost interest in ungulate hunting, the lower social echelons soon did likewise and the levels of wild ungulate exploitation gradually declined across the board. For England, this is reflected again in Figures 12.1 and 12.2: together they highlight the beginnings of the post-Medieval shift away from the wild ungulate exploitation, which has continued from this point to the present day.

12.2 Where are we now?

In modern times, management of wild ungulates in Europe shows a diversity of objectives and priorities, dependent in part on local issues and attitudes. Some taxa (whether species, or endemic subspecies) are considered threatened and management for those species may primarily be focused on conservation. At the same time, other species may, through local overabundance, cause conflict with other land-management objectives, through excessive impacts on agriculture or forestry, or damage to natural habitats (Reimoser and Putman, 2011), through problems as vectors of disease (e.g. Ferroglio *et al.*, 2011), through implication in collisions with vehicles (Langbein *et al.*, 2011) or through 'invasion' of our urban landscapes (Chapter 7); management focus here will be on control of those populations – or at least reduction of impacts. As a cross-cutting thread, throughout most of Europe, wild ungulates are valued as hunting quarry, though the main 'driver' is for recreational purposes rather than a subsistence economy reliant on wild game meat.

In most places, almost all these 'objectives' for management of wild ungulate populations coexist, albeit in different balance of priority for different species, or in different areas, within a given country. Thus in almost the whole of Europe, modern management must face similar problems (and often faces serious conflicts of interest between these various different management objectives). Yet management methods and approaches to management may vary significantly, as clearly illustrated in the problems of management of populations of ungulates which cross national or provincial boundaries (Chapter 9) where priorities and management actions may differ significantly in different parts of a common range.

These differences relate in part to differences in political history and socioeconomic history and thus differences in the legal status of game and the administrative structures developed to effect their management (and to control or facilitate recreational hunting). Among countries that have a more totalitarian or socialist history and where, in consequence, wild ungulates are legally considered the property of the state, or are deemed to belong to the public, there is a tendency to accept much more state control of, or at least state intervention in, management priority and in administration of management activities (Putman, 2011). In countries where game belong to no one, but where landownership confers the rights to take or manage wildlife populations, administrative and legal systems permit a far lesser level of state control over what management is carried out – if any at all – or control over what may be the primary objectives of such management.

But in addition to these 'political' differences', there are also clear cultural differences in the way wild game are managed and treated; indeed the very political differences we have just described are themselves at least in part a consequence of those same underlying cultural differences. We see totally different attitudes to hunting in, for example, the more Germanic countries, those of the former Austro-Hungarian Empire, or the former Polish Empire, than in, let us say, Scandinavian, or more Mediterranean countries (Putman *et al.*, 2011). These differences are reflected in administrative structures (the need in some countries to belong to a Hunters' Club or Hunters' 'family', for example), in hunting traditions and practice (the traditions preserved, for example, by membership of the St Hubert's Club), even in terms of the dress code and formality of costume: with very formal and traditional hunting costumes still widely worn as the norm, in places such as Germany, Austria, Hungary, the Czech Republic and Slovenia, amongst others. Attitudes are also changing in relation to ungulate populations more intensively managed by humans (thus wild species within hunting parks or recreational parks; the extent of intervention by humans even among wild populations through fenced enclosure, through selective hunting, or deliberate genetic 'improvement'; Chapters 2 and 6, this volume).

Thus, just as we have seen changes through time in both management objective and management practice, reflecting changes in attitudes to wild animals and changes in the social and economic importance of wild ungulate populations, we see persistent differences between modern European cultures in the cultural and economic importance of wild ungulate species reflected in different attitudes and management philosophy (Putman, 2011; Chapter 2 this volume). Whatever the differences in culture, legal and administrative system, or hunting practice, however, almost all European countries promote recreational hunting as a major component of the management (and maintenance) of wild ungulate populations; indeed this is sufficiently well-entrenched that in many European countries there have developed systems of compulsory compensation (whether paid directly by the state, or as a burden on the various Hunters' Clubs), to pay for the inevitable damage caused to agricultural or forestry interests as the result of maintaining ungulate populations at densities required to provide for this tradition of recreational hunting.

Indeed in the majority of European countries, management not only provides for recreational hunting but is itself largely dependent on such recreational hunting. While control of ungulate populations in national parks or national forests, for example, may be carried out by state-employed hunters, on private lands almost all management – whether carried out towards purely provate objectives, or to deliver public benefit – is carried out by volunteer, recreational hunters. This may be a concern for the future: a number of countries are experiencing a decline in the number of hunters within the human population and indeed an ageing population of hunters as fewer younger people take up hunting. This decline in the number of new hunters may in part be a reflection itself of the changing cultural attitudes of an increasingly urban population across Europe. Such formal surveys as have been undertaken confirm that one of the primary factors affecting individual attitudes to hunting is personal experience (as a hunter, or closely related to others who hunt; e.g. Floyd *et al.* 1986; Stokke, 2004; Burgin and Mattila, *in press*). But, whatever the cause, a significant change in demographics of human populations and the demographics of those who hunt has been recorded in a number of countries (e.g. Norway, Reidar Andersen, *pers. comm.*; Finland: Burgin and Mattila, *in press*). In such context, it is significant also to note that recent review by the US Fish and Wildlife Service of 'Fishing, Hunting and Wildlife-Associated Recreation' (US Fish and Wildlife, 2007, 2011) presents statistics to show a progressive decline in almost all States of the number of hunters (and hunters of large ungulates) from 1991–2001–2006–2011 (with a corresponding increase in numbers of Americans involved in non-consumptive pursuits (wildlife watching, etc.).[1]

With expanding ungulate populations (see Apollonio *et al.*, 2010) and a changing human demographic, there are significant concerns in many countries of a possible mismatch in the near future between management need and management capacity. We are facing a time when ungulate impacts may increase - impacts on agriculture, forestry, on our roads and in our towns and cities - because we simply do not have the capacity to control expanding populations; there may thus be a need to review current management models reliant in this way on recreational hunting.[2]

12.3 How far have we come?

The last ten thousand years have witnessed extreme transformations in human attitudes to wild ungulates and the natural world: from a situation when the wild did not exist because it was *all* that existed, to a situation where many 'wild' ungulate populations have been selectively managed to the point that they could rightly be classified as 'domestic'. The transformation that took place between these two extremes was not, however, a linear evolution. In this chapter we have shown that the dynamics and social significance of wild animal exploitation have followed the ebb and flow of culture change and wider attitudes to the natural world.

From an archaeological perspective, some of the most pivotal moments in human history have been characterised by shifts in the relationship between people and wild ungulates. The advent of farming and the gradual spread of domestic livestock across Europe during the Neolithic not only gave rise to the category

---

[1] For further comparison: a recent survey in Japan revealed an average age amongst recreational hunters of 65 years old (K. Kaji, *pers. comm.*).

[2] In Japan, the provincial authorities in some regions have already taken the step of recruiting paid hunters to compensate for the inability of recreational hunters to control ever-increasing numbers of sika deer and contain increasing damage reported to agriculture and forest crops.

'wild', but also meant that it was no longer necessary to kill wild animals because meat and other products from domestic cattle, sheep, pigs and goats were readily available. Yet people continued to exploit wild ungulates, albeit at low levels: the English zooarchaeological data suggest that, with the exception of the Mesolithic, wild ungulates cannot have made a significant contribution to human diet, their remains generally making up less than 4% of archaeological assemblages. Indeed, in the 6000 years since the hunter-gatherer culture of the Mesolithic, humans have spent approximately 5,200 years deliberately avoiding the killing of wild ungulates. This is not to suggest that they were deemed culturally insignificant, quite the opposite; there is good evidence that wild ungulates played important roles in the cosmology and religious practices of many of Europe's ancient societies, such as those of the Iron Age.

In other periods and places, such as the Bronze Age of the Mediterranean, the Roman Empire and Medieval Europe, wild ungulate killing was more common. It can be no coincidence that these cultures are all characterised by an anthropocentric worldview in which the natural world was conceptualised as something to be transformed and controlled by people. This is evidenced by the art and literature of the time which also highlight that hunting was less about sustenance and more a social and political performance, used as an expression of martial prowess, elite identity and masculinity. It was these cultures that brought about the most significant exploitation and translocation of wild ungulates. Reflecting upon their impact–such as the over-hunting of roe deer in England and the extirpation of fallow deer in Bulgaria–serves as a lesson from antiquity about the need for deer management strategies that are both sensitive and sustainable. However, legacies from the past also remind us that modern humans have a responsibility to manage the changes instigated by our ancestors, such as medieval importations of new species and the extinction of natural predators.

It seems paradoxical that today, when ungulate populations are expanding and there is a growing need for people to manage them, levels of exploitation are far lower than was the case in the past. This situation is itself is a product of the cultural, economic, environmental shifts that have occurred during the last 500 years. In many countries, in particular those that are highly urban, modern attitudes to the natural world show similarities to those of the Neolithic and Iron Age, with wild ungulates being seen as sacred icons of the wilderness and, therefore, above human exploitation. For some, however, the status of wild ungulates as icons of nature is the very reason why hunting has become attractive as a leisure pursuit. To be sure, recreational hunting is culturally and economically important across Europe and many of the groups involved share ideological traits with the ungulate hunters of the past, in particular the timeless connection between hunting, martial ability and masculinity. However, it is evident that in many European countries masculine identities are increasingly being negotiated in ways other than expressions of machismo, a situation reflected by the declining number of hunters in the human population. It seems likely that this trend, which arguably began at the end of the medieval period, will continue and future management models need to take this into account.

Whilst the outlook for ungulate management in the future is concerning, studies of the past provide some comfort. The zooarchaeological record demonstrates that attitudes to wild ungulates can change, have changed repeatedly and will certainly continue to do so. Importantly, what these variations reveal is that human responses to wild animals are shaped as much by social and cultural ideology as they are by environmental factors, if not more so, and this must be taken into consideration when developing ungulate management regimes.

References

Albarella, U. (2010) Wild boar. In T. O'Connor and N. Sykes (eds), *Extinctions and Invasions: A Social History of British Fauna*. Oxford: Windgather Press, pp. 59–67.

Albarella, U., Dobney, K. and Rowley-Conwy, P. (2009) Size and shape of the Eurasian wild boar (*Sus scrofa*), with a view to the reconstruction of its Holocene history. *Environmental Archaeology* **14**, 103–136.

Aldhouse-Green, M. (2004) *An Archaeology of Images: Iconology and Cosmology in Iron Age and Roman Europe*. London: Routledge.

Allen, M. (2010) *Animalscapes and empires: new perspectives on the Iron Age/Romano-British transition*. PhD thesis, Department of Archaeology, University of Nottingham, UK.

Allen, M. (in press) Chasing Sylvia's stag: placing deer in the countryside of Roman Britain. In K. Baker, R.F. Carden and R. Madgwick (eds), *Deer and People: Past, Present and Future*. Oxford: Windgather Press.

Anderson, J. K. (1985) *Hunting in the Ancient World*. Davis, CA: University of California Press.

Apollonio, M., Andersen, R. and Putman R. (2010) Present status and future challenges for European ungulate management. In M. Apollonio, R. Andersen and R.J. Putman (eds), *European Ungulates and their Management in the 21st Century*. Cambridge, UK: Cambridge University Press, pp. 578–604.

Arbuckle, B. S. (2012) Animals in the ancient world. In D.T. Potts (ed.), *A Companion to the Archaeology of the Ancient Near East*. Oxford: Wiley-Blackwell, pp. 201–219.

Baker, K. (2011) *Population genetic history of the British roe deer* (Capreolus capreolus) *and its implications for diversity and fitness*. PhD thesis, School of Biological and Biomedical Sciences, Durham University, UK.

Bartosiewicz, L. (2005) Plain talk: animals, environment and culture in the Neolithic of the Carpathian Basin and adjacent areas. In D.W. Bailey, A.W. Whittle and V. Cummings (eds), *(Un)settling the Neolithic*. Oxford: Oxbow, pp. 51–63.

Birrell, J. (1996) Peasant deer poachers in the medieval forest. In R. Britnell and J. Hatcher (eds), *Progress and Problems in Medieval England: Essays in Honour of Edward Miller*. Cambridge: Cambridge University Press, pp. 68–88.

Boyle, K. (in press) The Italian Neolithic red deer: Molino Casarotto. In K. Baker, R.F. Carden and R. Madgwick (eds), *Deer and People: Past, Present and Future*. Oxford: Windgather Press.

Burgin, S. and Mattila, M. (in press) The profile of male and female hunters in Finland.

Carandini, A. (1985) *Settefinestre: una villa schiavistica nell'Etruria Romana, III*. Modena, Italy: Panini Editore.

Carden, R. F. (2012) *Review of the natural history of wild boar* (Sus scrofa) *on the island of Ireland*. Report for the Northern Ireland Environment Agency, Northern Ireland, UK, National Parks & Wildlife Service, Department of Arts, Heritage and the Gaeltacht, Dublin, Ireland and the National Museum of Ireland–Education & Outreach Department.

Carden, R. F., McDevitt, A. D., Zachos, F. E., Woodman, P. C., O'Toole, P., Rose, H., and Edwards, C. J. (2012) Phylogeographic, ancient DNA, fossil and morphometric analyses reveal ancient and modern introductions of a large mammal: the complex case of red deer (*Cervus elaphus*) in Ireland. *Quaternary Science Reviews* **42**, 74–84.

Cartmill, M. (1993) *A View to a Death in the Morning: Hunting and Nature Through History.* Cambridge, MA: Harvard University Press.

Conneller, C. (2004) Becoming deer: corporeal transformation at Star Carr. *Archaeological Dialogues* **11**, 37–56.

Conneller, C. (2011) The Mesolithic. In T. Insoll (ed.), *The Oxford Handbook of the Archaeology of Ritual and Religion.* Oxford: Oxford University Press, pp. 358–370.

Cotton, J., Elsden, N., Pipe, A., and Rayner, L. (2006) Taming the wild: a final Neolithic/Earlier Bronze Age aurochs deposit from west London. In D. Serjeantson and D. Field (eds), *Animals in the Neolithic of Britain and Europe.* Oxford: Oxbow, Oxford, pp. 149–167.

Dalix, A.S. and Vila, E. (2007) Wild boar hunting in the Eastern Mediterranean from the 2nd to the 1st millennium BC. In U. Albarella, K. Dobney, A. Ervynck and P. Rowley-Conwy (eds), *Pigs and Humans: 10,000 Years of Interaction.* Oxford: Oxford University Press, pp. 359–372.

Daujat, J. (2013) *Ungulate invasion on a Mediterranean island: the Cypriot Mesopotamian fallow deer over the past 10,000 years.* PhD thesis University of Aberdeen and the Muséum National d'Histoire Naturelle.

Davis, S. and Payne, S. (1993) A barrow full of cattle skulls. *Antiquity* **67**, 12–22.

Dunbabin, K.M.D. (2004) *The Roman Banquet: Images of Conviviality.* Cambridge, UK: Cambridge University Press.

Fabiš, M. (2003) Troai and fallow deer. In G.A.Wagner, E. Pernicka and H.-P. Uerpmann (eds), *Troia and the Troad: Scientfic Approaches.* London: Springer, pp. 263–275.

Ferroglio, E., Gortazar, C. and Vicente, J. (2011) Wild ungulate diseases and the risk for livestock and public health. In R.J. Putman, M. Apollonio and R. Andersen (eds), *Ungulate Management in Europe: Problems and Practices.* Cambridge, UK: Cambridge University Press, pp. 192–214.

Fletcher, J. (2011) *Gardens of Earthly Delight: The History of Deer Parks.* Oxford: Windgather Press.

Floyd, H.H., Bankston, W.B., Burgesion, R.A. (1986) An examination of the effects of young adults' social experience on their attitudes toward hunting and hunters. *Journal of Sport Behavior* **9**, 116–130.

Gál, E. (in press) Evidence for the differentiate exploitation of cervids at the Early Bronze Age site of Kaposújlak–Várdomb (South Transdanubia, Hungary). In K. Baker, R.F Carden and R. Madgwick (eds), *Deer and People: Past, Present and Future.* Oxford: Windgather Press.

Goldberg, E.J. (2013) Louis the Pious and the Hunt. *Speculum,* **88**(3), 613–643.

Green, M. (1992) *Animals in Celtic Life and Myth.* London: Routledge.

Hambleton, E. (2008) *Review of Middle Bronze Age to Late Iron Age faunal Assemblages from Southern Britain.* Portsmouth, UK: English Heritage.

Hamilakis, Y. (2003) The sacred geography of hunting: wild animals, social power and gender in early farming societies. In E. Kotjabopoulou, Y. Hamilakis, P. Halstead, C. Gamble and V. Elafanti (eds), *Zooarchaeology in Greece: Recent Advances.* London: British School at Athens, pp. 239–247.

Hammon, A. (2010) The brown bear. In T. O'Connor and N. Sykes (eds), *Extinctions and Invasions: A Social History of British Fauna.* Oxford: Windgather Press, pp. 96–103.

Harding, A.F. (2000) *European Societies in the Bronze Age.* Cambridge, UK: Cambridge University Press.

Harris, K. (in press) Hunting, performance and incorporation: human–deer encounters in Late Bronze Age Crete. In K. Baker, R.F Carden and R. Madgwick (eds), *Deer and People: Past, Present and Future.* Oxford: Windgather Press.

Helms, M. (1993) *Craft and the Kingly Ideal: Art Trade and Power.* Austin, TX: University of Texas Press.

Hetherington, D. (2010) The lynx. In T. O'Connor and N. Sykes (eds), *Extinctions and Invasions: A Social History of British Fauna.* Oxford: Windgather Press, pp. 75–82.

Ingold, T. (2000) *The Perception of the Environment: Essays in Livelihood, Dwelling and Skill.* London: Routledge.

Jennbert, K. (2011) *Animals and Humans: Recurrent Symbiosis in Archaeology and Old Norse Religion.* Lund, Sweden: Nordic Academic Press.

Kothari, B. and Sharma, B.K. (2013) An anthropological account of bonhomie and opprobrium between communities and animals in Rajasthan. In B.K. Sharma (ed.), *Faunal Heritage of Rajasthan, India: Ecology and Conservation of Vertebrates.* New York: Springer, pp. 213–225.

Kroll, H. (2012). Animals in the Byzantine Empire: an overview of the archaeozoological evidence. *Archeologia Medievale* **39**, 93–121.

Langbein, J., Putman, R.J. and Pokorny, B. (2011) Road traffic accidents involving ungulates and available measures for mitigation. In R.J. Putman, M. Apollonio and R. Andersen (eds), *Ungulate Management in Europe: Problems and Practices.* Cambridge, UK: Cambridge University Press, pp. 215–259.

Leduc, C. (2012) New Mesolithic hunting evidence from bone injuries at Danish Maglemosian sites: Lundby Mose and Mullerup (Sjælland). *International Journal of Osteoarchaeology,* doi: 10.1002/oa.2234

Lindquist, G. (2000). The wolf, the Saami and the urban shaman: predator symbolism in Sweden. In J. Knight (ed.), *Natural Enemies. People–wildlife Conflicts in Anthropological Perspective.* London: Routledge, pp. 170–188.

Lister, A.M. (1996) The morphological distinction between bones and teeth of fallow deer (*Dama dama*) and red deer (*Cervus elaphus*). *International Journal of Osteoarchaeology* **6**(2), 119–143.

Marvin, W. P. (2006) *Hunting Law and Ritual in Medieval English Literature.* Woodbridge, UK: Brewer.

Masseti, M., Cavallaro, A., Pecchioli, E. And Vernesi, C. (2006) Artificial occurrence of the fallow deer, *Dama dama dama* (L., 1758), on the Island of Rhodes (Greece): Insight from mtDNA Analysis. *Human Evolution* **21**(2), 167–176.

Müldner, G.H. and Richards, M. P. (2007) Stable isotope evidence for 1500 years of human diet at the city of York, UK. *American Journal of Physical Anthropology* **133**(1), 682–697.

Mulville, J. (2010) Red deer on Scottish islands. In T. O'Connor and N. Sykes (eds), *Extinctions and Invasions: A Social History of British Fauna.* Oxford: Windgather Press, pp. 43–50.

Mysterud, A. (2010) Still walking on the wild side? Management actions as steps towards 'semi-domestication' of hunted ungulates. *Journal of Applied Ecology* **47**, 920–925.

Nadasdy, P. (2007) The gift in the animal: The ontology of hunting and human–animal sociality *American Ethnologist* **34**, 25–43.

Noe-Nygaard, N. (1974) Mesolithic hunting in Demark illustrated by bone injuries caused by human weapons. *Journal of Archaeological Science* **1**, 217–248.

Orton, D. (in press) False dichotomies? Balkan Neolithic hunting in archaeological context. In J. Mulville and A. Powell (eds), *A Walk on the Wild Side: Hunting in Farming Societies.* Oxford: Oxbow.

Pluskowski, A. (2007) Communicating through skin and bones: Appropriating animal bodies in medieval Western European seigneurial culture. In A. Pluskowski (ed.), *Breaking and Shaping Beastly Bodies: Animals as Material Culture in the Middle Ages.* Oxford: Oxbow, pp. 32–51.

Pluskowski, A. (2010a) Animal magic. In M. Carver, A. Sanmark and S. Semple (eds), *Signals of Belief in Early England: Anglo-Saxon Paganism Revisited.* Oxford: Oxbow, pp. 102–127.

Pluskowski, A. (2010b) The wolf. In T. O'Connor and N. Sykes (eds), *Extinctions and Invasions: A Social History of British Fauna.* Oxford: Windgather Press, pp. 68–74.

Pollard, J. (2006) A community of beings: animals and people in the Neolithic of southern Britain. In D. Serjeantson and D. Field (eds), *Animals in the Neolithic of Britain and Europe.* Oxford: Oxbow, pp. 135–148.

Politis, G. and Saunders, N. (2002) Archaeological correlates of ideological activity: Food taboos and spirit-animals in an Amazonian hunter-gatherer society. In P. Miracle and N. Milner (eds), *Consuming Patterns and Patterns of Consumption.* Cambridge: MacDonald Institute, pp. 113–130.

Poole, K. (2010) *The Nature of Society in England AD 410–1066.* PhD thesis, Department of Archaeology, University of Nottingham, UK.

Putman, R.J. (2011) A review of the legal and administrative systems governing management of large herbivores in Europe. In R.J. Putman, M. Apollonio and R. Andersen (eds), *Ungulate Management in Europe: Problems and Practices.* Cambridge, UK: Cambridge University Press, pp. 54–79.

Putman, R., Andersen, R. and Apollonio, M. (2011) Introduction. In R.J. Putman, M. Apollonio and R. Andersen (eds), *Ungulate Management in Europe: Problems and Practices.* Cambridge, UK: Cambridge University Press, pp. 1–11.

Reimoser, F. and Putman, R.J. (2011) Impact of large ungulates on agriculture, forestry and conservation habitats in Europe. In R.J. Putman, M. Apollonio and R. Andersen (eds), *Ungulate Management in Europe: Problems and Practices.* Cambridge, UK: Cambridge University Press, pp. 144–191.

Reinken, G. (1997) Wieder-verbreinung. Verwendung und namesgebungung des damhirsches *Cervus dama* L. in Europa, *Zeitchrift Jagdwissenshaft* **43**, 197–206.

Richards, M.P. and Schulting, R. (2006) Touch not the fish: The Mesolithic–Neolithic change of diet and its significance. *Antiquity* **80**, 444–56.

Richards, M.P., Schulting, R.J. and Hedges, R.E. (2003) Archaeology: Sharp shift in diet at onset of Neolithic. *Nature* **425**, 366.

Russell, N. (2012) *Social Zooarchaeology: Humans and Animals in Prehistory.* Cambridge, UK: Cambridge University Press.

Schmidt, K. (in press) Neolithic revolution anew. The way supplemental feeding changes our attitude towards red deer and natural mortality. In K. Baker, R.F Carden and R. Madgwick (eds), *Deer and People: Past, Present and Future.* Oxford: Windgather Press.

Schulting, R. (2013) On the northwestern fringes: Early Neolithic subsistence in Britain and Ireland as seen through faunal remains and stable isotopes. In S. Colledge, S. Connoly, K. Dobney, K. Manning and S. Shennan (eds), *The Origins and Spread of Domestic Animals in Southwest Asia and Europe.* Walnut Creek, CA: Left Coast Press, pp. 313–338.

Serjeantson, D. (2011) *Review of animal remains from the Neolithic and Early Bronze Age of Southern Britain.* Portsmouth, UK: English Heritage.

Ševčenko, N. P. (2002) Wild animals in the Byzantine park. In A.R. Littlewood, H. Magiure and J. Wolschke-Bulmahn, J. (eds), *Byzantine Garden Culture*. Washington DC: Dumbarton Oaks, pp. 69–86.

Stephan, E. (in press) Red deer hunting and exploitation in the Early Neolithic settlement of Rottenburg-Fröbelweg, South Germany. In K. Baker, R.F Carden and R. Madgwick (eds), *Deer and People: Past, Present and Future*. Oxford: Windgather Press.

Stokke, E. (2004) *Nordsmenns Holdninger til jakt: [Norwegians' attitudes to hunting]* Masters Thesis at the Department of Ecology and Natural Resource Management, The Norwegian University of Life Sciences.

Sykes, N.J. (2007a) *The Norman Conquest: A Zooarchaeological Perspective*. Oxford: Archaeopress.

Sykes, N.J. (2007b) Taking sides: the social life of venison in medieval England. In A. Pluskowski (ed.), *Breaking and Shaping Beastly Bodies: Animals as Material Culture in the Middle Ages*. Oxford: Oxbow, pp. 150–161.

Sykes, N.J. (2010) Worldviews in Transition: the impact of exotic plants and animals on Iron Age/Romano-British landscapes. *Landscapes* **10**, 19–36.

Sykes, N.J. (2011) Woods and the Wild. In Hamerow, H., Hinton, D. A., and Crawford, S. (eds), *The Oxford Handbook of Anglo-Saxon Archaeology*. Oxford: Oxford University Press, pp. 329–347.

Sykes, N.J. (in press) *Beastly Questions: Animal Answers to Archaeological Issues*. London: Bloomsbury Press.

Sykes, N.J. and Carden, R.F. (2011) Were fallow deer spotted in Anglo-Saxon England? Reviewing the evidence for *Dama dama dama* in early medieval Europe. *Medieval Archaeology* **55**(1), 139–162.

Sykes, N.J., White, J., Hayes, T. and Palmer, M. (2006) Tracking animals using strontium isotopes in teeth: the role of fallow deer (*Dama dama*) in Roman Britain *Antiquity* **80**, 848–959.

Sykes, N., Carden, R.F. and Harris, K. (2013) Changes in the size and shape of fallow deer: evidence for the movement and management of a species. *International Journal of Osteoarchaeology* **23**, 55–68.

Thomas, K. (1983), *Man and the Natural World: Changing Attitudes in England 1500–1800*. London: Penguin.

Tolan-Smith, C. (2008) Mesolithic Britain. In G. Bailey and P. Spikkins (eds), *Mesolithic Europe*. Cambridge, UK: Cambridge University Press, pp. 132–157.

Towers, J., Montgomery, J., Evans, J., Jay, M., and Parker Pearson, M. (2010) An investigation of the origins of cattle and aurochs deposited in the Early Bronze Age barrows at Gayhurst and Irthlingborough. *Journal of Archaeological Science* **37**(3), 508–515.

Toynbee, J. M. (1971) *Death and Burial in the Roman World*. Ithaca, NY: Cornell University Press.

Trantalidou, K. (in press) Archaeozoology of the red deer in the southern Balkan Peninsula and the Aegean region during the antiquity: confronting bones and paintings. In K. Baker, R. F. Carden and R. Madgwick (eds), In K. Baker, R.F Carden and R. Madgwick (eds), *Deer and People: Past, Present and Future*. Oxford: Windgather Press.

Treherne, P. (1995) The warrior's beauty: The masculine body and self-identity in Bronze-Age Europe. *Journal of European Archaeology* **3**(1), 105–44.

US Fish and Wildlife Service (2007) *National Survey of Hunting, Fishing and Wildlife-Associated Recreation*. Washington DC: US Fish and Wildlife Service.

US Fish and Wildlife Service (2011) *National Survey of Hunting, Fishing and Wildlife-Associated Recreation*. Washington DC: US Fish and Wildlife Service.

van Vuure, C. (2005) *Retracing the Aurochs: History, Morphology and Ecology of an Extinct Wild Ox*. Sofia/Moscow: Pensoft Publishers.

Vigne, J. D., Carrere, I., Briois, F. and Guilaine, J. (2011) The early process of mammal domestication in the Near East. *Current Anthropology* **52**, 255–271.

Webster, J. (1995) Sanctuaries and sacred places. In M. Green (ed.), *The Celtic World.* Oxford: Routledge, pp. 445–464.

Willerslev, R. (2007) *Soul Hunters: Hunting, Animism, and Personhood among the Siberian Yukaghirs.* Berkley, CA: University of California Press.

Yannouli, E. and Trantalidou, K. (1999) The fallow deer (*Dama dama* Linnaeus, 1758): Archaeological presence and representation in Greece. In N. Benecke (ed.), *The Holocene History of the European Vertebrate Fauna.* Berlin: Verlag, pp. 247–281.

Yvinec, J.-H. (1993) La part du gibier dans l'alimentation du Haut Moyen Âge. In J. Desse and F. Audoin-Rouzeau (eds), *Exploitation des Animaux Sauvages a Travers le Temps.* Juan-les-Pins, France: APDCA, pp. 491–504.

Zvelebil, M. (2008) Innovating hunter-gatherers: the Mesolithic in the Baltic. In G. Bailey and P. Spikkins (eds), *Mesolithic Europe.* Cambridge, UK: Cambridge University Press, pp. 18–59.

Index